Exploration and Science

Other titles in
ABC-CLIO's

Science and Society: Impact and Interaction
Series

The Environment and Science, Christian C. Young

Imperialism and Science, George N. Vlahakis, Isabel Maria Malaquias, Nathan M. Brooks, François Regourd, Feza Gunergun, and David Wright

Literature and Science, John H. Cartwright and Brian Baker

Race, Racism, and Science, John P. Jackson, Jr., and Nadine M. Weidman

Women and Science, Suzanne Le-May Sheffield

Exploration and Science

Social Impact and Interaction

Michael S. Reidy
Gary Kroll
Erik M. Conway

A B C ⬥ C L I O

Santa Barbara, California   Denver, Colorado   Oxford, England

Library of Congress Cataloging-in-Publication Data

Reidy, Michael S.
  Exploration and science : social impact and interaction / Michael S. Reidy,
Gary Kroll, and Erik M. Conway.
      p. cm. — (Science and society)
  Includes bibliographical references and index.
  ISBN-13: 978-1-57607-985-0 (hard cover : alk. paper)
  ISBN-10: 1-57607-985-6 (hard cover : alk. paper)
  ISBN-13: 978-1-57607-986-7 (ebook)
  ISBN-10: 1-57607-986-4 (ebook)
  1. Science—Social aspects. 2. Discoveries in geography—Social aspects.
3. Discoveries in science—Social aspects. 4. Science—History. I. Kroll, Gary R.
II. Conway, Erik M., 1965– III. Title.

  Q175.5.R44 2007
  303.48'3—dc22

                                                                    2006033258

  10 09 08 07 06   10 9 8 7 6 5 4 3 2 1

This book is also available on the World Wide Web as an eBook.
Visit abc-clio.com for details.

ABC-CLIO, Inc.
130 Cremona Drive, P.O. Box 1911
Santa Barbara, California 93116-1911

Media Editor	*Ellen Rasmussen*
Media Manager	*Caroline Price*
Editorial Assistant	*Alisha Martinez*
Production Manager	*Don Schmidt*
File Manager	*Paula Gerard*

Contents

Series Editor's Preface, vii

Preface, ix

1   *Navigating the Oceans, 1*

2   *Ordering Nature in the Age of Enlightenment, 39*

3   *Humboldt and the Rise of the Geophysical Sciences, 71*

4   *Natural History in the Nineteenth Century, 103*

5   *Scientific Exploration of a Manifest America, 135*

6   *The Exploratory Tradition in the Ocean Sciences, 161*

7   *Human Exploration under the Sea, 189*

8   *Human Exploration of the High Frontier, 219*

9   *Robotic Space Exploration, 245*

Chronology of Significant Events, 273

Glossary, 279

Primary Source Documents, 287

Notes, 335

Bibliography, 339

Index, 361

Series Editor's Preface

The discipline of the history of science emerged from the natural sciences with the founding of the journal *Isis* by George Sarton in 1912. Two and a half decades later in a lecture at Harvard Sarton explained, "We shall not be able to understand our own science of to-day (I do not say to use it, but to understand it) if we do not succeed in penetrating its genesis and evolution." Historians of science, many of the first trained by Sarton and then by his students, study how science developed during the sixteenth and seventeenth centuries and how the evolution of the physical, biological, and social sciences over the past 350 years has been powerfully influenced by various social and intellectual contexts. Throughout the twentieth century the new field of the history of science grew with the establishment of dozens of new journals, graduate programs, and eventually the emergence of undergraduate majors in the history, philosophy, and sociology of science, technology, and medicine. Sarton's call to understand the origins and development of modern science has been answered by the development of not simply one discipline, but several.

Despite their successes in training scholars and professionalizing the field, historians of science have not been particularly successful in getting their work, especially their depictions of the interactions between science and society, into history textbooks. Pick up any U.S. history textbook and examine some of the topics that have been well explored by historians of science, such as scientific racism, the Scopes trial, nuclear weapons, eugenics, industrialization, or the relationship between science and technology. The depictions of these topics offered by the average history textbook have remained unchanged over the last fifty years, while the professional literature related to them that historians of science produce has made considerable revision to basic assumptions about each of these subjects.

The large and growing gap between what historians of science say about certain scientific and technological subjects and the portrayal of these subjects in most survey courses led us to organize the Science and Society series. Obviously, the rich body of literature that historians of science have amassed is not regularly consulted in the production of history texts or lectures. The authors and editors of this series seek to overcome this disparity by offering a synthetic, readable, and chronological history of the physical, social, and biological sciences as they developed within particular social, political, institutional, intellectual, and economic contexts over the past 350 years. Each volume stresses the reciprocal relationship between science and context; that is, while various circumstances and perspectives have influenced the evolution of the sciences, scientific disciplines have conversely influenced the contexts within which they developed. Volumes within this series each begin with a chronological narrative of the evolution of the natural and social sciences that focuses on the particular ways in which contexts influenced and were influenced by the development of scientific explanations and institutions. Spread throughout the narrative, readers will encounter short biographies of significant and iconic individuals whose work demonstrates the ways in which the scientific enterprise has been pursued by men and women throughout the last three centuries. Each chapter includes a bibliographic essay that discusses significant primary documents and secondary literature, describes competing historical narratives, and explains the historiographical development in the field. Following the historical narratives, each book contains a glossary, timeline, and most importantly a bibliography of primary source materials to encourage readers to come into direct contact with the people, the problems, and the claims that demonstrate how science and society influence one another. Our hope is that students and instructors will use the series to introduce themselves to the large and growing field of the history of science and begin the work of integrating the history of science into history classrooms and literature.

—*Mark A. Largent*

Preface

The tendency to explore may be one of the actions that defines us as humans. Perhaps it is rooted in our ancestral movement out of Africa and our spread throughout the world's continents during the last million years. Indeed, the practice of exploration has had a peculiarly strong iconic status in the standard narrative of Western civilization. For instance, histories of Western civilization generally mark the oceanic voyages of the Portuguese, Spanish, and Dutch in the fifteenth and sixteenth centuries as the great age of exploration. It is no coincidence that as these European explorers were plying the world's oceans, Europe also witnessed the beginnings of a new formal way of learning about nature, now referred to as the scientific revolution. Since that time, the institution of Western science and the practice of exploration have moved apace, though never consistently, hand in hand. Exploration may be a human universal, but here we are interested particularly in the historical relationship between exploration and Western science.

There is probably no figure in the Western mind that trumps the iconic status of the heroic explorers. Their deeds and stories are the stuff of legends, and even today a voracious reading public snatches up the latest chronicle of this or that explorer at the local bookstore. The history of exploration has become encrusted in a kind of mythology that raises the social significance of the explorer to an almost sacrosanct status. While it was rarely the overt objective to knock these explorers from their pedestal, over the past thirty years historians and journalists have worked hard to present the history of exploration as a complex social, political, and cultural saga. As a result, we can now tell stories of a demythologized and truly human history of exploration.

One of the most enduring myths has been that the explorer boldly sets off alone. It is a myth that casts exploration as a solitary venture in which man sets out to conquer an untamed wilderness. This is seldom true for scientific exploration. Not only did explorers usually travel with large parties and rely heavily on the indigenous knowledge of the people they met in the wilderness, but explorers also moved into unknown territories with the social, economic, and technological backing of political states or empires. Historians often invoke the language of the core-periphery model of development, describing the scientific explorer as moving from the urban core (Paris, London, Philadelphia) into the unknown periphery (Lapland, Fiji, the Missouri River). Scientific exploration was thus a projection of political entities into regions that were relatively unknown to the Western mind. Perhaps best symbolized by the federal coins that Meriwether Lewis and William Clark distributed to Native Americans on their voyage up the Missouri River, the values and technology of the core traveled with the scientific explorer into the periphery.

Another myth is that the explorer moved into the periphery without any inherent interest other than a desire to experience and understand what lay beyond the horizon. This myth was bolstered by such statements as Sir Edmund Hillary's explanation that he scaled Mount Everest "because it was there." Yet, exploration has always been an inherently goal-oriented process. As representatives of the core, scientific explorers set out to map, catalog, survey, and envision remote territories for cultural and economic reasons. An explorer was considered to be successful, after all, only if he came back to the core with maps, journals, and other data. This information was used, in the parlance of geography, to fill in the blank spaces on the map. But maps are important and powerful social and political documents that help states to collate knowledge for utilization, whether it was for commercial extraction, settlement, or war. The desire to conserve or preserve nature was also an important use for exploration. Even when it appears that scientific exploration has had no direct social or economic benefit, it is crucial to put such exploration into context. For example, human space exploration by the United States, the former Soviet Union, and now China can only be understood within the context of guided-missile technologies during the Cold War.

The third myth suggests that explorers move into an unpeopled region that lay outside of human history. The ideal of exploration was historically predicated on the faulty assumption that the wilderness has no previous human presence. Clearly, scientific explorers usually had some knowledge of the people they would encounter on their journey into the periphery. But for much of the modern period, these people rarely seemed to count. The term "terra incognita," meaning unknown land, was always uttered by people coming from the core who rarely appreciated that the periphery was already known by other people. Historians sometimes use the term "contact zone" to refer to the territory where the explorer meets the indigenous "other." Such encounters factor into the scientific process in stunning ways.

While this volume does not pretend to make any headway in further demythologizing the scientific explorer, it does participate in the project of telling a history of scientific exploration as a multilayered social, cultural, political, and economic drama. To do so, we have relied heavily on scholarly literature in the history of science and technology, environmental history, general history, and historical geography. Moreover, we have drawn from the primary sources of scientific exploration in order to enrich and enliven our narrative. The narrative itself takes both a chronological and a geographical approach. The volume focuses primarily on European and American exploration in the modern period, from the seventeenth century to the present. The first part of the volume focuses on the horizontal exploration of our oceans and landmasses, including the Atlantic and Pacific Oceans, the polar regions, the Americas, and Australia. The second part focuses on vertical exploration below and above Earth, including deep-sea exploration and atmospheric and space exploration. This geographical approach coincides with a chronological approach, allowing us to focus on common themes that unite the history of exploration and science across time.

Although scientific exploration has been practiced by all cultures throughout history, we focus primarily on Western exploration, partly because there is so much to gain from a close study of the interweaving set of political, scientific, and economic forces that came together in the Western context. This includes the creation of large-scale economic developments that we term "capitalism," the formation of representative institutions of government, the foundation of our modern conceptions of science, and, of course, the heavy investment in exploration and colonization. From advances in astronomy motivated largely by overseas expansion in the seventeenth century to our modern advances in deep-ocean and deep-space exploration, science and exploration have gained from each other's exertions. This reciprocal relationship constitutes our first major theme.

The advances in both science and exploration, however, made major leaps at specific times, namely when institutions, including national governments, stepped in and funded exploring expeditions, whether for militaristic, imperialistic, economic, or nationalistic reasons. Thus, a second theme is the social and cultural forces that have sustained the exploration enterprise. This includes the institutionalization of science and the internationalization of the exploration process. The last theme attempts to answer the question "Why do we explore?" As is true for the scientific revolution, the eighteenth-century Enlightenment, the opening of the American West, and today's deep-space programs, exploration creates both personal and national identity. We explore outwardly in an attempt to understand ourselves inwardly. The narrative of the text follows these three themes: the relationship between science and exploration, the cultural forces sustaining exploration, and the creation of personal and national identity.

After a thirty-year hiatus, there has been a resurging interest in the history of exploration by both historians and the general public. The Vietnam War, the energy crisis, and the recession of the 1970s dampened much of the enthusiasm for the explorer that was typical in the immediate post–World War II years. Historians have cast a suspicious gaze on the often egocentric tendencies of Western exploration. As the twenty-first century progresses, it will be necessary for us to keep in mind that the history of exploration is a complex drama in which the fruits of scientific knowledge came at an all too real social, political, and economic cost.

Western culture has been peculiarly outwardly looking over the last 500 years, as is indicated in the spread of Western economic, cultural, and social values over the face of Earth. Political states invested heavily in the science and technology needed for colonial adventurism. States also patronized the exploratory efforts of geographers, hydrographers, naturalists, cartographers, and geologists whose natural knowledge lubricated the wheels of commerce and control. As European and, later, American powers incorporated far-flung territories into their empires, scientific explorers faced head-on the troublesome diversity of the world's natural environment, and they invented sophisticated theories of Earth and innovative explanations for the mutability and evolution of life.

The late nineteenth and twentieth centuries brought this social and cultural tradition of scientific exploration into the hostile regions of the world's oceans and outer space. Certainly, there were many changes to the practice of exploration: funding mechanisms, popular culture, an expanded mandate to explore by corporations, and massive technological advances, to name a few. But in many ways, twentieth-century exploration was just an elaboration of older traditions in new regions. We have not been able to cover the full scope of the field; our intent is to be more representative than encyclopedic. Exploration will continue—to deeper areas of the oceans and farther reaches of space—as humans continue to construct their place in the universe. As the explorations become grander and the science gets bigger, the process will continue to internationalize, national governments will continue to take a leading role, and the adventure of science and exploration will continue to fascinate, bewilder, and captivate the attention of the wider public.

Navigating the Oceans

Francis Bacon, one of the premier architects of modern science, maintained that three inventions—the compass, printing, and gunpowder—marked the transition from the ancient to the modern world. Roughly translated as ease of navigation, communication, and control, all three profoundly affected European exploration. The compass guided the early voyages of discovery, allowing the Spanish, Portuguese, and Dutch to sail beyond the limited confines of the Mediterranean region and explore the vast expanse of the world's oceans. Printing publicized these remarkable achievements. Extravagant folio picture books and exciting travel narratives brought new worlds and vivid experiences to the forefront of European consciousness. These bold voyages of exploration, although initiated partly for scientific and geographic curiosity, paved the way for European colonization and commercial expansion, a process in which gunpowder proved ever present, helping to transform consciousness into control.

Writing in the early seventeenth century, Bacon was also convinced that the early voyages of discovery necessitated a complete restructuring of knowledge. The geographical revolution of the preceding century, Bacon averred, required an equally profound scientific revolution. He quoted the prophet Daniel—"many shall run to and fro, and knowledge shall be increased"—as biblical support that oceanic exploration could shake the foundations of ancient learning.[1] While mariners traveled for glory and riches, the experiences they brought back to Europe, including descriptions of new and exotic flora and fauna, helped undermine faith in ancient forms of learning. When mariners traveled to new lands beyond the Mediterranean world, they discovered that far more land and ocean existed than had been previously believed, and that only personal experience could uncover these new additions to knowledge. These early voyages of exploration helped motivate a new approach to studying the natural world in which firsthand experience and detailed observations reigned over ancient authority.

Francis Bacon (1561–1626)

Francis Bacon was a scientist and politician. The son of the lord keeper of the great seal, England's most eminent judge, Bacon was born into the cradle of the court, spending much of his life as a political analyst and royal servant. He was minister to two British monarchs, Elizabeth I and James I. Educated at Trinity College, Cambridge, Bacon also became one of the premier advocates for a new natural philosophy. He combined these two lives by consistently linking science to the state, urging English monarchs to endow scientific research. Knowledge of the physical world, Bacon argued, conferred remarkable power on both the British kingdom and the Kingdom of God, leading to unprecedented advances in technology, medicine, and warfare.

Bacon's most influential works include *The Advancement of Learning*, published in 1605 and dedicated to King James I; the *Novum organum*, or *New System of Learning*, published in 1620; and the *New Atlantis*, published in 1627 a year after his death. In these monumental works, Bacon advanced a new approach to natural philosophy, one grounded in direct observation and experiment. He related his new vision for science directly to exploration, arguing that the discovery of new lands in the preceding century necessitated an equally profound reorganization of natural philosophy. He advocated cooperation among a large workforce of scientific observers and experimenters. In his *New Atlantis*, he outlined just this type of organized, collaborative enterprise based on a division of labor whereby philosophers in Salomon's House on the uncharted island of Bensalem equipped humans with power over, and thus control of, the natural world.

Bacon was knighted in 1603 for his extensive writings on politics, law, history, and natural philosophy. He rose within the English court to the position of lord chancellor and keeper of the great seal. Yet he remained until the end of his life a controversial figure in politics, dismissed from the chancellorship in 1621, for instance, after an extremely public trial for bribery. His influence rests with the role he envisioned for science in public affairs. His ideas were brought to fruition in the founding of scientific societies across Europe, particularly the Royal Academy in Paris and the Royal Society of London, both consciously based on Bacon's Salomon's House.

From the early sixteenth century to the Enlightenment, geographic and intellectual exploration advanced in unison, a process that links the remarkable rise of natural philosophy to broader cultural, economic, and political forces sweeping through Europe. Dreams of commercial success led European monarchies to invest heavily in oceanic exploration, leading to the discovery of new continents and cultures and ultimately to the establishment of economic policies that we now call modern capitalism. As voyages of discovery multiplied and moved into uncharted territory, the need for advanced navigational techniques directed the practice of astronomy, the science most associated with the scientific revolution. While European powers sought to span the globe, modern science developed.

After the first burst of European exploration by the Portuguese, Spanish, and Dutch in the late fifteenth and early sixteenth centuries, natural philosophers were increasingly called on to mitigate the risks of overseas exploration. They had to learn about the ocean itself—its tides, currents, and magnetism— and the means by which mariners could determine longitude at sea. Natural philosophers helped transform an inherently and notoriously dangerous space into a more secure place for European expansion by graphing rule and rationality onto an ostensibly chaotic region of the globe. This new learning occurred on the high seas and in distant lands, aboard ships in uncharted regions of the North Pacific, and in newly discovered tropical forests of South America.

The requirements of navigation also shifted the emphasis of natural philosophy in Europe, where the science of navigating the oceans also took place in the palatial observatories built by wealthy monarchs, in scientific societies in major metropolitan centers, and in the dark denizens of artisans' shops. Land-based natural philosophers solved significant problems concerning navigation, including Earth's magnetic properties and the determination of longitude at sea. Their quest required a faith in the ability of natural philosophers to understand the natural world and a faith in the natural world that it was unvarying and rational, subject to strict and understandable laws.

The desire to navigate the world's oceans contributed significantly to the advance of science not only by directing its questions but also by helping to define the manner in which natural philosophers created scientific knowledge. Oceanic navigation relied on extensive governmental patronage, the establishment of astronomical observatories, and increased precision in instrumentation and measurement. The sheer mass of the varied activities associated with the new natural philosophy reinforced the perception that the new form of learning could help advance the economic and political agendas of maritime nations, a perception that led to increased funding and institutional support for the burgeoning scientific process. Natural philosophy could be profitable for national governments, while national governments proved essential for the advance of modern science. This combination, played out in the form of European explorations to remote areas of the globe, proved to be a powerful instrument of European expansion.

When China Ruled the Waves

None of the three inventions that Bacon argued had changed the world were invented in Europe; all three came from China. After a momentous period of philosophical energy in ancient Greece, Europe lagged far behind China by the Middle Ages in both natural philosophy and technological sophistication. Owing to their interest in both astrological forecasting and oceanic exploration, emperors of the great Chinese dynasties placed special emphasis on observational astronomy. By the end of the Tang dynasty (circa 900), China had produced the most advanced star charts of any civilization. During the Sung dynasty (960–1279), China developed the largest navy and merchant marine in the world. Advances in navigational techniques, including an understanding of the difference between true north and magnetic north, combined with improved shipbuilding technology, including fore and aft rigging and watertight bulkheads, enabled the Chinese to assemble the largest fleet to sail the oceans until the massive convoys of World War I and World War II.

In the first quarter of the fifteenth century, Emperor Zhu Di of the Ming dynasty (1366–1644) sent fleets of hundreds of Chinese vessels into the Indian Ocean to help bring India, the Persian Gulf, and the east coast of Africa under Chinese economic and political influence. The leader of these massive fleets was Admiral Zheng He (1371–1433), a Muslim eunuch from the mountainous Yunnan province of Southwest China, the site of fierce battles between China and Mongols from the north. After the armies of the Ming dynasty conquered Yunnan in the late fourteenth century, Zheng He was taken captive, transported to Nanjing, castrated according to custom, and made a servant to the future emperor of China, Zhu Di.

Between 1405 and 1433, a half century before the voyage of Christopher Columbus, Zheng He commanded seven voyages of exploration and discovery throughout the China Sea and Indian Ocean. He sailed with a magnificent complement of vessels, reaching more than 250 ships and 25,000 men at its height. In comparison with European voyages of exploration later in the century, the size, shape, and number of vessels were nothing short of awe-inspiring. The largest vessels of the squadron were known as treasure ships. With nine masts and measuring more than 400 feet, these stupendous vessels were aptly named, carrying precious cargo of silks, fine art, and porcelain. The rest of the fleet was made up of smaller cavalry ships, warships, troop transports, patrol boats, supply ships, and water tankers. In addition to the sailors and soldiers, the fleet also embarked with astronomers, scholars, and 180 medical officers in charge of collecting herbs and other medicinal flora from distant lands.

Ready to depart in 1405, the first voyage set sail from Nanjing on 11 July, a day celebrated today in China as National Maritime Day. This and subsequent voyages helped open renewed trade with kingdoms throughout the periphery of the China Sea and Indian Ocean. Several of these voyages sailed far beyond India all the way to the East African coast. Indeed, some historians believe that

Zheng He's ships may have traveled beyond the Cape of Good Hope all the way to the Americas. Yet the death of Zhu Di in August 1424 signaled the beginning of the end of the treasure fleet. Several years earlier, he had moved the capital of China from Nanjing to Beijing, thus signaling a change in political focus from the southern and western seas to the northern borders on land. The emperors who followed Zhu Di focused their energies even more on China's northern borders to secure the area from resurgent Mongolian tribes, thus bringing the state-sponsored oceanic voyages of the treasure fleet to an end. Ironically, Chinese rulers turned their attention away from oceanic voyages of exploration just as European countries were venturing farther and farther through the Atlantic Ocean.

Europe's Rebirth

China's oceanic voyages resulted in an accumulation of expertise in the art of navigation, linked closely to technological progress and an intricate study of the heavens. These advances did not vanish with the great dynasties of early modern China. They made their way, through trading lanes carried by adventurous explorers, to the West through the Muslim world. Great East-West trading routes spanned the Middle East, and Islamic scholars served as a conduit for both commodities and ideas. Like China, they developed a sophisticated knowledge of observational astronomy for religious reasons, such as finding the direction of Mecca and the correct seasons for the holy month of Ramadan. As early as the ninth century, Baghdad had become a center of learning, where scholars at the Bait al-Hikmah, or House of Wisdom, performed daily translations and calculations. These Islamic scholars excelled at both translating ancient texts from all over the Northern Hemisphere, including Egyptian, Babylonian, and Chinese treatises, and preserving ancient Greek mathematical and scientific texts, such as those by Galen, Euclid, and Ptolemy.

In the process of translation, moreover, Islamic scholars made significant additions to these texts in the form of commentaries, including more accurate determinations of the path of the sun and moon. Words such as "azimuth," "zero," and "zenith," which were co-opted from these Arabic commentaries, also made their way into Western thought. The revival of Greek mathematical and astronomical learning in the Latin West emerged through these Arabic translations and commentaries, opening up a large, and largely forgotten, treasury of Greek scholarship, combined now with advances from India, the Middle East, and China. It was this cross-cultural contact and exchange that took place along the major trading routes of Europe and Asia that set the stage for the remarkable advances in Europe between 1300 and 1600, a period known as the European Renaissance.

The term "renaissance" means "rebirth," an intellectual movement with roots deeply entrenched in the ascent of autonomous city-states such as Venice, Florence, and Genoa, where a remarkable economic growth produced a wealthy aristocratic class interested in advancing new forms of learning. One text in particular, recovered in Florence in the early fifteenth century, excited mariners and traders alike and played a significant role in the rise of mathematics as a useful tool for the art of navigation. Ptolemy's *Geographia* was an instructional text for cartographers that described how to depict a three-dimensional space, such as Earth, onto a two-dimensional plane, such as a map, through lines of latitude and longitude. The power of visual representations of Earth linked the practical study of mathematics with European political and economic agendas.

The renewed interest in mathematics proceeded apace with the phenomenal advances in exploration and overseas trade emanating from Western Europe. Throughout the early Renaissance, however, natural philosophy and mathematics remained divided. John Wallis, one of the founding members of the Royal Society of London in the 1660s, insisted that mathematics was at that time still the purview of merchants, traders, land surveyors, and almanac makers. The connection between mathematics and natural philosophy occurred within practical fields such as navigation and cartography, those practices in which quantity, not quality, mattered. Setting one's course by the stars and Earth's magnetic field prompted the merging of mathematics and natural philosophy, a union that would become one of the defining features of modern science. Europeans turned this new union toward maritime trade, ushering in the first great age of European exploration and discovery.

Portugal, Spain, and the Atlantic World

In Europe, exploration of the world's oceans began with the Portuguese in the fifteenth century. They were the first of the European nations to search out the Atlantic, before the Spanish, Dutch, British and French became involved, all obsessed with finding new routes to the spices of the East. Prince Henry the Navigator, although he did little exploratory travel himself, spurred the early interest in oceanic exploration, setting up the School of Navigators and a new scientific approach toward sea-bound discovery. His followers refined the quadrant, an instrument for measuring latitude while at sea, allowing navigators to stretch ever farther, tentatively, down the west coast of Africa. Before Henry the Navigator's death in 1460, valiant adventurers had reached all the way down to the Ivory Coast.

King John II continued the exploratory fervor after his accession to the Portuguese throne in 1481. He sent Bartolomeu Dias farther down the African continent in the hopes of rounding the Cape of Africa. Dias died on his second journey in 1500, but only after proving the feasibility of voyaging around Africa's southern tip, a treacherous stretch of navigation christened the Cape of Good Hope by John II. Vasco da Gama followed Dias in 1497 and, after finally rounding the Cape, made it to the east coast of Africa before finding an Arab guide, Ibn Majied, to lead the way to the spices of India. The fate of da Gama's voyage, however, was representative of these early voyages of discovery: only 55 of the original 170 men returned alive, most falling prey to shipwrecks, scurvy, or dysentery.

Not everyone viewed the feasibility of an eastern route to India as a success. Columbus had been politicking for funds to attempt the same journey by traveling in the opposite direction, west rather than east, but he could no longer interest the monarchy of Portugal. He slyly petitioned other European monarchies and found interest from Queen Isabella and King Ferdinand, the rulers of Spain. Quite by accident and quite contrary to his own beliefs, he discovered a new continent, vehement to the bitter end that he had discovered no such thing. Amerigo Vespucci was less certain. He believed that Columbus had indeed discovered a new world and set sail to explore its outer rim. Vespucci sailed along the coast of Florida, then Brazil, and was the first European to sail south of the equator off the coast of South America.

Eventually, it became apparent that Columbus and Vespucci had discovered a landmass new to Europeans, yet the European nobility maintained their fascination with the spices in the East. Thus, the question became how to get around the Americas, a search that sparked several momentous oceanic voyages, including the epic voyage of Ferdinand Magellan. Like Columbus, Magellan was Portuguese by nationality, but after a confrontation with the king of Portugal, Magellan sought patronage from Spain's young King Charles I, grandson of Ferdinand and Isabella, the patrons of Columbus. With five ships and a crew of 250 men under his command, Magellan led the first successful circumnavigation of the globe.

Magellan's voyage was replete with hardships of all kinds. He was forced to imprison one mutineer even before the vessels had crossed the Atlantic, and by the time they had sailed down much of the South American coast, three of the five ships had mutinied. Magellan had two of the ringleaders drawn and quartered, and the third he marooned on the South American coast. The worst, however, was yet to come. Before Magellan sailed through the straits that would eventually bear his name, one of his vessels, the *Santiago*, was wrecked in a storm. A second ship, the *San Antonio*, carrying a large portion of the crew's food, became separated from the remaining ships as they crossed through the present-day Straits of Magellan. It retreated back to Europe, leaving only three understocked vessels to cross the vast Pacific Ocean. With barely 150 malnourished crewmen left, they managed to make it to the Philippines, where Magellan was killed during a skirmish with local tribes. Laden with tons of valuable spices, the remaining crew on board one beleaguered vessel limped back to Spain. Of the original crew of 250, only 18 returned alive.

These early Portuguese and Spanish expeditions utilized the rudiments of cartography and the increasingly mathematical science of navigation. But, in general, they relied on their seasoned intuition rather than on what we would today term science. Oceanic travel remained exceedingly dangerous, eventually leading the Portuguese and Spanish monarchs to rely more heavily on natural philosophers and expert instrument makers. Throughout the sixteenth century, Portuguese and Spanish monarchs founded professorships in astronomy to aid overseas expansion, offered monetary incentives for the construction of navigational instruments and maps, and began licensing pilots and cartographers. Phillip II of Spain, for instance, established the Casa de la Contratación (House of Trade) in Seville in 1503, a center of calculation for the art of navigation. Yet the science and technology they had at their disposal were still rather crude. Mariners used the compass for direction, an hourglass to determine time, the pole star and quadrant to find latitude, and not much to find longitude. Being lost at sea remained a strong possibility.

As voyages of exploration grew in number and their cargoes increased in value, problems surfaced that demanded solutions. If navigators traveled too far south, for instance, they could no longer use the pole star to find latitude. Mariners could use the sun, but the sun's motion is exceedingly complicated. Experience also showed that the mariner's compass pointed to magnetic north rather than geographic north, a variation known as magnetic declination. Magnetic declination, moreover, did not depend on longitude or latitude but rather seemed to vary through space and time. Competition among maritime nations for unexplored territories in the Atlantic, Indian, and Pacific Oceans stimulated European monarchs to have the stars studied more closely, their possessions mapped, and explorations funded to ever more distant areas of the globe. All this placed a new emphasis on direct observation through personal experience and accuracy through precise measurement, the foundations of the new scientific method that would sweep through Europe in the ensuing centuries. The harrowing dangers of navigating the world's oceans precipitated a new program for natural philosophers.

The Scientific Revolution

Astronomy, the science most associated with the scientific revolution, became a focus of study for natural philosophers, owing partly to the dangers and rigors of oceanic travel. Ironically, early voyages of discovery themselves played a decisive role in undermining confidence in the ancient authority of Ptolemy, whose geocentric system of the universe had held sway for more than 1,300 years. In particular, Ptolemy's geographical views, including the inhabitability of land near the equator, helped undermine the long-held faith in his model of an Earth-centered universe. In 1543, Nicolaus Copernicus published *De revolutionibus*, offering a model of the universe in which Earth was one of the many planets circling the sun. He offered little proof, however, of what was essentially a mathematical tool with which to determine the position of celestial bodies. It was left to others, most notably Tycho Brahe, Johannes Kepler, and Galileo Galilei, to refine the new system of the cosmos.

Claudius Ptolemy's chart of the new continents, particularly the West Indies, from his Geographia Universalis, *1540. The era of European oceanic exploration during the Renaissance undermined faith in Ptolemy and other ancient authorities, helping to usher in the scientific advances of the sixteenth and seventeenth centuries. (The New York Public Library/Art Resource)*

Brahe was the top observational astronomer before the advent of the telescope. Owing to his expertise, King Frederick II of Denmark charged him with building a palatial observatory on Hven, a 2,000-acre island off the coast of Denmark. Working over the course of twenty years, from 1576 to 1597, Brahe relentlessly observed the positions of the fixed stars, the planets, the moon, and the sun to improve knowledge of their motion. Although his model of the heavens differed from Copernicus's model, the accuracy of Brahe's observations facilitated others in refining the Copernican system. Brahe's observations of Mars, especially, enabled Kepler, one of his assistants, to postulate his now-famous three laws of planetary motion, publishing the first two, including the elliptical path of the planets, in his *Astronomia nova* (1609) and the third in his *Harmonices mundi libri V* (1619). Like Brahe, however, Kepler was supported by the state, and after Brahe's death, Kepler was charged with completing Brahe's catalog of the position of the planets and fixed stars, published as the *Rudolphine Tables* in 1627.

The *Rudolphine Tables* served as a foundation text not only for observational astronomy but also for the practice of navigation. It included a catalog of 1,000 fixed stars; planetary, lunar, and solar tables; and the means by which mariners could calculate the position of these celestial bodies at any time in the past or future. Owing to the importance of these celestial bodies for the practice of navigation, Kepler's *Tables* were copied, pirated, and dispersed throughout continental Europe and Britain. Despite his debt to Brahe, Kepler's astronomic work also convinced him of the truthfulness of the Copernican system, and he joined other astronomers, notably Galileo, in an effort to convince the world that Earth did indeed move.

Although he did not invent the telescope, Galileo was one of the first to point the new instrument to the starry heavens. He discovered innumerable and unforeseen wonders, including mountains on the moon, rings around Saturn, phases of the planet Venus, and dark spots on the sun, all of which lent strong support to the new Copernican system. By the time of Galileo's death in 1642, also the year of the birth of Isaac Newton, few practicing astronomers doubted the power of the Copernican system or its usefulness in helping to advance the safety of navigation. Within a century after the publication of *De revolutionibus*, Earth was no longer at the center of the universe, and astronomers had developed a working model to determine the position of the sun, moon, and planets, exactly what was needed to safely ply through the blue waters of the world's oceans.

While the Portuguese and Spanish attempted to form a monopoly on long-distance trade, other European nations with similar maritime interests had used the powerful new astronomy for similar ends. Wealthy merchants throughout Europe recognized the lucrative advantages of overseas exploration and overtly sought to convince their respective monarchs to engage in voyages of exploration and discovery. The Dutch, English, French, and Russians all pursued oceanic voyages from the mid-sixteenth century onward. Britain's island geography, in particular, led to a maritime orientation that rose steadily from the middle of the sixteenth century, and during the reign of Queen Elizabeth (1558–1603), England looked increasingly toward the oceans. As the commercial prospects of overseas exploration blossomed, the English nobility encouraged merchants to travel on long seaborne voyages of trade and exploration. In 1577 Sir Francis Drake circumnavigated the globe, only the second time such a feat had been accomplished. In 1595 Sir Walter Raleigh sought the legendary city of El Dorado. And prior to their journeys, the queen's astrologer, John Dee, published *General and Rare Memorials Pertayning to the Perfect Arte of Navigation* in 1570, a formative text that contained practical tables for mariners and contributed to the revival of interest in mathematics in England. It also laid down England's territorial claims to North America, along with hardy arguments for the continued expansion of England's possessions overseas.

Science also began to play a leading role in the exploration process. The queen's physician, William Gilbert, studied Earth's magnetism, partly in response to the need to improve the safety of navigation at sea. His text, *De magnete*, published in 1600, was an experimental treatment of the magnetic properties of Earth, demonstrating, among other things, how to use magnetic declination to determine longitude. He conversed with both navigators and instrument makers, and through experiments with a device called a terrella, a spherical magnet, he argued that Earth was subject to the same laws of magnetism. Gilbert's judicious use of the terrella also symbolized the burgeoning role of experimentation within the learned scientific community, one geared toward understanding the workings of nature.

It was within this atmosphere of heightened interest in oceanic discovery, combined with the increasing role of science within that exploratory process, that Bacon, minister to Elizabeth I and her successor James I, outlined a new vision of science and its role in colonial expansion. Bacon argued that science conferred a power and mastery over nature that could help advance both the Kingdom of God and the British kingdom by easing the physical suffering, disease, and death of its inhabitants. Nature, however, had to be obeyed before it was commanded, a preliminary process that required an organized set of procedures to study the intricacies of the natural world. He therefore delineated a new methodology for science based on the accumulation of observations and the performing of experiments, lambasting those who would solve problems by studying books and other forms of ancient learning. It was only through the classification and organization of experimental and observational results, Bacon argued, that useful knowledge could be attained. Thus, he is known as the father of induction, the process by which general conclusions are drawn from amassing a wide and varied range of observations of the natural world.

Bacon's scientific methodology required a fundamental restructuring of the scientific process and a reorganization of its followers. He advocated a collaborative workforce for science, incorporating what he termed a "brotherhood" of natural philosophers, with the desire and ability to distribute their results to one another. This is one reason that he placed such great stock in the discovery of both the compass and printing. The first allowed access to new sources of information, and the second enhanced the exchange of information among its practitioners from distant lands. Cooperation among natural philosophers and organization by the state were the keys to science's future success.

In his *New Atlantis*, published shortly after his death, Bacon laid out his utopian vision of just such an organized, cooperative scientific establishment. His fictitious story began as other utopian novels of the time began. A vessel was caught in a storm in the deep Pacific, finally making land on an uncharted island called Bensalem. The vessel was, significantly, a Spanish vessel; Spain held a monopoly on oceanic travel that Bacon was hoping to both emulate and undermine. Scientific research in the form of careful observation and experimentation held a prominent place in Bensalem. At the center of Bensalem stood Salomon's House, a great institution of scientific and technological learning where "Elders" oversaw the performance of experiments and organized collaborative research. Because Bensalem was isolated geographically from Europe, "merchants of light" were sent out on voyages of exploration and discovery every twelve years to spy on other civilizations, incorporating their most recent discoveries and inventions into Salomon's House.

Bacon presented the practice and process of science not as an individual effort but rather as a social institution dependent on collaborative and organized experimentation, the cataloging of facts, and the prudent exchange of information. Published some three-quarters of a century after Copernicus's major work outlining the heliocentric system of the universe, Bacon's *New Atlantis* fit squarely within the ferment of the new reform of learning that was taking hold throughout Europe. He was a contemporary of Galileo, Kepler, and René Descartes, all of whom advocated an authoritative position to matters of fact attained through observation and experiment.

The Royal Society and English Science

Within several decades of Bacon's death, similarly inclined thinkers began to act on his suggestions, resulting in the first formal societies specifically

The frontispiece to Sir Francis Bacon's Instauratio Magna, *1620, depicting an English vessel passing through the "Pillars of Hercules." The Latin phrase below the ship is from the Book of Daniel: "Many will pass through, and knowledge will be increased." (Image Select/Art Resource)*

focused on the organization and publication of scientific knowledge. The first such organizations began in Italy. The two sons of Cosimo de' Medici, the patron of Galileo, founded the Accademia del Cimento (Academy of Experiments) in Florence in 1657. It consisted of nine members who worked collaboratively on experiments and published their results as a cooperative venture, inspiring the formal organization of other learned groups across Europe. Those that formed in England and France increasingly cited as their inspirations both the precedents set by the Accademia del Cimento and Bacon's Salomon's House.

In England in the 1640s, several different groups began to meet to perform experiments and build instruments. England was on the brink of a civil war that eventually led to regicide in 1649, and these groups sought to introduce social stability amid the political upheaval. One group met at Gresham College in London, where they worked on problems of practical utility for the state, including questions of navigation. Following the suggestion of Christopher Wren, Gresham professor of astronomy and architect of St. Paul's Cathedral in London, they formalized their meetings based on similar organizations in Italy and France. The end of civil war and the restoration of the monarchy in 1660 brought King Charles II to the throne, eager to establish his rule, his authority, and his own glory. He offered a royal charter to the group, thereafter christened the Royal Society of London.

The creation of the Royal Society provided a central locale where natural philosophers could perform experiments and build instruments. It was consciously modeled after Bacon's *New Atlantis*, an empirical mission to catalog observations and organize the workings of nature. The process of gathering numerous facts was matched, moreover, by an equal emphasis on communicating those facts to other correspondences, both across the Channel and abroad. Through the exchange of erudite letters, certain learned men became the center of scientific communication: Henry Oldenberg held the position in England, Denis de Sallo in France, and Otto Mencke in the Germanic states. Their unofficial job was to receive letters, make copies (sometimes fifty or more), and send them to learned gentlemen in other parts of Europe. It is no surprise that two of these overburdened centers of communication, Oldenberg and de Sallo, published a collection of such letters to ease their dissemination. Thus, the scientific journal article was born.

The journal article emerged in 1665 as a new outlet with which natural philosophers could rapidly disseminate their ideas in print to a larger community of researchers who could then either accept or question the original claim. The brief content of the articles proved the perfect venue to describe recently performed experiments or new phenomena observed from far-off voyages of exploration. Moreover, the brief format also helped set the standards of what counted as correct argumentation in science.

Oldenberg stressed in the "Dedication" to the first issue of the *Philosophical Transactions of the Royal Society of London* that his journal aimed to "spread abroad Encouragements, Inquiries, Directions, and Patterns, that may animate, and draw on Universal Assistances."[2] A brief entry titled "Directions for Sea-Men, Bound for Far Voyages," published in the *Philosophical Transactions* in January 1666, makes these "directions" and "patterns" particularly apparent. The article included a list of inquiries that ocean-bound explorers should answer and send back to the Royal Society for publication. The next issue included "An Appendix to the Directions for Sea-Men Bound for Far Voyages," and the inquiries culminated in an entire issue in April 1667 dedicated to "Directions for Observations and Experiments to Be Made by Masters of Ships, Pilots, and Other Fit Persons in their Sea-Voyages." It stipulated in great detail how to observe the declination of the compass, chart ocean currents, and register all changes of wind and weather. Natural philosophers used these inquiries to direct research according to the initial aims of the Royal Society of London through a medium that could reach a public that the erudite letter could not. Scientific journals provided the ideal venue to catalog the diverse information acquired from seagoing philosophers traveling throughout the world's oceans.

Edmund Halley, the Last Philosopher-Commander

The period from the restoration of the monarchy in 1660 to the end of the seventeenth century represents one of the most productive times for science in England. Isaac Newton, Robert Boyle, Robert Hooke, and Edmund Halley, along with foreign visitors such as Christian Huygens and Dominique Cassini, pursued scientific investigations and reported their research to the Royal Society of London, often publishing them in the *Philosophical Transactions*. They also found a host of new observations reported from new territories dispersed around the globe that they could use for their own theories and to postulate more general scientific laws.

It was within this heightened intellectual context that Newton published perhaps the greatest mathematical text in the history of science, the *Principia mathematica*, in 1687. His brilliant insight was to relate the motions of objects in the terrestrial and celestial spheres under one all-embracing law of universal gravitation, thus establishing a new framework from which to study both the heavens and Earth. He became the most esteemed natural philosopher in the world, and his approach to mechanics, astronomy, and optics directed scientific

Edmund Halley (1656–1743)

Best remembered for the process of computing the motion of comets and predicting their orbits, Edmund Halley was one of England's most distinguished astronomers and the chief scientific analyst for the Royal Navy. At the age of twenty-one, he embarked on a voyage of scientific discovery to St. Helena, an island off the west coast of Africa, where he spent a year observing the southern stars and making detailed measurements of the transit of Mercury. It was during his sojourn in St. Helena that he hit upon the idea of using the transit of Venus to measure the dimensions of the solar system. He described his method in papers published in the *Philosophical Transactions* in 1691, 1694, and, yet again, in 1716. After his return from St. Helena, he took an avid interest in the study of astronomy to improve navigation, corresponding with other scholars throughout Europe and England, including Isaac Newton. It was Halley who financed and oversaw the publication of Newton's *Principia* in 1687.

After the publication of Newton's *Principia*, Halley spent the rest of his life engrossed in problems of physical astronomy and navigation, including commanding three voyages on the *Paramore* at the turn of the century. In his first two voyages, between November 1698 and June 1700, he undertook the first systematic magnetic survey of the southern Atlantic, producing a map that graphically represented the points of equal magnetic variation. On his third voyage, he made the first systematic hydrographic survey of the Thames estuary, and following his visual representation of terrestrial magnetism, he created a tidal map of the English Channel. Such visual representations seem commonplace today, but they were revolutionary in the late seventeenth century, and helped to promote the graphical representation of scientific data, one of the most powerful tools of modern science.

Halley was the official clerk of the Royal Society for fourteen years, from 1686 to 1700, and editor of the *Philosophical Transactions* from 1685 to 1693, where he published significant investigations concerning science and navigation, including "An Historical Account of the Trade Winds, and Monsoons" and "The True Theory of the Tides, Extracted from that Admired Treatise of Mr. Isaac Newton." In addition to his study of the actual ocean—its magnetism, tides, and currents—Halley worked for almost half a century on methods for finding longitude at sea. He published tables for the satellites of Jupiter in 1719. He worked most of his life on the motion of the moon, attempting to demonstrate the feasibility of the lunar-distance method. He was an initial member of the Board of Longitude and was visited on several occasions by John Harrison, the inventor of the marine timepiece, introducing him to other scientists and suggesting to George Graham, England's most renowned instrument maker, that he support Harrison financially.

Halley was offered the Savilian Chair of Geometry at Oxford in 1704 and the position as second astronomer royal in 1720, owing to his intellectual command of the new natural philosophy; his interest in oceanic winds, tides, and magnetism; and his indefatigable efforts in discovering a solution to the problem of finding longitude at sea.

research for more than two centuries. He had failed, however, to include much in the way of original data to help prove his theory of universal gravitation in the *Principia*, relying on data from voyages of exploration as his main source of observational support. He compared observations of the oceanic tides taken on the coasts of England with those taken by Charles Davenport during an exploration of the Bay of Tonkin in China to show that the tides were produced by the gravitational attraction of the sun and moon. And Newton compared pendulum experiments made in France with those made by the French academician Jean Richer on a voyage of discovery to Cayenne to demonstrate that the gravitational attraction was weaker at the equator than at the poles. It was left to other researchers, such as Edmund Halley, to extend Newton's analysis and fortify his place in history.

Halley was one of England's top astronomers, and he recognized immediately the revolutionary character of Newton's conclusions. It was Halley who prodded Newton into publishing the *Principia*, acting as both editor and financier, a risky proposition for a mathematical text that only a handful of natural philosophers in the world could understand. Once the *Principia* was published, Halley presented a special copy to King James II, along with a letter outlining Newton's theory of the oceanic tides, which Halley later reprinted in the *Philosophical Transactions* in March 1697. With the new, powerful theory of universal gravitation at his disposal, he then set out to expand on Newton's achievements by exploring the world's oceans.

After the publication of Newton's *Principia*, Halley spent the next decade engrossed in problems of astronomy and navigation, culminating in three voyages of discovery at the turn of the seventeenth century. Impressed with Halley's abilities and interests, James II appointed him master and commander of the *Paramore*, with orders to sail through the Atlantic Ocean to measure Earth's magnetic variation. The variations in Earth's magnetic force, Halley reasoned, if plotted on charts, could help mariners determine their correct heading without the need for astronomical observations. His orders, written by Halley himself, were open-ended, including attempts to discover land in the southern ocean. This gave him the opportunity to follow his own itinerary and complete command of a vessel to survey the entire Atlantic Ocean.

Although Halley began to speak like a sailor in his later years, he was not trained as a professional mariner, and he was assigned his position as master and commander partly to ward off any obstructions from envious mariners under his command. His lieutenant did not take the bait. He represented Halley to the ship's company as entirely insufficient for the task. After realizing that the crew had been directing the vessel on a course contrary to Halley's orders, Halley was forced to terminate his cruise early after reaching Barbados. He returned to the Atlantic three months later, again under orders from the king. Sixty years before the famous voyages of Captain James Cook to the southern seas, Halley traversed the southern latitudes farther than any European before him, on an explicitly scientific mission to enhance the navigation of English sailing vessels. His voyages represent an early link between science, the English state, and the perfection of navigation. Halley published a map of the magnetic variation over the Atlantic in 1701, using isogonic lines to represent lines of equal magnetic variation, a remarkably new and powerful graphical device that formed the basis of all subsequent studies in terrestrial magnetism. Alexander Humboldt adopted Halley's graphical techniques in the early nineteenth century, expanding their use to most other questions in physical astronomy.

For his third and final voyage, Halley scoured the English Channel for tide observations and soundings throughout the summer of 1701, and following his visual representation of terrestrial magnetism, he created a map to represent the direction of the tide and tidal streams in the Channel. In the legend of the tidal chart, he asked for the help of mariners voyaging to distant regions of the globe to communicate any information on the tides that they amassed on their journeys. It is no coincidence that the area he surveyed became the stage for some of the most notorious sea battles in the eighteenth century. As a master of an English vessel, he was working directly for the state. The state supplied the vessels, the crews, and the equipment, and the results were valuable for the state in terms of trade and national defense. Science was being solicited by the government to help with questions of navigation and warfare.

Halley's personal command of the *Paramore* was less successful. After the near mutiny during his stewardship, no longer were natural philosophers allowed to captain ships. They became supernumeraries on oceanic voyages of discovery, visitors who rarely had control of when and where they could take measurements. Perhaps this is why Halley never again went to sea. He did, however, retain his interest in the relationship between science and navigation throughout the rest of his life. His work on the sea earned him the Savilian Chair of Geometry at Oxford in 1704 and eventually the position as second astronomer royal, a Crown appointment he received in 1720. He was then officially working for the state, a paid position with the sole purpose of investigating the science of the sea for the protection of mariners.

The Royal Academy and French Science

The political atmosphere in Restoration England formed the backdrop to the initial formation of the Royal Society of London, but similar objectives were under way in France as well. A number of groups interested in the new experimental philosophy began to meet to perform experiments, build scientific instruments, and communicate their findings around Europe. One such group revolved around Marin Mersenne, a Minorite friar who corresponded with some of the greatest minds in Europe, including Descartes, Pierre Gassendi, and Blaise Pascal. Mersenne promoted the Baconian ideals of collaboration through his extensive correspondence and his group's intensive experimentation. Both formed the backdrop to the first national scientific organization in France, the Royal Academy of Sciences in Paris.

Within a decade after Mersenne's death in 1648, plans were under way to form a national organization, not unlike the precedent set by the Accademia del Cimento. The founding of the Royal Society of London in the early 1660s added a further nationalistic incentive. Unwilling to be outdone by the English, Jean-Baptiste Colbert, France's finance minister, convinced Louis XIV, the Sun King, to found the Royal Academy of Sciences in 1666. Although modeled on the Royal Society of London and explicitly designed under the banner of Baconianism, it was quite distinct from its British counterpart, an amalgamation of the previous Accademia del Cimento and the Royal Society. Its membership was limited to a handpicked set of academicians from different disciplines of science. The initial list of fifteen members included seven geometers and three astronomers, suggesting that the organization focused heavily on questions of mathematics and astronomy in relation to navigation.

Whereas the Royal Society received little financial encouragement from the Crown, irrespective of the word "Royal" in its title, the French academy was from the beginning an extension of the state. The academicians represented a professional group of natural philosophers on the king's payroll, and they focused their science on questions that interested the Crown. Colbert often personally directed the type of research pursued through the awarding of prizes for questions posed by the academy, a tradition that continued throughout the eighteenth century. Prizes by the academy included the most efficient arrangement of masts of ocean-plying vessels in 1727, the proper shape and installation of hourglasses in 1725, the best form of an anchor in 1737, the science of the tides in 1740, and problems of magnetism in relation to navigation in 1743 and 1746. Answers from philosophers arrived from all over Europe and helped solidify the feeling that philosophers living in widely distant places were part of a joint research venture.

Direct access to the state's treasury also had its advantages. With the support of Louis XIV, the greatest of monarchs in late seventeenth-century Europe, the French academy brought to Paris some of the top minds in Europe, including Christian Huygens from Holland and Dominique Cassini from Italy. Most

Jean-Baptiste Colbert, France's finance minister, convinced King Louis XIV to found the Royal Academy of Sciences in 1666. In this detail of a painting by Henri Testelin (1616–1695), Colbert presents the members of the Academy to the Sun King, suggesting the close connection between the new Academy, geographic exploration, and the glory of the French Crown. (Réunion des Musées Nationaux/ Art Resource)

significantly, the deep pockets of Louis XIV also allowed the academy to send out expeditions of scientific discovery. In 1675, the academy sponsored an expedition to Cayenne, an island about five degrees north of the equator, where the French academician Jean Richer made exact measurements of the swinging of a pendulum. From these observations, Richer reached the now famous conclusion that the pendulum did not swing at the same rate in Cayenne as it did in Paris, observations that Newton used to support his theory of universal gravitation. Indeed, Richer's pendulum experiments added to a growing controversy concerning the exact shape of Earth, a scientific question that occupied the academy for the rest of the seventeenth century and well into the eighteenth century.

Louis XIV's strategy of bringing glory to his reign through the coordination of scientists and the funding of exploratory voyages attracted a regal following. In Russia, Peter the Great sought to modernize the country by emulating his Western counterparts. He was foremost interested in a strong military, which led him to the study of geometry, ballistics, fortifications, and the sciences associated with navigation. Clad in Western clothes, he led a group of his Russian officials on a tour of Western Europe, particularly its manufacturing districts, in an attempt to learn about the cultural and intellectual life of the West. Upon his return, he corresponded with some of Europe's most famous philosophers, including Gottfried Wilhelm Leibniz, who had earlier helped establish the Berlin Academy of Sciences. With Leibniz's help, Peter the Great imported Western science and technology into Russia and founded the National Academy of Sciences in St. Petersburg in 1724, which he staffed largely with Western natural philosophers. Like Louis XIV, he then turned to exploration to help solidify and honor his absolutist rule.

Peter the Great chose Vitus Jonassen Bering to explore the farthest northern and eastern regions of his realm, partly to determine if the eastern reaches of Siberia were connected to the American mainland. Bering was born in Denmark, but he joined the Russian navy in 1703, at the age of twenty-two, where he received renown for his accomplishments as a sea captain and explorer. With support from Peter the Great, Bering departed St. Petersburg in 1725, reaching the Kamchatka Peninsula on the eastern coast of Siberia after a two-year journey by land. After building vessels to prepare for his sea-bound voyage, Bering sailed in the summer of 1728 along the eastern coast of Siberia, an area completely unmapped and unexplored by Europeans. By August, he had passed through the strait that today bears his name. Although he never actually saw the American coast, he determined that he had sailed far enough to assure that Siberia was not connected to the Americas. He returned to St. Petersburg in 1730, only to sail yet again under the Russian flag in an attempt to more thoroughly chart the Russian, Siberian, and North American coasts. It was during this second journey that he notoriously landed on the North American continent. Bering discovered numerous islands on the Aleutian chain and was the first to map large sections of the Alaskan coast. He paid a heavy price, however, for his successes. Overworked and malnourished, he succumbed to scurvy and died in December 1741 on Bering Island, near the Kamchatka Peninsula.

Maupertuis and the Shape of Earth

Voyages of discovery sponsored by the Royal Academy of Paris in the late seventeenth century focused almost exclusively on questions of astronomy, the one area in which the academy excelled. The scientific data brought back, moreover, proved essential in answering questions of navigation. Richer's experiments in Cayenne suggested to Newton that Earth produced a weaker gravitational force at the equator, a result, he argued, of Earth's oblong shape. Newton argued that the globe was flattened at the poles, resembling the shape of an onion. Several French philosophers disagreed, citing the results of massive cartographic ventures that Louis XIV had commissioned to survey the extent of his territory. Financial difficulties, partly occasioned by wars against the English, had interrupted these surveys, which were only brought to completion after the Sun King's death. Jacques Cassini then used these cartographic measurements to show that Earth was not flattened at the poles as Newton had surmised but rather was flattened at the equator. The terrestrial globe, he averred, was shaped more like a balloon than an onion.

The debate between Newton's theoretical conclusions and Cassini's observations became both a competition among nations and a test of Newton's theory, which had yet to be confirmed empirically and was still fighting for recognition outside of England. The shape of Earth, moreover, connected the king to his academy. The academicians in the Royal Academy of Sciences, always eager to make themselves indispensable to the Crown, insisted that the question could be solved through mathematical analysis and observational astronomy. King Louis XV, in turn, wanted the shape of Earth determined under his reign. Thus, the king used the academy to glorify his rule, just as the academicians used the king's interest to legitimize their own positions as the arbiters of expert knowledge for the good of the state. To determine the shape of Earth, measurements of a degree of longitude were required, made as close to the equator and as close to the pole as possible, no small task in the age of wood and sail but well within the bounds of an Enlightenment monarch and his group of distinguished philosophers. If Newton was correct, a degree of longitude measured near the equator would be shorter than a degree measured in Paris and shorter still than similar measurements made in the Arctic.

Enhancing its own reputation at home and that of Louis XV abroad, the Academy of Sciences organized two voyages of scientific exploration, one directed south to the jungles of Peru, the other north to the frozen tundra of the Arctic. Both groups were equipped with the latest surveying and astronomical instruments. The voyage to the equator, headed by Charles-Marie de La Condamine in the summer of 1735, was the first to set sail but the last to be completed. Filled with monumental trials of all kinds, it took the La Condamine expedition more than a decade to return to Paris; by then the question of the shape of Earth had long been solved. La Condamine, therefore, focused his work almost exclusively on the natural history of South America, publishing valuable information on the animals, plants, and cultures of the Amazon region.

Shortly after the departure of the South American expedition, Pierre-Louis Moreau de Maupertuis offered to lead the northern expedition to Lapland, a rugged, Arctic region that is now part of Finland. With the full support and financial patronage of the Crown, a team of researchers—which included several mathematicians and surveyors, twenty-one soldiers, a draftsman, and an interpreter—set out to the Arctic Circle to measure the length of one degree of longitude. Determining an arc of Earth's meridian was no simple matter, requiring painstaking astronomical observations and mathematical reductions combined with a detailed and extensive trigonometrical survey through rugged terrain. Carrying a large transit telescope over treacherous mountain passes and traveling by sleighs pulled by reindeer across the frozen terrain, the scientific expedition to Lapland proved equal to the most daunting of oceanic voyages in the eighteenth century.

Within two years, their terrestrial and celestial measurements complete, the party made its way back to Paris, and Maupertuis quickly put pen to paper to publish the results of the expedition. His text, *La figure de la terre*, published in 1738, was written as an adventure narrative, full of suspense and daring acts of courage in which natural philosophers barely escaped with their lives after confronting seemingly insurmountable obstacles placed before them by nature. Yet the physical accomplishments were outweighed in Maupertuis' narrative by the equally dramatic mathematical and astronomical accomplishments in determining an arc of longitude in the northernmost borders of the world. The descriptions of trigonometrical surveying, the means of calculation, and design of instruments all added support to Maupertuis' conclusions. Newton, he argued, was correct.

La figure de la terre was widely read and discussed in the cafés and salons of Paris, going through several editions and quickly traveling throughout

Pierre-Louis Moreau de Maupertuis (1698–1759)

Pierre-Louis Moreau de Maupertuis spent his formative years in Saint-Malo, a town on the English Channel in northern France with easy access to the Atlantic. Watching merchant vessels enter and exit the port, he yearned to travel from an early age. After being privately tutored and showing signs of excellence in mathematics, he went to Paris at the age of sixteen, where before long he was moving in polite circles among intellectuals associated with the Academy of Sciences. He was elected to the academy in 1723 when he was only twenty-five. A subsequent sojourn to London in 1728 made an indelible mark on the aspiring mathematician. He returned to Paris destined to become one of the premier champions of Newtonian mechanics in France, including Newton's views on the shape of Earth. Maupertuis completed his first mathematical text based on Newtonian mechanics in 1732, endearing him to some of the most influential actors in the burgeoning French Enlightenment, including Voltaire and the Marquise du Chatelet, both of whom he tutored in Newtonian mechanics.

In 1735, the Academy of Sciences in Paris sent two expeditions to measure the shape of Earth, one south to Peru, the other north to the Arctic Circle. King Louis XV funded both ventures, patronage that linked the acquisition of scientific knowledge symbolically to the glory of France and practically to questions of navigation and commerce. Maupertuis, hoping that the voyage would establish his career, volunteered to lead the northern expedition to the Swedish territory of Lapland, now part of Finland. The scientific party acquired the best and latest equipment, including a

the Enlightenment world of letters. The controversy over the shape of Earth, however, continued for almost three years after the publication of Maupertuis' text, owing partly to objections made by Cassini concerning the reduction and interpretation of data and partly to the still incognito state of the South American voyage. Although Maupertuis stressed that his conclusions supported Newton's theory of universal gravitation, he was also careful to link the acquisition of scientific knowledge to the glory of France and to the perfection of the practical art of navigation.

Questions of navigation linked the Sun King and the Royal Academy of Sciences, and both continued to support voyages of exploration to remote parts of the world to solve specific astronomical questions, including those in the 1760s to observe the transit of Venus. Thus, not only was France the center of the European Enlightenment in the mid-eighteenth century, but it was also the point from which some of the most consequential exploratory voyages disembarked. The academy, moreover, took an early lead in conquering the largest problem still left for mariners intent on navigating the world's oceans, one that transit telescope and a zenith sector made by George Graham that they carried over mountain passes and up treacherous peaks. After successfully measuring the length of a degree of longitude in the northernmost regions of Europe, the scientific party returned safely to Paris in mid-August 1837. In his *La figure de la terre*, published the following year, Maupertuis announced his conclusions that Earth was indeed flattened, as Newton had suggested. Written in the style of sensational travel literature, the text made Maupertuis' reputation. The salons and cafés in Paris were abuzz with the news of natural philosophers risking life and limb, overcoming both physical and emotional distress, to complete their mathematical and astronomical calculations in the wild environs of the Arctic.

His Lapland expedition and acclaimed text on the figure of Earth brought Maupertuis accolades from monarchs and jealousy from other academicians. He fought endless battles with both Cassini and Voltaire while attempting to solidify his reputation as the spokesman for Newtonianism in France. Maupertuis accepted Frederick the Great's invitation to preside over the Berlin Academy of Sciences in 1745, thus initiating the next stage of his intellectual career. That same year Maupertuis published *Venus physique*, in which he attacked the accepted argument of the preformation of the embryo. This was followed by *Systeme de la nature* in 1751, a study of heredity that foreshadowed modern theories. He postulated that hereditary particles was passed down through both parents and that new species might arise through variation and geographical isolation. Maupertuis is remembered, therefore, for both his exploration of Lapland and his pathbreaking work in both the physical and natural historical sciences.

directed the entirety of observational astronomy throughout the first half of the eighteenth century: the process of finding longitude at sea.

Determining Longitude

In the late seventeenth century, mariners used a combination of the stars, the moon, and the sun, along with an intimate practical knowledge of the ocean, to determine their location far from the sight of land. They learned their art either through apprenticeship on the ocean or from private teachers. London, Paris, Amsterdam, and other major metropolitan centers were replete with mathematical practitioners who taught the rudiments of scientific navigation. To find latitude, they instructed mariners on how to use the sun by day and the stars by night. For longitude, however, mariners were forced to rely on dead reckoning, whereby a piece of wood attached to a rope set off by knots was thrown overboard. Mariners then counted the knots and the time traveled to estimate the speed of their vessel. From the speed, mariners could then determine the distance traveled once they corrected for wind direction, ocean currents, and tidal streams. This was the art of navigation, and the seventeenth and eighteenth centuries witnessed a concerted effort to turn this art into an exact science.

An interlocking set of political, economic, and scientific concerns motivated astronomers and national governments to solve the problem of finding longitude at sea. As the politics of colonialism intensified throughout the eighteenth century, European nations looked increasingly to the ocean for its geopolitical ambitions. To control the sea meant to attain access to vast commercial possibilities on its outer rim. The problem of finding one's way in the tactless ocean, far from the sight of land, however, had no easy solutions. It alluded mariners and astronomers alike and became the most harrowing quest of the eighteenth century. Almost the entire science of astronomy was linked to the cause, as were the pocketbooks of kings and the lives of whalers, merchants, and naval men. Galileo, Cassini, Huygens, Newton, Halley, and all the major natural philosophers of the late seventeenth and early eighteenth centuries focused on the problem at some point in their careers. Often, as is the case with Nevil Maskelyne, it formed their entire life quest.

The establishment of national observatories, built primarily to solve the problem of finding longitude at sea, followed closely on the heels of the founding of scientific societies. In England, King Charles II had the Royal Greenwich Observatory built specifically to help determine longitude at sea, and he appointed John Flamsteed, an avid observer and indefatigable calculator, as his first astronomer royal. Construction of the Royal Observatory began in 1675 in Greenwich, miles down the river Thames from London, more than a decade after the founding of the Royal Society. In France, however, the Paris Observatory and the Royal Academy overlapped. Even before construction of the Paris Observatory, the academicians used the king's garden to make observations. And as soon as the Paris Observatory was completed, Louis XIV brought in the best astronomers money could buy, including Cassini, to run the observatory. Cassini and his successors focused on observational astronomy and made the Paris Observatory the foremost astronomical research center in the eighteenth century.

The search for longitude led to a host of mathematical, technological, and scientific advancements. The quest spurred nations to invest large sums of money in the pursuit of observational astronomy, established an important link between the artisan community and the leading natural philosophers of the day, and produced a new emphasis on precision measurement and equally precise craftsmanship. But scientific and technological accomplishments did not form the main concern of governments. A lack of knowledge of one's longitude resulted in all sorts of problems for vessels plying Earth's blue frontier. Because oceanic travel was measured in months and vessels could carry only limited supplies, scurvy was always a problem. Although able sea captains realized that a stable diet and sufficient exercise curbed the effects, extending those voyages owing to a lack of longitude could be deadly. The determination of zero longitude also held political significance. It marked the center of the world and ultimately the place from which standard time would be measured.

The solution to the longitude problem was rather simple in theory. As Earth rotates a full 360 degrees each day, or about 15 degrees every hour, each degree of longitude translated into a change of local time of about four minutes. Mariners could determine their longitude, therefore, if they knew how much their local time differed from local time at Greenwich or Paris, where the longitude was known with relative exactness. The trick, of course, was to know both one's own local time and the local time at Greenwich or Paris *at the very same time.* The most obvious solution was to carry a clock on board ship, but to be practical, such clocks could gain or lose no more than a few seconds each day. Because even the best timepieces of the seventeenth century gained or lost about fifteen minutes each day, the method appeared highly impractical.

The solution seemed to lie in the regular motions of the celestial bodies, especially the heavenly dance among the moon, sun, and fixed stars. In 1610, Galileo witnessed moons revolving around the planet Jupiter. He studied the satellites of Jupiter intensely for several years, noting the extreme regularity by which they eclipsed the planet—regular enough, indeed, to be used as a clock at sea. He measured the different periods of each of the satellites and calculated tables of their motions. He even devised and then tested an elaborate headgear mounted with a telescope with which a mariner could observe Jupiter's satellites while on board a moving ship.

The method of using Jupiter's satellites as an astronomical clock to find longitude, though impractical at sea, proved accurate enough to use on land, and tables of the eclipses provided Cassini with his main research agenda for his first year at the observatory. In 1676, the Danish astronomer Ole Roemer visited the Paris Observatory at the urging of Colbert and Louis XIV, where he also took up the study of Jupiter's satellites. He compared his own precise observations with Cassini's newly minted tables and found slight but distinct anomalies between the two. The eclipses occurred earlier than predicted when Earth was closer to Jupiter and later when it was farther away. He realized that the difference in time could be accounted for if light had velocity. He used the incongruity between his observed eclipses and Cassini's predictions to measure, for the first time in history, the speed of light. Cassini and Roemer were searching for a solution to the problem of longitude, but they ended up uncovering one of the fundamental constants in nature.

The remarkable achievement of Cassini and Roemer did not solve the problem of longitude, and astronomers proposed all sorts of radical methods in the years to come. The one with the most promise entailed measuring the distance between the fixed stars and the moon as it traveled on its path around Earth. By observing the distance of the moon when it passed a specific star and then comparing that distance with printed timetables of when the moon passed the same star at Greenwich or Paris, mariners could compute the difference in time and thus their longitude. This became known as the lunar-distance method, and its perfection occupied astronomers throughout the eighteenth century.

If there were obvious problems with the construction of an accurate mechanical clock, there were equally pressing problems with using the moon as an astronomical clock. The lunar-distance method was simple in theory but extremely complicated in practice. The printed lists of the distance from the moon to the stars or the sun, publications known as ephemerides, had to be published several years in advance for the convenience of navigators preparing for lengthy voyages to distant regions. That is, ephemerides had to be predicted, which meant that astronomers required a precise theory of lunar motion. The moon, however, does not follow a simple circular path around Earth. Rather, it sometimes moves closer, sometimes farther away, causing its apparent position to vary considerably with relation to the starry vault, a motion that proved to be one of the most recalcitrant problems in celestial mechanics. Finding longitude by the lunar-distance method also required intensive calculations and several corrections, and it could take several astronomers working in unison upwards of four hours to compute. Moreover, even if a workable theory of the motion of the moon had existed, astronomers had yet to observe accurately the positions of the fixed stars. Thus, no accurate map of the stars and no workable theory of the moon's motion yet existed. National governments and natural philosophers alike realized that the problem could only be solved through painstaking and time-consuming research of the celestial realm.

Mapping the Moon's Motion

The perfection of the lunar theory was one of the great feats of eighteenth-century celestial mechanics, irrespective of its worth in helping navigators determine their longitude at sea. It was accomplished through institutional means, through prodigious funding by governments of several countries, and by the indefatigable efforts of those mathematicians and astronomers who devoted their lives to the quest. The starting point was Newton's *Principia*.

In his monumental text, Newton explained that an inverse square law was responsible for the motion of Earth and the planets around the sun, and most importantly for finding longitude, the same law accounted for the motion of the moon around Earth. Although his theory explained the actual cause of this motion, his solution was general in nature and failed to account for the precise position of the moon, a motion so extremely complex that after struggling with it for years, even Newton admitted that it made his head hurt.

In the late seventeenth and early eighteenth centuries, when astronomers were still attempting to digest Newton's difficult text, the inconsistencies between the actual position of the moon and those predicted according to Newton's theory cast some doubt on his conclusions. Along with the practical question of finding longitude, therefore, an equally important motivation for astronomers entailed refining Newton's lunar theory to prove that it could account for the motion of the moon. Both aims, the solution to longitude and the power of Newtonian mechanics, complemented one another.

The problem of finding longitude became so acute, and its solution so important, that in 1714 the British government intervened, passing the first of several Longitude Acts, offering a reward of £20,000—equivalent to almost $15 million today—to anyone who could solve the longitude problem. The Longitude Prize was significant for several reasons. Today, governments offer incentives for scientific and technological research, including agencies within government whose sole function is to supply grants for research and development. In England at the beginning of the eighteenth century, no such inducements existed, and the Longitude Act filled this niche. It stipulated that smaller grants could be offered for promising solutions, thereby stimulating research and focusing the best minds in the world on the problem. The act also called for the formation of a prize committee, known as the Board of Longitude, to judge the many proposals that arrived from all over Europe. The board included the top minds in England concerned with astronomical matters: the astronomer royal, the president of the Royal Society, the premier professors at Oxford and Cambridge, and the first lord of the Admiralty, just to name a few. All believed that the lunar-distance method held the key to unlocking the secret of the longitude calculation.

For the next half a century following the Longitude Act, researchers pursued the lunar-distance method without interruption. Although London had taken the lead in science in the late seventeenth century with the formation of the Royal Society, the highly successful *Philosophical Transactions*, and the construction of the Royal Greenwich Observatory, after the turn of the eighteenth century much of the painstaking work of determining the longitude through astronomical means moved across the Channel to continental Europe. Several distinguished mathematicians, including the renowned Leonard Euler, worked throughout the second quarter of the eighteenth century to find the minute inequalities affecting the moon's motion.

After attending the University of Basel, Euler had risen to become one of the premier mathematicians on the Continent, one of those responsible for the steady advance of mathematical physics. His mathematics, though difficult to penetrate, was always closely linked to practical problems of technology and navigation. He was one of the main recipients of prizes offered by the Paris Academy, receiving twelve between 1738 and 1772, and as his reputation grew, national governments throughout Europe sought his services. He was invited to teach at the St. Petersburg Academy, where he worked on problems of shipbuilding and navigation, helping Russia rise as an important sea power throughout the second half of the eighteenth century. He was also drawn, as most mathematicians were, to the longitude problem. After studying the moon's motion for almost half a century, Euler finally reduced its motion to a set of simple equations that accounted for all its inequalities. He published his first lunar theory in 1753 with funding from the St. Petersburg Academy. All that remained was to compare his theory with careful observations, correct any deficiencies, and publish predictions of the moon's future path.

Johann Tobias Mayer accomplished these feats in the mid-eighteenth century. Born near Stuttgart, Germany, Mayer took an early interest in architecture, fortifications, and cartography, working a stint in the Cartographic Bureau in Nuremberg. In 1747 he turned his energies to the lunar-distance method, making accurate and detailed measurements of the times of the moon's meridian transits. In November 1750, he accepted a position at the academy in Göttingen, where he taught applied mathematics. While there, he compared Euler's tables with his own precise observations, producing tables of the moon's position in 1753. Through additional observations of the moon and refinements to Euler's theory, Mayer corrected his tables and sent them in 1755 to the Board of Longitude in London to claim the prize.

The year after Mayer sent his tables to the Board of Longitude, the Seven Years' War (1756–1763) broke out between France and England, preventing the proper trials of his method and thus delaying their recognition. Mayer died in 1762 at the age of thirty-nine, too young to see his tables put into practice. The original Longitude Act had called for the publication of a national ephemeris that listed the positions of all major stars and positions of the moon, exactly what was needed for the lunar-distance method to succeed. The *Nautical Almanac and Astronomical Ephemeris* finally appeared in 1766, the work of Nevil Maskelyne, astronomer royal from 1765 to 1811. Maskelyne had devoted his entire career to observational astronomy, and he was responsible for printing and publishing Mayer's solar and lunar tables in 1770. He used these publications as the basis for the *Nautical Almanac*, which he published several years in advance in forty-nine editions between 1765 and 1813. The *Nautical Almanac*, the pride of the Royal Navy, was made possible through the unflagging efforts of Euler and Mayer.

By the mid-1760s, the lunar-distance method finally seemed practical enough to find longitude at sea based on, and therefore in support of, Newton's theory of universal gravitation. It provided the premier method of finding longitude in the second great age of discovery that was about to commence in Europe in the second half of the eighteenth century, when it was slowly superseded by an entirely different method based on a very different set of principles. Just as the lunar-distance method was reaching perfection, Europe witnessed one of the greatest technological feats in history.

The Mariner's Watch

While natural philosophers and national governments trudged onward in their time-consuming and expensive efforts to perfect the lunar-distance method, another process was being pursued with the same goal in mind: a method to show the time simultaneously at both the home port and at sea. As the mathematically adroit astronomers looked to the stars, the artisan community in the major maritime centers attempted to build extremely sophisticated timepieces that could be used on board a swaying ship. At the time of the founding of the major European observatories, such a technical solution seemed impossible in practice. The best clocks gained or lost as much as fifteen minutes each day, and even the most accurate did not fare well on a moving, rocking, ocean-bound vessel where, in addition to its erratic motion, changes in temperature affected lubricating oils, causing gears to expand or to contract and thus to run inconsistently.

The artisan community was not completely outside of the burgeoning scientific establishment. Newton's *Principia* was just as much theirs as it was the astronomers. Although Newton had not emphasized mechanical operations in his *Principia*, European artisans expanded Newtonian mechanics in just this direction. Itinerant lecturers such as John T. Desaguilier promoted the practical side of Newton's work, helping to bridge the gap among artisans, natural philosophers, and England's industrializing economy. England's maritime focus catapulted London to the status of the largest port in the the world, and the nautical instrument trade became a competitive and lucrative business. As the science of studying the heavens intensified in eighteenth-century England, so too did the craft knowledge needed to build ever more sophisticated instruments, from massive telescopes requiring fine brass fixtures and precisely ground lenses to barometers and thermometers, sextants and quadrants. London instrument makers carried on a trade that brought them renown even within the nascent scientific community. Several of the best were fellows of the Royal Society, including George Graham, London's finest scientific instrument maker, while others received financial and moral encouragement from the learned body. Natural philosophers could never have constructed a mechanical, clockwork universe without mechanics and clock makers.

John Harrison, the most gifted clock maker in the eighteenth century, was part of the artisan community that thrived in industrializing England. In contrast to the university-educated mathematical astronomers, Harrison had very little formal education. He was trained as a carpenter and was comfortable working with different types of woods. He also understood the motions of gears and pulleys and was well versed in the art of precision measurement. Most importantly, he was gifted with a drive that allowed him to devote his entire life to the production of a clock that could be carried across the ocean without missing a beat.

Harrison began working on a marine timepiece in the early 1720s, and his training as a carpenter paid off handsomely. Rather than using iron, which required lubrication, he experimented with different types of wood, eventually using a hardwood that exuded its own oils. For intricate pieces of the instrument that required metal, he studied how each type of metal responded to changes in temperature. He then matched different types of metal that compensated for the other's reaction, one contracting while the other expanded. Harrison thereby eliminated the problem of the effects of temperature change, producing metal parts that did not change size under varying conditions. He also dispensed with the use of a pendulum, substituting a set of seesaws operated by springs. The judicious use of trial and error, his experiments with the behaviors of different types of wood and metals, and his obvious command of technical and mechanical processes proved both meticulous and time-consuming. It took him the better part of a decade before he was ready to show his ideas to London's instrument makers.

As many as forty different trades were needed to construct Harrison's timepiece, a division of labor that one could find in the bustling capital city of London, the workshop of the world by the mid-eighteenth century. He visited the shop of George Graham, a fellow of the Royal Society, in the summer of 1730. Harrison found not only help with technical problems and liberal financial support in the form of monetary grants but also a connection to the Royal Society of London, including men such as Halley and Newton. Within five years of first meeting Graham, Harrison finished his first mariner's timepiece, known as H1. It was an amazing piece of mechanical ingenuity.

Everyone seemed satisfied with Harrison's design except Harrison himself. While others were confident that H1 deserved the Longitude Prize, Harrison was confident that he could do better. He believed that H1 was too big and clunky, so he went back to his shop and worked for another twenty years. He produced two other clocks similar in design to H1 before hitting on his finished product, a small pocket watch version, known as H4, completed in 1759. All the while he received grants from the Royal Society, mostly through the perseverance of Halley, and further assistance from the many expert artisans found in Graham's shop in London. By every horological standard, H4 was an incredible timepiece, one that required no lubrication and ticked consistently irrespective of temperature change or the erratic movement of a swaying vessel.

Thus, as history would have it, after centuries of failure, two solutions to the longitude problem arose at about the same time: the lunar-distance method and Harrison's H4 timepiece. The actual tests of the different methods went on simultaneously in the 1760s and early 1770s, filled with mischievousness and acrimonious debate of the worst kind. Maskelyne, having spent his life working on the lunar-distance method, was biased in its favor, and he struggled at every turn to promote the astronomer's method over the lowly artisan's timepiece. When the Board of Longitude met to award prizes, Euler received £300 for his elegant equations reducing the motion of the moon to a set of rules, while Mayer's widow received £3,000 for the tables based on Euler's equations. Yet Maskelyne failed to elevate the lunar-distance method above the mechanical timepiece. After years of wrangling for funds, and being resolutely ill-treated in the process, Harrison finally received £10,000 in 1765 and an additional £8,750 in 1773 for the most elegant mariner's watch the world had ever seen.

Of the many directives given to Captain Cook on his voyages of discovery, testing the lunar-distance method and the reliability of the new marine timepieces were two of the most important. Both seemed to meet the challenges of finding longitude. By the end of the eighteenth century, the lunar-distance method was still laborious in practice, requiring an immense number of calculations, though the entire process had been whittled down to about an hour. The mariner's watch, likewise, was touch and go. It was too expensive, and although it often worked to perfection, even the best timepieces still fell prey to inconsistencies. The lunar-distance method was still needed to check its validity and to reset the timepieces if errors did occur. The rise of the mariner's watch for finding longitude, however, was not far off. When Charles Darwin circumnavigated the globe in the 1830s, the vessel in which he sailed carried twenty-two chronometers.

In the end, all three techniques for finding longitude—eclipses of the satellites of Jupiter, the lunar-distance method, and the chronometer—were all pursued simultaneously, especially the latter two from the 1720s through the 1760s. More instructively, they were pursued through the support of national governments, through the construction of observatories, within the halls of scientific societies, and through the awarding of grants of research and lavish prizes. The search for longitude defined the practice of observational astronomy throughout the eighteenth century, producing an array of scientific and technological advances. Philosophers for the first time measured the velocity of light, mapped the starry vault, and accounted for the motion of the moon. A closer link was also established between the scientific and artisan communities, the defining feature of industrialization in England, leading to, among other things, an increase in precision measurement in the sciences.

Conclusion

Bacon stressed that knowledge of the physical world conferred remarkable power on aspiring nations, and he placed great weight on the importance of inventions such as the mariner's compass. Those living after the invention of Harrison's timepiece would certainly have agreed. Both the compass and the timepiece represent the mariner's increasing reliance on scientific and technological advances, allowing them to travel more confidently over the world's oceans. Providing useful knowledge for navigators was one of the premier aims of the new natural philosophy, part of the larger extension of scientific knowledge in general occurring between the sixteenth and the eighteenth centuries, one that helped define the scientific revolution and usher in the Enlightenment.

Social, political, and economic transformations set the stage for attempts by these nations and their explorers and natural philosophers to conceptualize the world's oceans in a manner fit for long-distance travel. European countries spanned the globe looking for new lands and new souls, relying increasingly on sophisticated instruments, precise measurement, and exact knowledge of the motion of the sun and moon and positions of the fixed stars. Correct theories of Earth regarding its size, shape, magnetism, and ocean currents were of immediate practical value in the field of navigation, while outlining the ocean's territory through latitude and longitude as well as with surveys and charts proved valuable for commerce and geopolitics. The rise of Europe from the early sixteenth through the mid-eighteenth centuries rested on the use, understanding, and control of the seas and oceans.

The growth of modern science, in turn, profited from the willingness of European monarchs to commit financial resources to the study of the world's blue frontier. The longitude problem formed an early and important realm of investigation for the fledgling scientific societies and the newly constructed observatories in Paris and Greenwich. Within these institutions, astronomers demonstrated the power of observational astronomy to answer questions for the good of the state and thus their own positions as the purveyors of expert knowledge. The actual voyages of exploration themselves also helped advance physical theories of Earth and its environment. They provided observations, for instance, for Newton to support his theory of universal gravitation through tide and pendulum observations while also providing the main arena for testing the new theory's validity through the shape of Earth and the perfection of lunar theory. Advances in natural philosophy, in turn, spurred further exploration of the globe, a reciprocal relationship that led to a sophisticated understanding of Earth's oceans and financial patronage for natural philosophers in Europe. From the early sixteenth century through the mid-eighteenth century, physical and intellectual exploration advanced together in mutually supporting roles.

Bibliographic Essay

The early sections of this chapter follow David Goodman and Colin A. Russell, eds., *The Rise of Scientific Europe, 1500–1800* (1991), which has its origins in an Open University course of the same name. It contains highly informative chapters on Chinese and Arabic science, Portuguese and Spanish exploration, and British and French seventeenth-century developments. All work on Chinese science must begin with Joseph Needham's exceedingly comprehensive six-volume *Science and Civilization in China* (1956–2006). Recent scholarship on China's maritime strength also includes the short introductory text by Lynda Shaffer, *Maritime Southeast Asia to 1500* (1996). For the voyages of Zheng He, see Louise Levathes, *When China Ruled the Seas: The Treasure Fleet of the Dragon Throne, 1405–1433* (1997). Gavin Menzies, in his *1421: The Year China Discovered America* (2004), argues that Zhen He sailed past the Cape of Good Hope all the way to the Americas. Dick Teresi, *Lost Discoveries: The Ancient Roots of Modern Science—from the Babylonians to the Maya* (2002), offers a positive, if polemical, portrayal of non-Western science before the European Renaissance. The literature on early European voyages of discovery is immense, including more picturesque approaches such as Eric Flaum's *Discovery: Exploration through the Centuries* (1992) to more sophisticated studies such as Anthony Pagden's *European Encounters with the New World:*

From Renaissance to Romanticism (1994) and Georges Van Den Abbeele's *Travel as Metaphor: From Montaigne to Rousseau* (1992).

The rise of science from the Renaissance through the scientific revolution was one of the first areas in the history of science to be opened to historical analysis, and the literature is now vast. Classics include H. Butterfield, *The Origins of Modern Science, 1300–1800* (1957), and A. R. Hall, *The Revolution in Science, 1500–1750* (1983). For the specific relationship between navigation and science, see R. Hooykaas, "The Rise of Modern Science: When and Why?" (1987), and D. W. Waters, "Science and the Techniques of Navigation in the Renaissance" (1967). The role of navigation as a nexus connecting mathematics to natural philosophy can be found in two influential articles by Jim Bennet, "The Challenge of Practical Mathematics" (1991) and "Projection and the Ubiquitous Virtue of Geometry in the Renaissance" (1998).

Francis Bacon's original works have been collected in J. Spedding et al., eds., *The Works of Francis Bacon* (1857–1859), with the English translation of his scientific works in volumes 4 and 5. For a listing of Bacon's publications, along with a summary of his influence, see Mary Hesse's entry in the *Dictionary of Scientific Biography*, 1:372–377 (1970). Perez Zagorin's biography, *Francis Bacon* (1999), provides valuable information on Bacon's vision of a collaborative, cooperative program for scientific research. A particularly informative source is Rose-Mary Sargent, "Bacon As an Advocate for Cooperative Scientific Research" (1996).

For the role of scientific societies and national observatories, see the dated but highly informative study by M. Ornstein, *The Role of Scientific Societies in the Seventeenth Century* (1975). Most of the work on the role of argument and the rise of the scientific journal has focused on the scientific revolution in England. See, for example, the work by Steven Shapin, particularly "The House of Experiment in Seventeenth-Century England" (1988), and the work of Peter Dear, particularly "*Totius in verba*': Rhetoric and Authority in the Early Royal Society" (1985). For the role of the Greenwich Observatory throughout the late seventeenth and eighteenth centuries, see Eric G. Forbes, *Greenwich Observatory*, Vol. 1, *Origins and Early History (1675–1835)* (1975). The best sources for Edmund Halley's voyages on the *Paramore* are Alan H. Cook, *Edmond Halley: Charting the Heavens and the Seas* (1998), and Colin A. Ronan's entry on Halley in the *Dictionary of Scientific Biography*, 6:67–72 (1972). Eugene Fairfield MacPike has edited selections of Halley's correspondence in *Correspondence and Papers of Edmond Halley* (1975). The articles referred to in this chapter include Halley's "An Historical Account of the Trade Winds, and Monsoons, Observable in the Seas between and near the Tropicks, with an Attempt to Assign the Physical Cause of the Said Winds" (1686–1692), and "The

True Theory of the Tides, Extracted from That Admired Treatise of Mr. Isaac Newton" (1696).

For French science in the seventeenth and eighteenth centuries, see Roger Hahn, *The Anatomy of a Scientific Institution: The Paris Academy of Sciences, 1666–1803* (1971), and Josef W. Konvitz, *Cartography in France, 1660–1848: Science, Engineering, and Statecraft* (1987). The definitive historical treatment of the expeditions to find the shape of Earth is John Leonard Greenberg, *The Problem of the Shape of the Earth from Newton to Clairaut* (1995). It is extremely dense and highly technical and is best reserved for graduate-level courses. The works of Maupertuis are collected in his *Oeuvres* (1756), but they have not been translated into English. An award-winning biography of Maupertuis is Mary Terrall's *The Man Who Flattened the Earth: Maupertuis and the Sciences in the Enlightenment* (2002). For Russian science and exploration in the time of Peter the Great, see Orcutt Frost, *Bering: The Russian Discovery of America* (2003).

The above texts also deal with the problem of longitude, as all texts that cover eighteenth-century science must. Dava Sobel's recent best-selling *Longitude: The True Story of a Lone Genius Who Solved the Greatest Scientific Problem of His Time* (1996) is a popular account and an enjoyable read, but as its title suggests, it is a bit ahistorical. It is based on the highly informative and richly illustrated work by William J. H. Andrewes, ed., *The Quest for Longitude* (1996). Stemming from a Longitude Symposium held at Harvard in 1993, Andrewes's volume includes twenty-three entries, four appendices, and a rich bibliography.

2

Ordering Nature in the Age of Enlightenment

While on a voyage bound for the Americas, the professor of mathematics at Cambridge, Isaac Greenwood, was struck by the comprehensiveness of the meteorological journal kept by the master of his vessel. In a "New Method for Composing a Natural History of Meteors," published in the *Philosophical Transactions* (1727), Greenwood suggested that the Royal Society of London and the Royal Academy of Sciences in Paris collect all such journals to produce a comprehensive physical account of the ocean, a project that he believed would be "of no inconsiderable Service to Navigation" and equally productive for science.[1] Greenwood's hope of linking science and navigation, of course, was well under way, having made great strides in the previous century. Oceanic exploration directed much of early natural philosophy, advances that in turn allowed mariners to travel ever farther to distant and exotic lands.

For the next half century, European powers embarked on one of the greatest periods of exploration in history, known as the second great age of discovery. From the expeditions to determine the shape of Earth in the mid-1730s to the international effort to ascertain the size of the solar system in the late 1760s, European voyages of exploration attempted to complete the mapping of the world's oceans and coastlines. Ostensibly scientific in nature, each voyage had obvious political agendas: to survey islands and harbors that could serve as strategic ports for colonization and trade. Establishing overseas bases, however, required minute studies of a new land's local resources. After astronomers and artisans perfected the lunar-distance method and the marine timepiece, the scientific emphasis shifted ever inland, to the plants, animals, and peoples of the newly discovered lands, and naturalists played increasingly visible roles as agents of the colonial process.

As naturalists stepped tentatively beyond the ocean's rim, they amassed immense stores of material from North and South America, Africa, India, and the Pacific Islands that they sent back to European centers of analysis. From the mid-sixteenth century onward, private natural history collections and Royal Society botanical gardens arose in numerous universities and European metropolitan centers. Exotic plants and animals previously unknown to Europe, along with descriptions of new peoples and cultures, overwhelmed the centuries-old exonomic systems of ancient authorities. These new discoveries sparked a classification frenzy, leading to *Systema naturae* by Carl Linnaeus and *Histoire naturelle* by Georges Louis Leclerc, comte de Buffon, in the mid-eighteenth century—publications that in turn directed subsequent data collection and analysis. Naturalists stretched ever farther afield looking for exotic species to fit within their newly ordered systems of nature. Before the creation of the modern scientist, in either name or profession, natural history emerged as a scientific discipline, an enterprise that sought to bring intelligibility to the world's extreme diversity revealed by European voyages of exploration.

Carl Linnaeus (1707–1778)

Carl Linnaeus was born in Stenbrohult, a provincial town in Sweden, and educated at the nearby city of Vaxjo. He was later tutored by Johan Rothman, who introduced him to the study of natural history and the classification system of Joseph Pitton de Tournefort. Beginning in 1727, Linnaeus studied medicine first at the University of Lund and then at the University of Uppsala in Sweden. He then traveled to the University of Harderwijk in Holland to attain a medical degree, where an influential group of naturalists convinced him to stay to practice botany. He was drawn to the botanical collection at the University of Leiden and to Hermann Boerhaave, the greatest botanist in Europe.

Upon his arrival at the University of Leiden in 1735, Linnaeus and Boerhaave gained a mutual affinity for one another. Boerhaave convinced George Clifford, the director of the Dutch East India Company, to hire Linnaeus to tend his garden. During his first year, Linnaeus published his *System of Nature*, a twelve-page document that outlined his sexual system of plant classification, still the basis of modern plant taxonomy. Written in the vernacular, it was translated into the major European languages, including French, English, Spanish, German, and Russian. He expanded his system the next year in his *Fundamenta botanica* of 1736 and used it in his *Flora lapponica*, a study of the botanical treasures of Lapland, as well as in his catalog of Clifford's garden, *Hortus ciffortianus* of 1737, a study focusing largely on tropical flora. These texts combined to offer a definitive method of naming, classifying, and exhibiting plants. Linnaeus produced his fullest treatment of his hierarchical system of taxonomy in two subsequent works, *Philosophia botanica* (1751) and *Species plantarum* (1753), the latter a record of thousands of known plants from around the world.

As naturalists studied the flora and fauna of distant lands, their published reports also described the initial encounters between European and non-European cultures. From adventure narratives such as Charles-Marie de La Condamine's *Voyage Made within the Inland Parts of South America* to Louis-Antoine de Bougainville's and Captain James Cook's vivid accounts of the Pacific, travel literature influenced how Europeans conceptualized newly discovered lands and people. The reports described a space separate from European culture, a liminal zone from which philosophers could question the foundations of their own society. A new genre of writing appeared that used the narrative of oceanic travel to critique European systems of values, exemplified, for instance, by Oliver Goldsmith's *Gulliver's Travels* and Robert Louis Stevenson's *Robinson Crusoe* in England and by Voltaire's *Candide* and Jean-Jacques Rousseau's *Emile* in France. These authors questioned whether noble virtues could still be found in savages untainted by European systems

---

After several years of intense work in Holland, Linnaeus returned to his native Sweden at the height of his career. He worked at the intersection of science and the state, serving as both a natural philosopher and a political economist. Within a year of his return, he helped found the Swedish Royal Academy of Sciences, serving as its first president. Under his leadership, the academy grew into one of the premier scientific societies in Europe, and one that is still prominent in scientific circles today in offering Nobel Prizes. Two years later, he was appointed physician to the Royal Navy and subsequently physician to the Swedish Court. His importance within government also helped him to secure positions at the University of Uppsala, first as professor of medicine, then as professor of botany. At Uppsala, he met Olaf Celsius, a renowned experimenter and astronomer, and the two worked together to enhance Uppsala's already notorious reputation in the sciences. Linnaeus tended to the botanical gardens while Celsius supervised the new observatory.

The professorship in botany at the University of Uppsala solidified Linnaeus's reputation, and he became a world-renowned lecturer and teacher. Students traveled from all over Europe to study under the great Linnaeus. Throughout his later years, he toiled away at the center of an increasingly global network of observers, collectors, and classifiers who spanned the world's trading routes in search of useful and exotic plants. His system of plant nomenclature and taxonomy was relatively quickly adopted, especially in England and Scandinavia, and he became renowned for his work in plant and animal husbandry, including plans for a pearl plantation in Lapland and a silk industry in Sweden. Not long after his death in 1778, his entire natural history collection was sold to the English naturalist James Edward Smith, who founded the Linnaean Society of London to catalog its massive holdings.

of government and religion. A new critical age blossomed, directly influenced by voyages of exploration and discovery that returned from distant and exotic lands.

The critique of European culture—its politics, economics, and systems of ethics—formed the basis of a new way to view the world known as the Enlightenment. The writers who contributed to the reform movement came to be known as philosophes, best represented by Denis Diderot and Jean d'Alembert. Between 1750 and 1780, these two indefatigable scholars edited a thirty-six-volume encyclopedia that included entries by the greatest thinkers in France, incorporating all the subjects that today we would group under the term "sciences." Following the discovery of rational laws governing the physical world, Enlightenment thinkers looked to nature to support their theories of political, economic, and cultural reform. They sought to extend their study of the natural world to help advance human society, including everything that they viewed as wrong with advanced civilization. This included the abuses of the ancient régime and the heightened power of the church. Enlightenment natural philosophers began with the natural world, but they sought to reform all of European society.

The Stimulus to Natural History

The new and powerful attitude of reform that swept through Europe and the American colonies in the eighteenth century had its roots in the exchange of curios and natural wonders in the preceding two centuries. The first great age of discovery by the Portuguese, Spanish, and Dutch opened the trade routes from Western Europe to the East and West Indies. These early maritime nations studied the plants, animals, and minerals of their new colonies overseas to exploit their commercial potential. The Spanish physician Garcia d'Orta was one of the first Europeans to describe the medicinal plants and tropical diseases brought back by Portuguese explorers in the sixteenth century. He joined several expeditions bound for India in the 1530s as the personal physician to Captain Martin Alfonso de Sousa and established a far-reaching trade in medicinal plants from all over India, Persia, and the Muslim world. In his authoritative text *Dialogues on the Simples, Drugs, and Materia Medica of India*, published in 1563, d'Orta disparaged ancient authorities for their lack of adequate descriptions of medicinal plants from outside of Europe, confident that he had received more knowledge from Portuguese explorers in one day than was known by the Romans after a hundred years. His text contained the first descriptions for Europeans of several useful species, including nutmeg, the durian tree, and mangoes. By the time of his death in 1568, the collection and distribution of medicinal plants from voyages of exploration had expanded into a profitable business.

Nicolas Monardes achieved similar fame for his collection of American flora. He was the most renowned physician in Seville, and although he never traveled from his home country of Spain, he traded in medicinal plants from all over the world, planting American plants in his personal herbarium in Seville. His principal text, *Historia medicinal* (1565–1574), introduced several new commercial and medicinal plants to Europe, including sarsaparilla, sassafras, coca, and the "holy herb" tobacco, then regarded as a potent medicinal plant. The translation into English of his text *Joyful Newes Out of the Newe Found World* (1577) exemplified the enthusiasm with which Europeans digested his new discoveries.

Monardes and d'Orta represent just two examples of a host of Portuguese and Spanish physicians who collected and exchanged medicinal plants from around the globe. With the blossoming of Renaissance culture throughout Europe, a network of collectors consisting of local physicians and provincial clergy helped establish the study of plants as a subject worthy of analysis for the European urban elite. Princes and nobles, moreover, began to support the commercial trade of valuable specimens, transforming the study of natural history from the realm of medicine to the realm of the court in the late sixteenth and seventeenth centuries. Collections linked princely culture in one area of the globe with the exotic from distant lands, demonstrating the power of the nobility to collect extravagant products from around the world. The study of nature became a way to demonstrate wealth and privileged status.

The fashionable interest in the exotic from around the globe fit squarely within the growing culture of curiosity on the rise in the seventeenth century. Large private collections, or cabinets, instilled a feeling of wonder in those who gazed upon them, demonstrating the world's diversity and God's munificence. Indeed, the inducement of wonder formed the main function of cabinets of curiosities, where radically different objects, such as an alligator from Florida or an exotic plant from Brazil, could be seen by Europeans for the first time. The sheer variety of natural objects intensified the feeling of awe, inspiring a desire to speculate on their larger purpose and place within the natural world.

Cabinets of curiosity, while demonstrating reverence, munificence, and wealth, also simultaneously offered European naturalists unprecedented collections with which to study nature firsthand. Wealthy princes helped advance natural history by supporting full-time naturalists as the curators of their cabinets and gardens. Massive botanical gardens in Amsterdam and Leiden housed plants from all over the world, and both attained renown in the late seventeenth century as centers of botanical research and exchange. Their arrangement helped undermine the confidence in ancient authority while necessitating new taxonomic schemes that could account for the seemingly chaotic diversity found in nature. The growth of local herbariums, private gardens, and state-funded botanical research stations were as important for the growth of natural history as were the traveling naturalists, physicians, and missionaries who collected the specimens.

The Jardin du Roi and Royal Academy of Sciences

As courtly and princely culture began to decline toward the end of the seventeenth century, natural history moved increasingly into the universities and newly formed scientific societies, where more formal networks of exchange crystallized. Philosophers in the early scientific societies thrived on international collaborations, exchanging seeds, plants, and methods of cultivation. Fellows of the Royal Society of London established correspondence with far-flung individuals from widely dispersed colonies overseas and visited famous collectors and collections on the European continent. The Royal Academy of Sciences in Paris also established networks of correspondence and exchange, working from the precedent set by the royal botanical garden in Paris, known as the Jardin du Roi, established in 1626 during the reign of King Louis XIII. The Jardin du Roi held plants from all over the world, offering early academicians unprecedented access to plants and resources for their acclimatization.

Joseph Pitton de Tournefort proved to be the most active plant collector of the early academicians and enjoyed a brief but distinguished career as the premier botanist in France. After studying medicine and botany at the University of Montpellier, he traveled each year on botanical expeditions, mountaineering and collecting throughout the higher elevations of the Pyrenees and Alps. After his appointment to the Jardin du Roi in 1683, he methodically expanded the garden's initial holdings. By extending his earlier botanical expeditions into Holland, England, and Spain, along with several sojourns throughout the Levant region, he established an extensive correspondence with contacts in private gardens throughout Europe. His publications, including *Elemens de botanique* published in 1694 and the account of his travels to the Levant published posthumously in 1717, advanced the study of plant taxonomy by focusing on the groupings of plants according to their genus, work that Linnaeus and other taxonomists later refined.

After his election to the Royal Academy of Sciences in 1691, Tournefort helped establish the academy's early interest in botany. With direct financial

The Jardin du Roi established in 1626 during the reign of Louis XIII held plants from all over the world, offering early academicians unprecedented access to seeds, plants, and resources for their acclimatization. (Giraudon/Art Resource)

support from the Crown, the academy supplied pensions, scientific equipment, and a laboratory for academicians to work together to catalog animals and plants along Baconian lines, publishing extravagant folio books replete with vivid illustrations. The academy also sent out voyages of discovery with orders to collect plants to enhance the holdings of the Jardin du Roi. The French academician La Condamine took such orders with him when he embarked to South America as part of the academy's obsession to determine the shape of Earth.

The Royal Academy sent two expeditions in different directions to measure Earth's shape, one south to the jungles of Peru, then a Spanish colony, and the other north to the frozen tundra of the Arctic. The South American expedition departed first, with extensive fanfare, in May 1735, with La Condamine and Pierre Bouger serving as the official astronomers. The king of Spain allowed passage through Spain's formerly closed territories, assigning two Spanish officers, Jorge Juan and Antonio de Ulloa, to join the voyage. It took the unfortunate party ten years to return to Paris.

Measuring the arc of Earth's meridian in a distant country filled with dangerous topography and uncompromising weather took its toll on the adventurous band of philosophers. Bouts of sickness, pervasive misinformation from Spanish officials, and internal rivalry tore the party apart. At first, major disagreements arose between the scientific party and colonial officials under the authority of the Spanish Crown. The local officials continually accused the traveling naturalists of espionage at every turn. But this quickly degraded into interparty rivalry. After managing to complete their work in 1743, each took separate journeys home. La Condamine resolved to travel the most harrowing route possible, down the Amazon River, to chart its course and collect ethnographical, botanical, and astronomical observations along the way. His eventual return to Paris, after many years of wandering inland through South America, and his published account, sixteen years after his departure from Paris, afforded Europeans a dramatic tale of scientific exploration in a foreign and dangerous land. It was La Condamine's seven-year odyssey tromping through the Amazon basin that fired the romantic imagination of Alexander von Humboldt and set the stage for his subsequent inland treks into the heart of the South American continent. La Condamine was heralded as a national hero upon his return, and his *Voyage Made within the Inland Parts of South America* (1747) was published, pirated, and copied and has since endured as a foundation text for scientific travel.

La Condamine's astronomical results came as confirmation rather than discovery; the shape of Earth had already been determined by the time of his belated return to France. He focused most of his narrative, therefore, on his extraordinary journey down the Amazon River, where he encountered "new plants, new animals, and new men."[2] He searched in vain for the lost golden city of El Dorado, pursued the famed Amazonian tribe of women warriors, and described all sorts of useful and exotic plants, including those that the natives took "to intoxicate themselves for the space of twenty four hours, during which time they have strange visions."[3] Although he failed to secure the precious cinchona plant, used as a medicine against malaria, his vivid description of South American flora occupied European botanists throughout the eighteenth century.

La Condamine's emphasis on the botanical treasures of the Amazon basis was eclipsed only by his equal fascination with the different tribes of American Indians that he encountered on his journey. He described them as "pusillanimous and cowardly to the 1st degree," and "incapable of foresight or reflection" owing at least partly from "the small number of their ideas, which extend no farther than their necessities."[4] His account highlights his own cultural predilections, a set of biases that provided the backdrop for similar questions posed by philosophes back in Paris concerned with man in a state of nature. His account of his travels set the stage for European ethnographic studies of native populations for more than a century. The Academy of Sciences in Paris and the Royal Society of London, moreover, used the fruits of his labors to incessantly argue that naturalists should always accompany voyages of discovery. The wealth of information brought back from these voyages necessitated new systems of nomenclature and taxonomy that could account for nature's increasing diversity.

The Rise of Natural History

Two Europeans, both born in the year 1707, set the course for the practice of natural history in the modern era. Linnaeus, a Swedish provincial botanist, and Buffon, a French naturalist, developed the methods by which observers, whether amateur collectors in Europe or seasoned explorers in new lands, could order the animal, vegetable, and mineral kingdoms under rational categories of scientific analysis. It was through the work of these two individuals, and their many followers and correspondents, that natural history evolved into a concrete subject in the second half of the eighteenth century. Although the two differed in their approach and ultimate goals, each in their own way was a product of a pervasive Newtonian worldview, responding to the chaotic upwelling of new flora and fauna brought back to Europe from voyages of discovery with an obsessive turn toward rationality and order.

Born into wealth and privilege, Buffon was educated at a Jesuit College in Dijon, where he showed early promise in mathematics. He moved to Paris in 1732 and, like Maupertuis and La Condamine, entered the scientific scene in France by expanding on the work of Isaac Newton. Buffon translated Newton's work on mechanics into French and undertook several studies in probability theory, for which he was elected a member of the Royal Academy of Sciences in 1733. In 1739, King Louis XV appointed Buffon director of the Jardin du Roi, one of the most powerful scientific positions in France.

During his long tenure at the Jardin du Roi from 1739 to 1788, Buffon added volumes of plants, transforming it into the largest repository of living plants in the world. With this collection at his disposal and with a far-flung network of correspondents stationed around the globe, he sought to formulate a rational and systematic history of Earth's natural wonders. Working over a period of half a century, Buffon produced the thirty-six-volume *Histoire naturelle*, vividly describing much that was then known about Earth's minerals, plants, and animals, including different types of humans. He argued fervently for the separation of the study of the natural world from theological and metaphysical influences, and his novel approach to the study of natural history complemented the work of Diderot and the encyclopedists.

While Buffon worked at the Jardin du Roi in Paris, the capital of one of the wealthiest and most powerful nations in Europe, his contemporary, Linnaeus, spent most of his life in the Netherlands, then a confederation of independent states. He also became associated with some of the oldest and most renowned universities in Europe, including the University of Leiden in Holland and the University of Uppsala in Sweden, both bastions of medical and botanical education. Philosophers at the University of Uppsala, following the growing trend in Italy, France, and England, organized themselves into the College of the Curios in the early eighteenth century, receiving a royal charter in 1728 to become the Royal Academy of Uppsala. It was this group of learned scholars who sponsored Linnaeus's first systematic foray into botanical research, sending him on an expedition of discovery to Lapland in the spring of 1732. Departing with the hopes of making his career, he traveled for five months through the rugged Arctic region, studying its exotic flora and fauna and the customs of its nomadic inhabitants. Lapland formed a possible area of Swedish geographical expansion, and Linnaeus linked his own interest in plants to the state's interest in the region's viable natural resources.

The requirements to attain a medical degree at the University of Uppsala were stringent, and Linnaeus opted to travel to the University of Harderwijk in Holland, where a degree could be attained in a very short time with the submission of a thesis. He brought with him a prodigious amount of scientific work, although little of it was published. With this work, a degree in hand, and experience as a field naturalist, Linnaeus endeared himself to a small group of Dutch naturalists who successfully convinced him to stay in Holland and practice botany at the University of Leiden.

Early eighteenth-century Holland was home to some of the wealthiest merchants in Europe, many of whom promoted trade in exotic plants. Several of these merchants, such as the wealthy Amsterdam banker George Clifford, director of the Dutch East India Company, kept exotic gardens replete with a rich variety of plants from all over the world. Clifford hired Linnaeus to name and classify his botanical holdings and oversee the acquisition of additional plants for his exquisite private herbarium. Although earlier systems of plant classification existed, such as Tournefort's, none was practiced systematically throughout Europe. Botanists used their own means of naming and classifying plants, often a mixture of Latin and the local dialect of the discoverer, including references to a plant's outward appearance, its internal characteristics, its place of origin, and perhaps its medicinal use.

While serving as Clifford's personal gardener, Linnaeus matured into the premier botanist in the world, publishing an extraordinary amount of material, including the earlier work that he had brought with him to Holland. His most influential work appeared in 1735, a twelve-page document titled *The System of Nature*. It introduced the world to his binomial sexual system of plant classification. Linnaeus advocated a system of nomenclature whereby naturalists could identify a plant by only two names, one for its genus, the other for its species. He based his system of classification on the sexual parts of plants. He grouped various plants into classes based on the number of their stamens, the male sexual organ, and divided each class into orders based on the number of pistils, the female sexual organ. After only a quick perusal, naturalists could place plants in their proper class and order and then could name them according to their genus and species. Both his method of nomenclature and his system of classification were simpler than previous methods, allowing the study of natural history to expand beyond the limited confines of the urban elite. It also systematized the process, simplifying the comparisons among herbaria of different universities, nations, or continents. New plants, moreover, fit easily into the hierarchical system, placing the ever growing number of new species into an ordered whole.

After three years of prodigious intellectual output in Holland, Linnaeus returned to Sweden at the height of his career. He secured positions at the University of Uppsala, first as a professor of medicine and then as a professor of botany, where students from all over Europe came to work under his guidance. He led botanical excursions into various Swedish provinces throughout the 1740s in search of plants to strengthen the university's botanical holdings, and he extended his correspondence with famous naturalists throughout Europe. He never again found the need to travel beyond Sweden. Rather, he worked to enhance the trade and cultivation of exotic plants, relying on the material coming into Sweden from voyages of exploration and trade. By the time of his death, he had transformed the botanical gardens in Uppsala into one of the largest collections in the world and a center of European botany.

Linnaeus's "Apostles" and Economic Botany

After a spate of unsuccessful military campaigns, by the mid-eighteenth century Sweden had lost most of its colonial territory, considered under the mercantile system of trade to be the means of attaining national wealth. Under Linnaeus's guidance, therefore, it turned inward to the importation and cultivation of useful plants. "Nature has arranged itself in such a way that each country produces something especially useful," Linnaeus wrote to the Swedish Academy of Sciences. "The task of economics is to collect from other places and cultivate such things that don't want to grow [in Sweden] but can grow [abroad]."[5] He experimented with different ways to get plants to grow in soils far different from their original environment, an acclimatization process that required an intense study of soil, climate, and weather conditions. The study and cultivation of colonial plants, Linnaeus reasoned, replaced the need for establishing colonial empires.

Linnaeus, however, could not do this alone. He needed plants from beyond Sweden, and he sent his students, whom he called his "apostles," on extensive explorations of the world's coastlines and landmasses. They traveled to Asia, India, South Africa, Japan, North and South America, and China; throughout the Ottoman Empire and North Africa; and to numerous islands in the Atlantic and Pacific. Linnaeus politicked for funds from all corners and found eager assistance from the Swedish Bureau of Manufacturers, the Swedish Academy of Sciences, the University of Lund, and the University of Uppsala. Some of his apostles traveled as members of private companies, such as the Swedish East India Company, while others were invited by foreign governments and scientific societies that recognized both the potential of Linnaeus's system and the commercial possibilities of acclimatizing foreign species.

From 1745 until his death in 1778, Linnaeus arranged passage for at least nineteen students, crafting special instructions for each based on their intended route. Most never made it back to Sweden alive. The life of a naturalist traveling to exotic and foreign countries was exceedingly dangerous. Those who did return were often mortally ill from their journeys. Linnaeus's first student, Christopher Tärnström, traveled widely in the Far East as an employee of the Swedish East India Company but succumbed to fever on his way back from China. Frederick Hasselquist collected plants throughout the Middle East before dying in Smyrna. Pehr Löfling sailed in the name of the Spanish Crown to South America but died in Venezuela at the age of twenty-seven. Pehr Falck slid into opium addiction and killed himself in Kazan.

A select few actually returned to Sweden in good health. Carl Peter Thunberd served as naturalist on a Dutch East India Company vessel, and after several attempts, he finally succeeded in gaining entry to collect botanical treasures in Japan, one of the first Westerners to do so. Pehr Kalm traveled widely throughout North America from 1747 until 1751, with special instructions from Linnaeus to send back red mulberry trees in a failed attempt to begin a silk industry in Sweden. Anders Sparrman, likewise, traveled to China and South Africa between 1765 and 1767 in search of useful plants that could grow in Sweden. He also served as an assistant to Johann Reinhold Forster on Cook's second voyage to the South Seas.

Linnaeus's most celebrated pupil, Daniel Carl Solander, who sailed as an assistant for Joseph Banks on Cook's first voyage, best represents the interest of nations outside of Sweden in Linnaeus's system of plant classification. Linnaeus was full of hope that the adventurous Solander would become his successor and perhaps even his son-in-law. After entering the University of Uppsala in 1750, Solander worked closely with Linnaeus on numerous projects, including the classification and indexing of several noteworthy natural history collections in

Sweden. Like Linnaeus, Solander also traveled widely throughout Swedish Lapland, gathering plants for Linnaeus and experience as a traveling naturalist.

In 1760, several prominent amateur botanists in London asked Linnaeus to send one of his pupils to England to familiarize them with his new system of plant classification and methods of cultivation. Linnaeus chose Solander. At the age of twenty-four, Solander departed to England to plant his and his master's knowledge of the sexual classification of plants on English soil. Linnaeus, in turn, hoped to attain valuable botanical specimens from the public and private herbaria in the British Isles. The Swedish government, moreover, also instructed Solander on the intricacies of technological and commercial espionage. He was to travel throughout the English manufacturing districts and, if possible, tempt artisans to leave England and work in Sweden. He thus served as Linnaeus's prized pupil, the popularizer of Linnaeus's ideas in England, and an industrial spy for the Swedish state.

As soon as Solander arrived in England in 1760, he was welcomed into the burgeoning scientific community. He attended meetings of the Royal Society of London and established a close working relationship with Joseph Banks, future president of the Royal Society and one of England's most respected natural philosophers. In 1763, Solander attained a position at the newly constructed British Museum and set to work organizing its natural history collection. He became Banks's personal assistant, and the two worked together to plan for a voyage along with Cook to the southern seas. Solander proved the most successful of Linnaeus's students in part because of the highly successful voyage of Cook.

During Cook's voyage, Banks and Solander collected more than 1,300 species of plants, assuring both a hero's welcome from the horticulturists in England and continued work for Solander in London. Linnaeus, seemingly forgotten by Solander, attempted to entice his favorite student back to Sweden by establishing paid positions for Solander in botany and even inviting him to succeed him at the University of Uppsala as professor of botany. Solander, however, increasingly distanced himself from his former mentor. His words of affection to Linnaeus diminished, and his last letter to Linnaeus was written while voyaging with Cook and Banks. Solander decided to stay in England, much to the chagrin of his former mentor.

The name Linnaeus gave to his traveling apostles was especially fitting. In the biblical account of creation, God allowed Adam to name the plants and animals of the world, conferring on him command over nature. Linnaeus viewed himself as a resurrected Adam, relating his system of naming and classifying the products of Creation to the account found in Genesis. The obligation to make use of God's creations also explains both the close connection of Linnaeus's work to political economy and his explicitly static view of nature, a system perfectly and intricately balanced by God's power and munificence. According to Linnaeus, each species was fixed by God and was unchangeable. Yet through his unprecedented success in ordering nature, Linnaeus guided naturalists to the close examination and comparison of plants, work that would eventually lend support to the highly unorthodox views of evolution, already gaining some footing in the work of Jean Baptiste Lamarck during Linnaeus's lifetime.

The Linnaean system of plant classification was quickly adopted throughout Europe, particularly in Scandinavia and England, and his program of sending students on scientific voyages of discovery helped establish the tradition in England, France, and elsewhere. After the mid-eighteenth century, vessels specifically equipped for scientific purposes spanned the globe in search of exotic and useful plants to enhance their economies back at home. This process, referred to by historians as economic botany, was one of the pivotal forces sustaining the early development of natural history as a discipline. It became a prominent feature of the second great age of discovery about to commence in Europe.

The Second Great Age of Discovery

From the mid-eighteenth century onward, European nations sent voyages to chart the largely unexplored expanse of the Pacific Ocean, often specifically outfitted for scientific reconnaissance. The voyages of Bougainville and Cook set the initial pattern; both carried naturalists on board equipped with Linnaeus's new system of classification and methods of plant acclimatization. Bougainville was the most renowned European mariner before Cook, a seasoned navy veteran who set the precedent for all future French explorations into the Pacific. Bougainville was known primarily as a mathematician before becoming a great explorer, publishing a treatise on calculus in 1752, which, among other accolades, earned his election to the Royal Society of London in 1756, the year of the outbreak of the Seven Years' War between France and England.

The Seven Years' War signaled the end to France's colonial territories in Canada and India, both allotted to England as a pillage of war. Bougainville was a member of the overthrown French forces in Quebec, and after the war he gathered other French Canadians to settle strategic islands in the South and East Pacific, partly to play havoc on English foreign trade. Overstepping his orders, he founded a small colony on the Falkland Islands in 1764. The Falkland Islands, a Spanish possession, formed a strategic set of islands from which European maritime powers replenished voyages bound to the South Pacific and Atlantic Oceans. The English Crown thus intervened, demanding that King Louis XV hand the colony back to Spain. Louis XV complied, but not before compensating Bougainville by offering him the frigate *Boudeuse*, equipped with enough provisions to sail in 1766 on the first French circumnavigation of the globe.

Bougainville's voyage took more than three years to complete and is remembered, at least in part, for the precedent it set for the systematic study of natural history on oceanic voyages of discovery. After duly returning the Falklands to Spain, Bougainville joined his supply ship, the *Étoile*, with the academician and botanist Philibert Commerson on board. Commerson began collecting plants for his own herbarium at an early age and was even caught stealing from the botanical gardens at Montpellier as a medical student. He was renowned for his hearty sojourns into the Pyrenees and Alps in search of botanical specimens, and his skills as a field naturalist led directly to his successful bid to join Bougainville on the tour around the world. Commerson wrote the natural history instructions for the voyage, a guide that subsequent French voyages of exploration used throughout the eighteenth century, including the ill-fated voyages of Jean-Francois Galaup, comte de La Pérouse, and Antoine Raymond Jesopeh Bruni d'Entrecasteaux.

After crossing the wide expanse of the Atlantic, the twin vessels passed through the Straits of Magellan heading to the Pacific. The first land they sighted was an uncharted island that would thrill all of Europe, becoming one of the most discussed places in the world by Enlightenment thinkers in the salons of Paris. They had sighted Tahiti. Although Samuel Wallis had discovered and claimed the island in the name of King George III only months earlier, Bougainville's subsequent description of his exotic welcome transformed it into an island paradise. In his *Voyage Round the World*, published upon his return, he described how, for the price of a nail, his crew could have sex with Tahitian women. The only one who did not rush to dislodge every nail holding the vessel together was Commerson's supposed male servant, Jean Baret, probably the first woman ever to circumnavigate the globe.

Commerson botanized throughout the Falkland Islands and Madagascar but published very little, one of the reasons he is not better known in the history of botany. Both Bougainville's description of native peoples and Commerson's collections of plants provided the raw material for thinkers interested in cultures and places decidedly different from Europe. The voyagers even returned with Ahutoru, the brother of a local chieftain in Tahiti, who resided in Paris upon their return. Bougainville's account, moreover, provided a scathing critique of social arrangements in France, from sexual relations and marriage to property rights and the role of religion. The first French circumnavigation of the globe documented the incredible wonders to be found in the South Pacific, and future explorers, particularly Cook, read him closely.

Bougainville's voyage was primarily economic, enhancing the geopolitical reach of France; scientific observation and natural history collecting remained secondary. The three voyages of Cook reversed this trend, at least ostensibly. His instructions were primarily scientific, with geopolitical agendas tucked conspicuously out of sight. One of the main objectives of Cook's first voyage was to ferry astronomers to the island of St. Helena to view the most significant astronomical occurrences of the eighteenth century: the transit of Venus.

The Transit of Venus

The expeditions sent to observe the transit of Venus encapsulate all that was heroic about the second great age of discovery, with its fantastic stories of ship-wrecks, war, death, privation, and, above all, perseverance. Yet beneath the drama of the expeditions themselves, one finds the usual backdrop of geopolitical motives, commercial interests, the role of scientific societies, the rudiments of international collaboration, and most of all the ceaseless desire to understand Earth and its environment. The result, moreover, was nothing short of momentous. After the observations of the transits of Venus in 1761 and 1769, natural philosophers understood the dimensions of the solar system.

The primary motive behind these voyages was scientific: to determine what even then stood as the standard measure of the universe, the distance between Earth and the sun. Its importance for astronomers cannot be overestimated. The leading maritime nations of Europe, especially England and France but also including astronomers from Prussia, Italy, Denmark, Sweden, and Russia, dispatched voyages to the middle of the South Pacific, Atlantic, and Indian Oceans; to Central America and South Africa; to Greenland and Siberia; and into the wilds of North America. Much of the European astronomical community prepared for the first transit of Venus in 1761, with more than 120 devoted astronomers fanning out to sixty-two locations around the globe in quiet anticipation of the most exalted seven hours in the history of astronomy.

The quest to measure the transit of Venus began, not surprisingly, with one of the most famous astronomers of the late seventeenth century. During his first sojourn into the Atlantic in 1677, Edmund Halley observed the transit of Mercury across the sun at the remote island of St. Helena. He reasoned that if astronomers observed the transit from widely dispersed areas of the globe, they could observe a phenomenon known as parallax, the apparent change in position of an object when measured from two different locations. The measurement of parallax, in turn, would allow astronomers to calculate the distance of Mercury from Earth through triangulation, the art of measuring distances through geometry.

By Kepler's third law, which gave the relative distances of all the planets from the sun, astronomers could then determine the distance from the sun to all the planets in the solar system, including Earth.

In practice, however, the process was not that simple. The measurements had to be extremely precise, and Mercury proved too small and traversed the sun too quickly to give accurate results. Venus, however, was closer to Earth than Mercury, producing a far larger parallax, and because it traversed the sun more slowly, taking approximately seven hours, its transit could be measured more precisely. Theoretically, only two positions on Earth were needed, but practically, the more measuring points the better in order to guard against bad weather and mistakes in measurement. Venus, Halley predicted, passed directly between Earth and the sun, eclipsing the latter, in pairs twice every 120 years. The next pair of eclipses was set to occur in 1761 and 1769, long after his own death, Halley realized, but allowing plenty of time for the future generation of astronomers to prepare. To add a flare of danger to an already daunting task, by the time the first transit took place in 1761, the two principal maritime nations involved, England and France, were immersed in the Seven Years' War. And increasingly, such wars were fought not only in Europe but also in European possessions overseas, in the waters of the southern Atlantic and Pacific, in India and South Africa, and on the North American continent.

The Frenchman Joseph Delisle took the lead in organizing the international venture, having visited Halley in London before Halley's death in 1743. Emboldened by the sheer immensity of the project, he spent decades preparing for the transit observations, gathering instruments, testing observational techniques, and making travel arrangements, all with a clear notion that nature and geopolitical rivalries could be fickle participants in scientific experiments. Cloudy weather, faulty equipment, hostile local inhabitants, tropical diseases, and geopolitical rivalries all colluded to turn the first international scientific expedition into a spectacle of adventure that was followed avidly by the public press, eclipsing even the war among nations. Astronomers were allowed free passage (letters of transit, so to speak) to their destinations, a scientific-political arrangement often overlooked by warring vessels.

The British sent several expeditions to observe the first transit of Venus of 1761. Nevil Maskelyne, the future astronomer royal, traveled to the island of St. Helena, the same island on which Halley had first observed the transit of Mercury. Unfortunately, cloudy weather on the day of the transit stifled Maskelyne's contributions to the international venture. Other British attempts, however, proved more successful. John Winthrop, a professor of natural philosophy at Harvard College in the American colonies, arranged to depart on a ship bound for Saint John's, Newfoundland, one of the few places in North America that offered good sightings. His successful observations formed the main narrative to his *Two Lectures on the Parallax and Distance of the Sun, as Deducible from the Transit of Venus*, which he published in 1769. Benjamin Franklin nominated him to the Royal Society of London a few years later.

Yet another British expedition first paired Charles Mason, an avid astronomer at the Greenwich Observatory, with Jeremiah Dixon, a surveyor and amateur astronomer, to observe the transit on the East Indian island of Sumatra. Only three days out of port, however, their vessel was attacked by a much larger French man-of-war, aptly named *Le Grande*. With eleven sailors killed and much of their scientific equipment in ruins, their vessel sailed back to England. Repairs took longer than expected, and Mason and Dixon then asked the government to allow them to change their destination. The Royal Society, however, filled with armchair philosophers intent on seeing the original course through, insisted that they depart once again to the East Indies, this time as part of a convoy that included a British man-of-war. After a three-month oceanic voyage, they arrived at the Cape of Good Hope in South Africa en route to the East Indies, only to hear that their final destination, the town of Bengkulu, was under siege by French forces. They remained in South Africa and took what turned out to be the best measurements in the Southern Hemisphere. Owing to their success, two years later the British government once again paired the two to survey the contested line between Pennsylvania and Maryland, known today as the Mason-Dixon Line.

Under the organizing leadership of Delisle, the French also had successes, near misses, outright failures, and several adventures that verge on the apocryphal. The Royal Academy of Sciences sent Alexandre-Gui Pingre to Rodrigues Island off the coast of Madagascar, where he managed to make several useful observations despite cloudy weather and the sacking of the island by British forces. Jean-Baptiste Chappe d'Auteroche traveled in the other direction, to Tobol'sk in Siberia, by horse and on foot through the rugged tundra of the Arctic. The spring floods that complicated his own travel also severely troubled the peasants of Siberia, who viewed d'Auteroche and his numerous instruments aimed at the sun with extreme apprehension. Under the protection of guards, d'Auteroche barely escaped with both his life and several good observations of the transit.

Pingre's and d'Auteroche's escapades paled in comparison to those of the French nobleman Guillaume Joseph Hyacynthe Jean-Baptiste Le Gentil de la Galaisiere. Dispatched by the Royal Academy of Sciences to Pondicherry in eastern India, he traveled into the heart of the Seven Years' War. While Le Gentil was en route, Pondicherry fell to the British, and his naval vessel could not make land. On a clear, sunny day on board a rocking ship, Le Gentil could only stare at the sun with disappointment. He had missed the transit. Undeterred and having already made his way halfway around the world, he decided to stay at Mauritius, a French colony in the Indian Ocean, and wait for eight years for the transit of 1769. He divided his time among botany, zoology, and geology, traveling to Manila, then Macao, and finally to Pondicherry, his original destination, safely in the hands of the French after the conclusion of the Seven Years' War.

After spending more than a year in Pondicherry setting up his observatory, calibrating his instruments, and otherwise waiting patiently for the transit, when the day finally came, it was cloudy and he could do nothing but stare at a grey sky. He again missed the transit. The hapless Le Gentil made his way back to France, eleven and a half years after his initial departure. Even then, however, his ill luck was to offer one last blow. Having heard a rumor that Le Gentil had died off the coast of South Africa, his family divided up his estate, and the academy filled his position. He returned to France, therefore, with no possessions, no job, and no astronomical measurements. However, the scientific account of his journeys describing his vast botanical and zoological research in India and Madagascar proved a scientific and financial success, and he died a rich and respected astronomer, one who had never witnessed the transit of Venus.

The adventures of Le Gentil perhaps best typify the trials and tribulations that astronomers encountered in their attempts to measure the transits of Venus. After years of preparation and travel, many never made a single observation. The values for the parallax measurements that were made, moreover, varied appreciably. The maritime powers of Europe thus once again planned a similarly organized venture for the transit of 1769. More than 150 astronomers dispersed throughout Europe and the world's oceans and coastlines to seventy-seven different locations. The extent of the global project, the collaboration among different nations, and the acknowledged significance of the results all combined to reduce the amount of error in the previous transit readings. Astronomers calculated the distance between Earth and the sun to be somewhere between 93 and 97 million miles, within range of the modern value of about 93 million miles. The most renowned voyage of the second transit, moreover, paired Cook, a former coal trader, with the young and adventurous botanist Banks. Both had unprecedented impact on the nature of eighteenth-century exploration and science.

The Three Voyages of Captain James Cook

Captain Cook is rightfully remembered as the epitome of the adventurous explorer, and his three voyages between 1768 and 1779 into the heart of the

James Cook (1728–1779)

Born in the Yorkshire village of Marton on 27 October 1728, Cook went to sea at the age of eighteen as an apprentice on a collier fleet in the North Sea. Owing to the notoriously dangerous wind, weather, and tidal streams on the east coast of England and Scotland, the North Sea collier fleet became renowned as the most gifted mariners in the world. There, Cook learned the principles of scientific navigation, including the practical use of mathematics and observational astronomy. When the Seven Years' War began in 1756, Cook joined the Royal Navy and was stationed in North America. Between engagements with the French, he charted the course of the lower St. Lawrence River, carried out a detailed survey of Newfoundland, and undertook precise observations of a solar eclipse. Upon his return to England, his meticulous surveying and astronomical work brought him to the notice of members of the Royal Society of London who recommended him to lead a voyage of scientific discovery into the largely uncharted waters of the Pacific Ocean.

With an assortment of astronomical instruments and material for transporting botanical and zoological specimens, the *Endeavour* departed England in 1768. Cook's orders were ostensibly scientific and geographic: to ferry astronomers to the island of Tahiti to view the transit of Venus and then to search the South Pacific for the hypothetical Southern Continent. He carried with him the naturalist Joseph Banks, future president of the Royal Society and Britain's top economic botanist. They returned to England after a circumnavigation of the globe that lasted almost three years. On his second voyage, Cook focused on the search for the Southern Continent, making three large sweeps of the South Pacific with stops at Tahiti and New Zealand. Cook had tested the feasibility of the lunar-distance method during his first voyage, and on his second voyage he took with him John Harrison's timepiece. Cook's favorable reports to the Admiralty played a significant role in the chronometer's future acceptance and eventually earned Harrison the Longitude Prize. On his third voyage, Cook surveyed the northwest Pacific, from present-day Canada to Alaska, before being killed when a dispute broke out between his crew and the inhabitants on the Hawaiian Islands.

Cook discovered numerous islands in the Pacific, surveyed the east coast of Australia and the islands of New Zealand, and searched in vain for the Northwest Passage and the famed Southern Continent. His geographical accomplishments, combined with his ethnographic studies of native populations, set the precedent for future voyages bound for the Pacific. The information he accumulated throughout his voyages made it easier for the trading nations to expand their reach, linking Europe through trade routes with the Pacific world.

Pacific, Arctic, and Antarctic regions are now legendary. His geographical accomplishments alone would have secured his place in history, yet they constitute only the most apparent of his lasting achievements. He was a meticulous observer of nature, examining everything from tides and currents of the ocean to the flora and fauna on its outer rim. He was especially fascinated with the traditions and customs of native populations, proving to be as good an ethnographer as he was a navigator.

After Cook's successful early career as both a coal trader and a military surveyor, the Royal Society of London recommended him to lead a voyage of scientific discovery to the South Pacific. He was to ferry astronomers to view the transit of Venus on the island of Tahiti, to search for any signs of a Northwest Passage between Europe and Asia, and to cross and recross the Pacific Ocean in search of the legendary Southern Continent. If he encountered new lands, he was to claim them in the name of England, especially key strategic islands that could serve as bases for further oceanic exploration. Commercial motives, however, also served alongside Cook's scientific and geopolitical instructions. In each of his voyages, he carried naturalists on board to collect and catalog commercially viable products from the island and coastlines that they surveyed.

Accompanying Cook's first voyage was Banks, in many respects a personality in direct contrast to Cook's. Banks was born into privilege and was privately tutored in many of the sciences, taking a special interest in the subject of botany. He traveled as a young man on a botanical tour of Labrador and Newfoundland, leading to his election to the Royal Society. Upon his return, he mixed freely with the social elite in London, establishing close ties to the first lord of the Admiralty and ministers to the Crown. Through these connections and backed by the Royal Society, he attained an appointment on Cook's first voyage, transforming Cook's vessel to accommodate his own scientific equipment, storage for his specimens, and room for his assistants, including Solander, the popularizer of the Linnaean system in England. According to John Ellis, one of the naturalists responsible for bringing Solander to England, no vessel ever sailed better equipped for the purposes of natural history than Cook's, a result Ellis attributed to the influential writings of Linnaeus.

Cook and Banks set sail from Plymouth, England, on 26 August 1768, accompanied by Charles Green, the astronomer in charge of observing the transit of Venus. After a grueling seven-month voyage, they arrived in Matavia Bay in Tahiti, and the crew set to work establishing an observatory at a place still named Point Venus. Green did not live to see his meticulous astronomical observations put to use; he died on the return journey. During their stay, Cook established good relations with the Tahitians, agreeing with both Wallis's and Bougainville's previous assessment of the island as a "terrestrial paradise." Cook's experiences in Tahiti began his fascination with islanders in the Pacific. He would return several more times to Tahiti, learning the rudiments of their customs and social organizations. For now, however, he had an entire ocean to explore. The *Endeavour* pushed out of Matavia Bay after a full three months'

During Captain James Cook's first voyage, Joseph Banks and Daniel Solander collected more than 1,300 species of plants, many from the east coast of Australia, depicted here. Cook's ship, the Endeavour, *can be seen in the background. Partly owing to the botanical bounty the crew amassed, the British government decided to settle New South Wales in 1786 as a penal colony. (Hulton Archive/Getty Images)*

stay with two new recruits, a Tahitian named Tupaia and his servant, Taiata. Like Green, they too died of disease before reaching England.

From Tahiti, Cook charted a course due south in search of a great southern continent, known as Terra Australis Incognita. But with pack ice as far as the eye could see, the *Endeavour* changed course, heading back north and west, and reached New Zealand in October 1769. For the next five months, Cook undertook a full survey of its coastlines while the naturalists on board surveyed the exotic flora and fauna on its shores. After charting and collecting in New Zealand, the *Endeavour* headed west in hopes of entering the Indian Ocean. Blown off course by winter squalls, the crew was instead set upon the east coast of Australia, which Cook claimed as New South Wales in the name of England. The naturalists were again overwhelmed by the sheer number of new plants, naming the area Botany Bay (just south of present-day Sydney). Partly owing to the botanical bounty that the crew amassed, the British government decided to settle New South Wales in 1786 as a penal colony.

Cook's first voyage produced a whirlwind of scientific information, including advances in geography, botany, zoology, and significant ethnological details on the inhabitants of Pacific cultures. Successful transit measurements, the charting of New Zealand and the east coast of Australia, and the botanical and zoological rarities amassed during the voyage assured Cook's heroic welcome back in England. Yet he had left one geographic mission unfulfilled. There were vast landmasses in the Northern Hemisphere, including Europe, Asia, China, Africa, and North America. Geographers contemplated a massive southern continent as a matter of balance, to keep the world from falling over. There, no doubt, would be found vast riches and unsaved souls.

Cook departed on a second voyage in July 1772, intent on ending the controversy over the existence of Terra Australis. He commanded the flagship *Resolution,* and Tobias Furneaux served as senior officer on the *Adventure* on what is often referred to as the greatest of all scientific voyages of discovery. Cook again circumnavigated the globe, making three large sweeps of the South Pacific, with intermittent stops at lush islands in the Pacific, including both Tahiti and New Zealand, to replenish his supplies and rest his crew. The *Resolution* sailed farther south than any previous European vessel, carrying the first Europeans below the Antarctic Circle. At one point, during his third sweep, Cook came within 120 miles of Antarctica. Each time he guided his vessel southward, however, thick pack ice thwarted his attempts to find land. His third retreat brought his search for Terra Australis to an end, and he returned to England after having covered more than 65,000 miles at sea.

Just as Cook had carried Banks as a naturalist on his first voyage, this time he was accompanied by the émigrés Johann Reinhold Forster and his son Georg Forster. The elder Forster arrived in England from St. Petersburg in 1766 with ample experience as a field naturalist. He drew the attention of the British government with his translation of Bougainville's *A Voyage Round the World,* which appeared just as Cook was preparing for his second voyage. Conscious of the role that Cook's first voyage had played in the careers of both Banks and Solander, Forster accepted a position as naturalist ten days before Cook's departure. Forster brought his son along as an assistant, and during the journey the *Resolution* stopped at Cape Town, where Forster was joined by Anders Sparrman, another of Linnaeus's prized pupils. Together, they returned with a copious supply of valuable botanical, zoological, and ethnological material, including thousands of plant species, many new to Europe.

Forster focused heavily on the humans he encountered on his voyages, integrating discussions of the moral sciences into the more familiar disputations on the flora, fauna, and geography of distant lands. He studied their customs, languages, and social systems, noting particularly their interrelationship with the physical and organic environments. He wrote knowledgeably on the Tahitians and Maori and was especially disparaging of the inhabitants he encountered in Tierra del Fuego. Caught in the Enlightenment fervor, he used his experiences of cross-cultural contact to comment on European cultural practices, including the status of women, material culture, private property, law, and education. Along with the travel narratives of Bougainville and Cook, Forster's published account served as fodder for the Enlightenment flame back in Paris and London.

Because Forster had signed on to the expedition as a naturalist working for the Admiralty, he had no control over the publication of his results. They belonged to the Admiralty, and Cook was given the exclusive rights to publish the official account. Forster's son Georg, however, was under no such restrictions. Immediately upon their return, he published *A Voyage Round the World*, a romantic travel narrative that, like his father's subsequent account, freely mixed a discussion of foreign plants with equally exotic native cultures. It was Georg Forster's account that had such a profound impact on the young naturalist Humboldt. In the 1790s, Forster accompanied Humboldt on a tour through Europe, and Forster's enthusiasm for natural history, combined with his romantic flair for writing about the natural world, set the stage for Humboldt's South American travels and writings. Humboldt, Charles Darwin, and others after him viewed indigenous peoples largely through the eyes of the Forsters.

Londoners, however, did not have to wait for Humboldt. Omai, a young Polynesian, accompanied the voyagers back to England and stayed for two years, the toast of the capital, under the charge of Banks. Omai's dramatic visit symbolized the ethnographic and botanical work of Cook, the Forsters, and Sparrman. Cook, moreover, became a national hero. He became a member of the Royal Society of London, the Admiralty promoted him, and he secured a well-paid sinecure at Greenwich Hospital. But Cook had lost his taste for land, and the endless formal functions did not mix well with his salty blood.

Cook sailed yet again in 1776, on his third and last voyage of discovery, this time in search of the Northwest Passage, with the added responsibility of returning Omai to Tahiti. Cook again sailed the *Resolution* into the Pacific, with extended stays in New Zealand, the Tonga Islands, and, of course, Tahiti. Instead of heading south as he had done on his previous two voyages, this time he headed north, reaching the islands of Hawaii in January 1778. From there, he led his crew on a search for the Northwest Passage. Surveying throughout the inlets and islands in present-day Alaska, Cook convinced himself that no such northern connection between Asia and Europe existed. After more than seven months in the bitter cold of the northern Arctic, he steered the *Resolution* back to the Hawaiian Islands, where the crew enjoyed the heat of the tropical sun throughout the winter. They disembarked in early February, but a broken foremast on the *Resolution* forced their quick retreat. Once back in Hawaii, a skirmish between Hawaiians and Cook's crew led to the death of Cook. Thus ended the life of the greatest navigator in history.

In the Wake of Cook

Cook's three voyages to the Pacific, Antarctica, and Arctic regions between 1768 and 1779 offered him a valuable knowledge of the flora, fauna, and peoples of the Pacific. While others were intent on bringing a civilizing mission to non-European cultures, Cook realized the necessity of pause. The negative aspects of European progress, Cook averred, were already apparent: "we debauch their Morals already too prone to vice and we interduce among them wants and perhaps diseases which they never before knew and which serves only to disturb the happy tranquillity they and their fore Fathers had injoy'd."[6] His rather nuanced approach to ethnography, especially the "happy tranquillity" supposedly reserved for those close to the state of nature, must be read in light of the ideals circulating at that time throughout Europe. As with Rousseau and other Enlightenment thinkers, Cook, when speaking of native cultures, had one eye fixed on European values.

Cook's three voyages also gave Britain an advantage in Pacific exploration and trade, which they sustained throughout the following century. Cook's voyages had covered large sections of both the Altantic and Pacific, and his published accounts contained information of importance to whalers, sealers, and the fur traders. His first voyage, moreover, carrying the botanist Banks, set the precedent for future naturalists to accompany British voyages of exploration and discovery. Once safely back in England, Banks established an extensive correspondence with naturalists throughout Europe, exchanging plants, seeds, and information about their cultivation. He also sent young naturalists on voyages of exploration, spanning out from London with orders to send back useful plants. Francis Masson traveled throughout North Africa and North America, David Nelson collected during Cook's third voyage, and Peter Good traveled in India and Australia. Like Linnaeus in Sweden, Banks established himself as the central node in an expansive network of botanical exchange.

Whereas Linnaeus had focused on similar objectives—to acclimatize exotic plants to his native Sweden—Banks attempted to study plants first in England, at the Royal Gardens in Kew, so that they could then be cultivated in English colonies overseas. Banks helped advise governments in their efforts to establish several botanical industries in the colonies, including sugarcane, coffee, tea, cocoa, and rubber, for both the good of England and the economic vitality of its colonies. He was, for instance, one of the masterminds behind the mutinous voyage of the *Bounty*, commanded by Captain William Bligh, that set sail in 1787 to transport breadfruit seedlings from the South Pacific to the West Indies in order to establish a cheap and healthy source of nutrition for its slaves.

Banks deliberately linked the advance of botanical research with the economics and politics of the state, helping to establish natural history as an

English naturalist Sir Joseph Banks (1743–1820). His voyage as naturalist on James Cook's first voyage set the precedent for future naturalists to accompany British voyages of exploration. Banks later served as president of the Royal Society of London for over forty years, establishing an extensive correspondence with naturalists throughout Europe. (Hulton Archive/Getty Images)

indispensable endeavor for the success of British imperial possessions overseas. He viewed his interests in the cultivation and distribution of plants as a way to provide for the betterment of European society. His Enlightenment views of natural history extended to his work as president of the Royal Society of London, a position he held for more than forty years until his death in 1820. Similar to the post held by Linnaeus in Sweden, Banks became the most powerful natural historian in England, and as a cultivator of plants for economic gain, he was an equally influential political economist. In effect, he was the scientific advisor to the Admiralty, one who always seemed to give equal time to the practice of science and the advance of empire.

The British sent several other expeditions to the North Pacific in the last years of the eighteenth century, including the search for the Northwest Passage by George Vancouver in the early 1790s that included a survey of more than 10,000 miles along the North American West Coast. With the ascension of Carlos III to the throne in 1759, Spain also began to encourage broader scientific exploration of their colonies in South America. Carlos III founded the Royal Botanical Garden, the Museum of Natural History, and the Astronomical Observatory in Madrid. In the 1780s, he also sponsored two major voyages of exploration and discovery, partly to index the flora and fauna of its colonial possessions, still the largest of any country in Europe. The first departed in 1785 to collect plants in New Spain, and the second and larger expedition departed in 1789, headed by Alejandro Malaspina, who explored throughout South America, Mexico, California, and up the Pacific Northwest to Alaska.

The most intensive voyages of discovery in the last quarter of the century, however, were those sponsored by the French government. In response to the successful voyages of Captain Cook in the 1770s, the French mounted their own voyages into the Pacific with an emphasis on the botanical treasures found along its islands and coastlines. Rear Admiral Jean-Francois Galaup, comte de La Pérouse, was chosen to lead one such voyage and, like Cook, combined natural history with geopolitical and commercial motives. Two frigates, the *Boussole* and *Astrolabe*, departed Brest in August 1785 better equipped with natural philosophers and scientific instructions than any previous European voyage. After two years of surveying the Pacific Ocean, La Pérouse arrived at Botany Bay in Australia in late January 1788. The crew rested for more than six weeks, set sail, and then vanished without a trace.

While La Pérouse was on his fateful voyage, the French Revolution erupted, first in Paris and then in the provinces and beyond, precipitated by the ideals of the Enlightenment—the same ideals that had swept through Britain's American colonies fifteen years earlier. La Pérouse's views likewise followed those of the philosophes. He owned busts of both Rousseau and Voltaire that

Jean-Francois de Galaup, Comte de La Pérouse (1741–1788)

Born near Albi, France, Jean-Francois de Galaup, comte de La Pérouse, entered the navy at the age of fifteen. He participated in several naval campaigns against the English throughout the late eighteenth century, distinguishing himself as both an able mariner and an acute observer of nature. In response to the successful voyages of Captain Cook, the Paris Academy of Sciences chose La Pérouse to lead a voyage of scientific discovery into the Pacific. He sailed with a full complement of natural philosophers specializing in a wide range of the sciences, from astronomy and physics to botany and geography. His staff included Andre Thouin, the chief horticulturalist at the Jardin du Roi, who took his instruction directly from Georges-Louis Leclerc, comte de Buffon.

The Academy Royal, as it had done previously in the century, played a significant role in the voyage's scientific aims. They imparted to La Pérouse more than sixteen pages of instructions specifying the exact information of interest to both the academy and the Crown. Geographically, he was to complete the voyages of Cook, focusing particularly on those regions Cook had left uncharted. Along with the actual route of the voyage and the geographical areas of special interest, La Pérouse's orders also included detailed instructions for establishing contact with aboriginal cultures.

La Pérouse departed Brest in August 1785. He voyaged through the Straits of Magellan and, instead of cruising through the South Pacific as his official instructions dictated, he headed up the South American continent, following more clandestine orders. As with Cook's final voyage, La Pérouse searched in vain for the Northwest Passage, all the while surveying and charting the inlets and islands from present-day Canada to Alaska. This was also the heart of the Northwest fur trade, and his accounts added to Cook's equally promising descriptions of the commercial possibilities in the Pacific Northwest. Receiving new orders from the French Crown relating British aims of establishing a colony in Australia, La Pérouse changed his course and headed south. He arrived at Botany Bay in late January 1788, rested his crew for six weeks, and then vanished without a trace.

The disappearance of La Pérouse, similar to the disappearance of John Franklin in the next century, led to further French exploration to determine his fate. The first voyage, captained by Antoine Raymond Jesopeh Bruni d'Entrecasteaux, departed in 1791, and two years later a second, privately funded expedition followed. They found no trace of La Pérouse. It was well into the nineteenth century before British explorers found evidence that he had been shipwrecked on the coral reefs off the islands of Santa Cruz. As was standard practice, he had forwarded his journals, notebooks, and researches back to France, and they formed a constant source of insight for both the philosophes in the cafés in Paris and the naturalists in the Jardin du Roi.

he kept above his desk as he contemplated his voyage to the Pacific. He was disparaging of despotic rule and equally skeptical of religious influences on native populations. As with other reforming philosophes of the time, however, his Enlightenment leanings did not reach as far as regicide. With the revolution in France taking on a life of its own, the king was beheaded in January 1793, inquiring to the last about the fate of his favorite explorer, La Pérouse.

Conclusion

That La Pérouse was last seen departing from Botany Bay is perhaps fitting. Botanical research formed a crucial component to many of the voyages of the eighteenth century. The study of plants began as a field subordinate to medicine but increasingly expanded into the courtly culture of seventeenth-century Europe, popularized through the creation of natural history cabinets and botanical gardens. European missionaries, traders, mariners, and naturalists encountered, described, and then attempted to order the daunting variety of exotic flora and fauna they amassed from voyages of exploration. The collections brought back from these native lands and their systematic organization under modern taxanomic schemes helped synthesize natural history into a coherent discipline. The collection, importation, cultivation, and ultimate domestication of plants allowed for new economies to develop while simultaneously forming a part of the colonial expansion of Western Europe throughout the seventeenth and eighteenth centuries.

The Pacific Ocean, virtually uncharted before the eighteenth century, became fully ingrained into the European sphere of influence. Its islands were charted, its wind and ocean currents studied, and its peoples incorporated into European consciousness. The voyages of Bougainville and Cook not only set the pattern for further exploration of the Pacific throughout the nineteenth century, but their ethnographic encounters formed the basis of how Europeans conceptualized other places and people. The philosophes were preoccupied with the corruption they saw in their own civilized society, and they continually questioned whether it was man or civilization that was corrupted. The Enlightenment's critique of the evils of society and ideas of the noble savage was represented by the discovery of Tahiti in midcentury; a paradise did exist, brought to the reading public by explorers and naturalists.

The Enlightenment notions of progress, reform, and improvement were encapsulated in the study of the world's natural diversity. As European mariners outlined the ocean's rim, the naturalists on board ships who were armed with the new liberalism of Enlightenment thought stepped ever tentatively inland, extending their scientific reconnaissance to the coastlines of foreign lands. Their emphasis on the power of science to better the plight of humans fostered a sense of dominion over the natural world. They represented this power geographically by naming the different places they visited, including Point Venus and Botany Bay, and botanically through the emphasis placed on European systems of nomenclature. Voyaging naturalists replaced odd-sounding aboriginal names for plants with Latin, the language of the European learned elite. By the mid-eighteenth century, the Royal Gardens in Paris and London served as the focal point of plant research and cultivation, demonstrating the role of national governments in linking the rise of natural history to the expansion of empire.

Yet, an understanding of the Pacific Ocean remained limited to the coast, in sharp contrast to the knowledge of the large landmasses beckoning from its outer rim. It was left largely to naturalists in the nineteenth century to explore the inland parts of the world's continents, initiated to a large extent by the Prussian mining engineer Humboldt. In his work, as with his predecessors, scientific data collection and analysis connected seamlessly with political and strategic motives to link Europe to the rest of the world.

Bibliographic Essay

Paul Farber's *Finding Order in Nature: The Naturalist Tradition from Linnaeus to E. O. Wilson* (2000) offers a wonderful introduction to the rise of natural history in the eighteenth century. Several recent edited volumes focus more specifically on the connection between the Enlightenment and science. These include Nicholas Jardine, Emma Spary, and J. A. Secord, eds., *Cultures of Natural History* (1996); William Clark, Jan Golinski, and Simon Schaffer, eds., *The Sciences in Enlightened Europe* (1999); Margarette Lincoln, ed., *Science and the Exploration of the Pacific: European Voyages to the Southern Oceans in the 18th Century* (1998); and Stephen Haycox, James K. Barnett, and Caedmon A. Liburd, eds., *Enlightenment and Exploration in the North Pacific, 1741–1805* (1997). See also the excellent studies in G. S. Rousseau and Roy Porter, eds., *The Ferment of Knowledge: Studies in the Historiography of Eighteenth Century Science* (1980), especially Roy Porter's entry "The Terraqueous Globe," which covers the changing perceptions of nature owing to European exploration in the eighteenth century. For the earlier period, see Paula Findlen, "Courting Nature," pp. 57–74, and Katie Whitaker, "The Culture of Curiosity," pp. 75–90, in *Cultures of Natural History* (1996) cited above.

David Goodman and Colin A. Russell, eds., *The Rise of Scientific Europe, 1500–1800* (1991) serves as an indispensable source for scientific developments in the sixteenth through the eighteenth centuries, particularly for the early medicinal study of plants in Portugal and Spain. For the work of Garcia d'Orta and Nicolas Monardes in particular, see C. R. Boxer, *Two Pioneers of Tropical Medicine* (1963). For seventeenth-century developments, see Alice Stroup, *A Company of Scientists: Botany, Patronage, and Community at the Seventeenth Century Parisian Royal Academy of Sciences* (1990). Anita McConnell covers the journey of La Condamine in "La Condamine's Scientific Journey Down the River Amazon, 1743–4" (1991), as does Mary Louis Pratt in *Imperial Eyes: Travel Writing and Transculturation* (1992).

Secondary material on Linnaeus is vast. A highly informative account is given in Londa Schiebinger, *Nature's Body: Gender in the Making of Modern Science* (1993). See also Lisbet Koerner, *Linnaeus: Nature and Nation* (1999) and "Linnaeus' Floral Transplants" (1994), which both focus on Linnaeus's approach to economic botany. Wilfred Blunt, *The Compleat Naturalist: A Life of Linnaeus* (1971), offers a more standard approach to Linnaeus's life and work. Frans A. Stafleu, *Linnaeus and the Linnaeans: The Spreading of Their Ideas in Systematic Botany, 1735–1789* (1971), covers the exploits of Linnaeus's apostles. Edward Duyker has focused specifically on the life and work of Daniel Solander in *Nature's Argonaut: Daniel Solander, 1733–1782* (1998).

Recent studies have also celebrated the accomplishments of the Forsters mostly through reprints of their original works. All contain informative introductory essays. They include Johann Reinhold Forster, *Observations Made during a Voyage Round the World*, edited by Nicholas Thomas, Harriet Guest, and Michael Dettelbach, with a linguistics appendix by Karl H. Rensch (1996), and Georg Forster, *A Voyage Round the World*, 2 vols., edited by Nicholas Thomas and Oliver Berghof, assisted by Jennifer Newell (2000). See also Michael E. Hoare, ed., *The Resolution Journal of Johann Reinhold Forster, 1772–1775* (1982).

For exploration in eighteenth-century France, see John Dunmore, *French Explorers in the Pacific, I: The Eighteenth Century* (1965). Roger L. Williams covers the rise of botanophilia in France throughout the eighteenth century in his *French Botany in the Enlightenment: The Ill Fated Voyages of La Pérouse and His Rescuers* (2003). Williams argues against the emphasis by historians on geopolitical ambitions in French voyages of discovery. See also the reprint editions of several primary sources, including *Voyages and Adventures of La Pérouse*, translated from the French by Julius S. Gassner (1969); John Dunmore, trans. and ed., *The Journal of Jean-Francios de Galaup de la Pérouse, 1785–7*, (1994); and Louis de Bougainville, *A Voyage Round the World* (1967).

For exploration in eighteenth-century England, see Alan Frost, *The Voyage of the Endeavour: Captain Cook and the Discovery of the Pacific* (1998), which, despite its title, covers all three of Cook's voyages of exploration and offers an exciting account of his accomplishments, including his scientific and ethnographic achievements. Alan Frost's "Science for Political Purposes: European Exploration of the Pacific Ocean, 1764–1806" (1996), focuses more specifically on Cook's geopolitical ambitions. See also G. M. Badger, ed., *Captain Cook: Navigator and Scientist* (1970), and Lynne Withey, *Voyages of Discovery: Captain Cook and the Exploration of the Pacific* (1987), along with the article by J. C. Beaglehole in the *Dictionary of Scientific Biography*, 3:396–397 (1971). Cook's journals and travel narratives are readily available in reprint. For the role of the Royal Botanical Gardens in the eighteenth and early nineteenth centuries, see the early chapters in Lucille Brockway, *Science and Colonial Expansion: The Role of the British Royal Botanical Gardens* (1979), and Richard Drayton, *Nature's Government: Science, Imperial Britain, and the 'Improvement' of the World* (2000). Patrick O'Brian, *Joseph Banks: A Life* (1987), also focuses on Banks's work as an economic botanist, as does John Gascoigne, "Joseph Banks and the Expansion of Empire" (1998). For the Spanish voyages after Cook, see Iris Wilson Engstrand, *Spanish Scientists in the New World: The Eighteenth-Century Expeditions* (1981).

The definitive work on the transits of Venus is Harry Woolf, *The Transits of Venus: A Study of Eighteenth-Century Science* (1959), a technical but thorough account of the science involved in the voyages. See also Steven J. Dick, *Sky and Ocean Joined: U.S. Naval Observatory, 1830–2000* (Cambridge: Cambridge University Press, 2003). A particularly informative and entertaining account can be found in the website run by Richard W. Pogge, Professor of Astronomy at The Ohio State University.

The original works by eighteenth-century explorers are extant in libraries and still make fascinating reading. These include Charles Marie de la Condamine, *Succinct Abridgment of a Voyage Made within the Inland Parts of South-America* (1747); Joseph Pitton de Tournefort, *A Voyage into the Levant . . . Perform'd by Command of the Late French King* (1741); Jean-Baptiste Chappe d'Auteroche, *A Voyage to California, to Observe the Transit of Venus by Mons. Chappe d'Auteroche; with an Historical Description of the Author's Route through Mexico, and the Natural History of That Province* (1778); Carl Linnaeus, *A General System of Nature, though the Three Grand Kingdoms of Animals, Vegetables, and Minerals . . .* , translated by William Turton (1802–1806); and Carl Linnaeus, *Lachesis Lapponica, or a Tour in Lapland, in Two Volumes*, edited by James Edward Smith (1811).

3

Humboldt and the Rise of the Geophysical Sciences

Advancing on the technological and scientific accomplishments of the late eighteenth century, from the perfection of the chronometer to the new systems of taxonomy, voyages of discovery in the nineteenth century extended outward to explore Earth's oceans and landmasses. Hearty explorers broke through the iced confines of the north and south polar regions, opened the "dark continent" of Africa, and braved the fearsome heat of the Australian outback. The nineteenth century was the age of imperialism, and European powers, especially Britain, traveled the oceans to establish trade routes and find raw materials for their industrializing economies. Advances in trade compelled national governments to provide correct soundings, tide tables, and sailing directions to mariners, both military and commercial. And where military or economic advantage arose, science profited. Set within the politics of imperialism and the economics of world trade, natural philosophers and governmental officials cemented an increasingly fruitful relationship.

All European nations contributed to this process, whether for militaristic, imperialistic, economic, nationalistic, or purely scientific purposes. The French, following the tradition set by Charles Marie de la Condamine, Louis de Bougainville, and Jean-Francois de Galaup, comte de La Pérouse, continued their active interest in voyages of scientific discovery, and by midcentury Germany, the United States, and several of the Scandinavian countries were actively involved in similar scientific voyages. But Britain, among all other nations, had emerged from the Napoleonic Wars as the dominant naval power, extending its reach and raising its flag over the four corners of the globe. As Britain extended its geographic operations throughout the oceans and as the Royal Navy became the protector of maritime commerce, the British Admiralty turned to science and the advice of the scientific community in a progressively more systematic manner.

The increasingly close relationship between national governments and natural philosophers was strengthened under the foment of imperialism. The imperialistic enterprise heightened the political and economic status of science as European governments repeatedly turned to science to aid in the imperial process of overseas expansion. In the second quarter of the nineteenth century, natural philosophers began to use the government's interest in the promotion of science to further scientific projects of a special nature: research in terrestrial science in which observational data was required from around the world. A variety of incentives encouraged governments to establish global networks of observations and observatories in terrestrial magnetism, meteorology, tidal observations, and the study of the ocean itself. The result was a coordinated interest in the physics of Earth viewed from a broad geographical perspective. The move from science practiced in the laboratory to science practiced on the whole of the globe was a defining feature of nineteenth-century science and exploration.

The new approach received much of its initial impetus from Alexander Humboldt, a Prussian traveling naturalist who forever changed both science and exploration. His inland trek through the heart of the South American continent at the turn of the nineteenth century set him apart from the numerous voyages of exploration of the eighteenth century and set the standard for the nineteenth-century explorer-scientists. Humboldt combined natural history, which emphasized data collection and networks of observers, with the physical sciences, which emphasized exact measurement, instrumentation, and mathematics, to advance a revolutionary approach to the study of Earth and its environment.

Two consequences for science followed Humboldt's contributions. First, science expanded from a relatively isolated, individual undertaking to large-scale research involving teams of scientists supported by national governments. Second, the organization of global geophysical initiatives, including the oceanic vessels, the men, and the equipment, all vastly exceeded the scientific budgets of individual nations. The execution and coordination of such endeavors passed beyond the arbitrary lines of nation-states, helping to internationalize the scientific process.

Scientific institutions played an intermediary role between scientific explorers and national governments. In Britain, the Royal Geographical Society was formed in 1830 from two previous organizations, the African Association and the Raleigh Club, both filled with plenty of armchair explorers. Its members were obsessed with filling in the blank spots on the world's map, transforming unknown spaces to known places, terra incognito into mapped and thus controllable outposts. The founding of the British Association for the Advancement of Science (BA) a year later fostered the participation of not only the British and foreign scientific elite but also the most humble of

British Association for the Advancement of Science (1831–present)

This biographical sketch of an institution rather than an explorer underscores the importance of scientific societies for both science and exploration. In 1830, Charles Babbage published *Reflections on the Decline of Science in England.* He was angered by the lack of professional scientists in Britain, which he blamed on the lack of direct governmental support for research. The "Decline of Science" debate raged through the halls of the Royal Society, a scientific organization that was in many respects the brunt of Babbage's onslaught. The Royal Society was increasingly run not by eminent scientists but by former navy officials and esteemed lords and dukes, persons who had become fellows more for their social standing than for their scientific accomplishments. It was in the midst of this debate, and one year after Babbage's scathing critique, that the British Association for the Advancement of Science (BA) was founded. Historians often note that the BA first met during the reform movement in British science. What is not stressed is the increasingly Humboldtian nature of science during its early years. The new association proved exceedingly valuable for such initiatives, based as they were on contributions to science from local and foreign observers located around the globe and on their insistence on the role of international cooperation.

The BA was also founded amid the extreme political and social unrest of the early 1830s and played an important role as an instrument of public order. By placing science above politics and religion, the BA became increasingly associated with social stability. Members of the BA also stressed the ability of science to become an agent of national prosperity and international harmony. Furthering the national interests of Britain as well as the particular interests of the British scientific elite, the BA quickly became a powerful lobbyist in the halls of government.

With a direct link to government, the BA played an important role in the patronage of science. The allocation of money went first to projects that were of benefit to the British Admiralty and were commercially significant in one way or another. The BA lobbied the government to fund many of the great seaborne voyages of the nineteenth century, including James Clark Ross's Antarctic voyage. It also supported and organized many of the global, geophysical initiatives in mid-century. Committees within the BA drafted instructions for voyages, printed off sheets to be used by naturalists and sea captains to record scientific data, and paid to have the results reduced once received from far-flung areas of the globe. The BA still exists as one of the most important scientific organizations in science.

researchers dispersed throughout the British Isles and its possessions. The BA was a peripatetic scientific organization, traveling to all the major cities in the British Isles and eventually to Australia and Canada. Its widely geographic leanings matched the type of science it fostered, where researchers from all classes of society and all areas of the globe could accumulate data and add to global geophysical research. The Royal Society of London also continued its role behind the scenes. The government requested its advice on topics ranging from copper sheathing for vessels to the cure for scurvy. Most importantly, the Royal Geographical Society, the BA, and the Royal Society provided the instructions, the methods, and often the scientific instruments for the voyages of discovery.

The focus on institutions such as the British Admiralty or the BA should not cloud the fact that to the Victorians themselves, the nineteenth century was a time of not only great explorations but also of great explorers. Based more perhaps on myth than reality, European culture became fascinated by the great nineteenth-century explorer who persevered through unrelenting hardship, facing danger at every swath through the Amazon or every iceberg in the northern Arctic. Humboldt's marriage of the undaunted explorer with the professionalizing scientist ushered in an exciting new era of the scientific voyager.

Eighteenth-century explorers often served as natural philosophers; part of their official orders included taking measurements and observing natural phenomena of interest to seafarers. In the nineteenth century, this observing impulse was expanded and systemized as naturalists became permanent members on voyages of exploration, accompanying vessels to the farthest reaches of Earth. Humboldt represented the intense focus on accurate descriptions and precise measurement of these new scientist-explorers. Captain James Cook traversed the South Seas, and Meriwether Lewis and William Clark opened up the American West, but it was Humboldt who set the tenor of science and exploration research and writing for the entire nineteenth century.

Alexander von Humboldt (1769–1859)

Two men born in 1769 altered the course of Western history. Napoleon Bonaparte became famous for parading his army and Enlightenment ideas throughout the continent of Europe. Alexander von Humboldt became equally famous to European intellectuals for taking his instruments and romantic ideas throughout the jungles of South America. Napoleon's run ended under house arrest, alone and broken on a small island in the middle of the Pacific. Humboldt's life ended in the middle of a worldwide network of scientific observers, a network that still forms a major part of modern geophysics.

Humboldt was born near Berlin, then the capital of Prussia, into Prussian high society during the middle of the second great age of discovery. He was privately tutored, first under J. H. Campe, a noted geographer and translator of *Robinson Crusoe*, and then under Karl Ludwig Willdonow, one of the greatest botanists

Alexander von Humboldt (1769–1859)

Polymath, world traveler, and influential statesman, Alexander Humboldt set the tenor of scientific travel in the nineteenth century. Born near Berlin, then the capital of Prussia, into high society, his father served in the Prussian court under Frederick II, an Enlightenment absolutist who corresponded with the philosophes of Europe and took an interest in the sciences. Humboldt was educated first by private tutors and then at the University of Göttingen. After a year of traveling throughout England and France with Georg Forster, the son of the naturalist on Cook's second voyage, Humboldt entered the Freiberg School of Mines before taking a position in the Prussian Department of Mines. His real passion, however, was to travel widely.

As a capable natural philosopher, a respected mining inspector, and a learned diplomat, Humboldt persuaded Charles IV of Spain to allow him to tour Spain's American colonies. Humboldt and Aimé Bonpland spent five years, from 1799 to 1804, traveling through Venezuela, Colombia, Ecuador, Peru, Cuba, and Mexico. Whether on the side of a mountain or in the heart of the Amazon jungle, Humboldt took scientific measurements of all types. He collected massive amounts of dried plants and accumulated information on plant geography, zoology, geology, mineralogy, and climatology as well as a host of writings on the political economy, archaeology, and ethnography of South America. He returned to Europe and published his findings over the next thirty years, including his widely read *Personal Narrative.*

After more than a quarter of a century living and working in Paris, Humboldt traveled to Siberia and the Urals, then served as a chamberlain in the royal court in Prussia. During his later years, he published *Cosmos,* a book that attempted to encompass the entire scope of human knowledge of the natural world. As with his other writings, he aimed not only at a description of nature but also at how to enjoy the practice of science. He died in 1859 at the age of ninety after living a long life that had transformed the process of scientific exploration.

in the Germanic states. Both tutors introduced the young and impressionable Humboldt to libraries overflowing with books filled with pictures and descriptions of exotic flora and fauna from far-off lands. He read with inspiration Bougainville's *Voyage Round the World* and other scientific travel narratives that stressed the excitement of exploration and the personal satisfaction of scientific discovery. Books about exotic plants excited the young Prussian's imagination, and he could hardly wait to travel to see them with his own eyes. But above all, it was Cook's three voyages that inspired Humboldt's sense of adventure. As he matured, so did his fascination with oceanic travel and the study of the natural sciences.

For those interested in scientific pursuits, the University of Göttingen was the place to be, the center of scientific scholarship in the Germanic states. Humboldt's own studies were wide-ranging, foreshadowing his own all-encompassing view of knowledge. During his year at Göttingen, he met Georg Forster, the son of the famous naturalist on Cook's second voyage. The two became close friends and traveled together in 1790, first to England and then to revolutionary France. In England, Humboldt met Sir Joseph Banks, a figure Humboldt had known through his reading as a companion of Cook and who by then served as president of the Royal Society.

Upon their return, Humboldt entered the Freiberg School of Mines, headed by one of the most esteemed geologists of his day, Abraham Werner. Humboldt learned the many skills he would use with such dexterity in his extended voyages, including mineralogy, geology, and chemistry. He also received practical training as a mining inspector, and the next year he accepted a position in the Prussian Department of Mines, where he put his scientific training to use. He thus became educated not only in the sciences but also in the use of practical and scientific instruments. It was in these years that he also befriended Johann Goethe and corresponded with Friedrich Schiller, one of the leaders and the epitome of German romanticism. Their seminal influence on Humboldt was extensive. Throughout his travels and scientific writings, he focused equally on the scientific observations that he made and the relationship of those observations to the totality of nature. This sensitivity always complemented his outright empirical investigations. His all-encompassing view of nature, however, could not be accomplished from studying mines in Europe. He decided that travel was the only path to comprehensive knowledge.

Humboldt's plans to explore the world met with disappointment at every turn. A trip to the West Indies fell through, as did his planned exploration of the Nile, this time owing to Napoleon's invasion of Egypt. Humboldt, however, was both a capable naturalist and a learned diplomat, and he managed to persuade Charles IV of Spain to give him open run in Spain's American colonies along with important assistance from Spanish and Creole officials. Humboldt's political acumen, combined, no doubt, with his experience in mining, endeared him to the king. Charles IV was eager to use Humboldt's scientific talents for his own colonial ambitions, thus linking Humboldt's scientific quest to European expansionism then taking hold in the Western world. Humboldt was to report back on everything he saw, especially concerning the areas' extractable natural resources. The significance of his South American journey for science cannot be overstated. The Spanish colonies of America had been all but closed to explorers for more than half a century, since the celebrated expedition of La Condamine. The continent of South America was relatively unexplored, and thousands of plant and animal species beckoned to be classified. Humboldt would combine all of his scientific and technical training to explore South America.

Humboldt was well prepared. He had spent his youth contemplating just such a voyage, amassing expertise in a wide range of scientific knowledge and equipment. More importantly still, he also traveled with a new and all-encompassing methodology of how to study the totality of nature. Describing his proposed exploration of South America, Humboldt wrote: "I shall collect plants and fossils and make astronomic observations. But that's not the main purpose of my expedition—I shall try to find out how the forces of nature interact upon one another and how the geographic environment influences plant and animal life. In other words, I must find out about the unity of nature."[1] At the age of twenty-nine, Humboldt was finally set to make a journey of scientific discovery. In the process, he created a new type of natural philosopher: the scientific explorer. He did not just describe South America; he invented an image of the continent and of the scientific explorer for all those in Europe to read and relish.

Humboldt and his traveling partner, Aimé Bonpland, arrived in South America in 1799. They traversed the jungles and highlands throughout Venezuela, Colombia, Ecuador, Peru, Cuba, and Mexico to measure, collect, survey, and describe. They spent more than a year traveling the waters of the Orinoco, a river more than 1,500 miles long that runs through the heart of the Venezuelan jungle. After the rivers and plains of Venezuela, they headed for the highest mountains of the Andes, where they planned to climb one of the largest peaks in South America. They crossed the Cordillera by land to Quito, Ecuador, where Humboldt attempted to scale Chimborazo, Ecuador's highest peak and long considered the highest mountain in the world. A thousand feet from the summit, Humboldt was forced back by altitude sickness, though he would hold the world altitude record for more than thirty years for this climb. It is a wonder that he was able to climb so close to the summit with only leather boots and cotton garments and even more miraculous that he was able to take scientific measurements throughout the trek, including barometric readings at different heights and air and soil samples for chemical analysis.

On their return journey to Europe, Humboldt and Bonpland traveled to Philadelphia to meet with Thomas Jefferson before returning to Paris in August 1804 to Europe's delight. Humboldt had become a celebrity, a cosmopolitan man of letters and the most famous scientist in Europe. During his travels, he had collected massive amounts of dried plants and specimens of all kinds, introduced hundreds of new species for European taxonomists, and accumulated information on plant geography, zoology, geology, mineralogy, and climatology as well as a host of writings on the political economy, archaeology, and ethnography of South America and its inhabitants. Except for short travels to Russia and Central Asia in his later years, Humboldt spent the rest of his life publishing the

A painting by Friedrich Weitsch of the renowned naturalist and scientific traveler, Alexander von Humboldt. Humboldt's inland trek through the heart of the South American continent at the turn of the nineteenth century set the standard for nineteenth-century explorer-scientists. (Bettmann/Corbis)

results of his explorations. He published on a myriad of subjects in what today would be called political economy, the social sciences, and science. The wealthy baron spent his entire fortune on its production and publication, and though scientifically a success, financially it left him broke. The cost of his completed works was so high that even Humboldt could not afford a copy.

Humboldt quite consciously attempted to transform scientific travel writing and the image of the scientific traveler. Surprisingly, his volumes are not filled with personal drama and struggle, though he and Bonpland certainly went through horrific ordeals. He caught malaria and recovered, only to live for weeks on muddy river water, animals that they had shot, and ants. He returned to Europe with a paralyzed right hand from rheumatism acquired during the voyage and was even presumed dead before his return, or so it was reported in the popular press. But his was not to be a heroic travel narrative. Rather, in all his writings, whether on the geographical distribution of plants or his own exploration narrative, he always balanced a sometimes overwhelming empiricism with the harmonies and occult forces underlying the aesthetic beauty of nature.

Humboldt owed his overwhelmingly holistic view of nature both to the German romantics, such as Goethe, and to his own study of the interconnectedness of the natural world. Nature in all its sublimity was for Humboldt a delicate balance of forces that interacted with each other. Careful quantitative measurements formed the basis of Humboldt's science and also served as a foundation for his emotional appreciation of nature. This holistic approach to uncover the interconnectedness of the forces of nature met with great success, and his scientific accomplishments are as diverse as they are important. He incorporated South America into European geography, and he began the modern treatment of physical geography, using the new science's methods to determine the structure of the Andes and the distribution of its volcanoes. He is also the founder of systematic meteorology through which he sought to determine the causes of large-scale phenomena, such as the origins of trade winds and magnetic storms. His study of the geographical distribution of plants and animals and their relation to altitude, temperature, and magnetic and electrical phenomena formed the basis of the new discipline of biogeography.

More important than specific findings or the introduction of new fields of research was the manner in which Humboldt attempted to study those fields. His scientific methodology emphasized the careful and continued measurement of interconnected phenomenon. In the distribution of plants, for instance, Humboldt noted not only where plants thrived but also their relation to sunlight, latitude and longitude, height above the sea, the chemical content of the soil, and the moisture, electricity, and magnetic properties of the air. For this, sophisticated and numerous instruments were needed, and he carried more than fifty different instruments on his travels, often accompanied by a host of servants and horses to carry his gear. His emphasis on massive amounts of data collection and relations between forces led in turn to an advance in the manner in which he represented that data. Humboldt, more than any other researcher, popularized the graphical method of data analysis and demonstrated its essential character in reproducing, in pictorial form, the interconnectedness of nature's forces.

Humboldt's inland trek through the heart of the South American continent set the standard for inland exploration of Earth's landmasses. While much of the South American coastline had been sounded and mapped, few scientific explorers had traveled inland where, Humboldt argued, his methods would bear the most fruit. Thus, for Humboldt, along with the measurement of interconnected phenomenon was the extension of that process to the gathering of large amounts of observational data over spatially distributed areas. "Amidst the apparent disorder which seems to result from the influence of a multitude of local causes," Humboldt noted, "the unchanging laws of nature become evident as soon as one surveys an extensive territory."[2] An emphasis on data collection and mapping over large geographical areas made an organized network of observers or corresponding physical observatories a necessity. Once combined with the practice of imperialism, this geographical distribution of correlated measurements became the most prolific and widespread facet of nineteenth-century science and exploration.

Humboldt and British Imperialism

Humboldt opened South America for those back in Europe. Missionaries, traders, soldiers, and scientists all clamored to the New World in increasingly large numbers. Not surprisingly, many of these travelers were British, looking for commercial opportunities, souls to save, or undiscovered flora and fauna. Humboldt's holistic view of nature, however, was a bit broad for most Europeans and too romantic for the British. Scientific travelers used Humboldt's methodological framework to focus on certain branches of science and to find specific quantitative laws in increasingly specialized disciplines. Scientists followed Humboldt's approach in the physical sciences to its reductionist conclusion, stripping it clean of what increasingly appeared to be its holistic appendages.

Humboldt had a profound effect on all of Europe and the New World, but it was in Britain that Humboldt's methodology fell on especially fruitful soil. Britain depended on the ability to maneuver through the oceans and dominate the coasts and channels around the world for its security and prosperity. At the end of the Napoleonic Wars, Britain exercised a larger influence in maritime operations than any previous power in history. It controlled the oceans and, increasingly, much of the world's colonial possessions on land. The British used their military might to ensure an open system of trade and commerce, both limiting in numerical terms and extending in geographical terms their naval operations to a less provocative policing force stationed throughout the world. This policy resulted in almost one hundred years of European peace, now known as Pax Britannica. And Pax Britannica proved extremely fruitful for science.

Open and safe travel throughout the world's oceans complemented Humboldt's scientific methodology. His science was geographic science; the greater the number of places on Earth's surface in which data was collected, the better the chance of finding order in nature. With British missionaries, sailors, seamen, and scientists disbursed around the globe, a coordinated network of observers or observatories could finally answer Humboldt's call. The Royal Navy and mercantile marine spanned the needed geographic territory, equipped with the men and the financial resources to make large-scale geophysical initiatives possible.

The expansion of empire and Britain's self-defined position as liberal leader within Europe allowed scientists, for the first time in history, to answer questions in the physical sciences that demanded observations on a global scale. Often referred to as Humboldtian science, this type of science relied upon simultaneous and interconnected observations over dispersed areas of Earth's surface to produce mathematical laws that explained the interrelationship of physical forces. The research was made possible by the increasingly strong relationship between the Admiralty and scientists in the second quarter of the nineteenth century.

Although the Admiralty had commissioned many surveys prior to the end of the eighteenth century, the captain's only obligation was to deposit a report to the Admiralty upon completion of the survey. The Admiralty was unwilling to pay for the publication of the results of the survey or the soundings, tidal observations, and sailing directions. It also declined the responsibility of distributing these publications to the vessels of the Royal Navy. The surveyor was expected to publish the results of his efforts privately, at his own expense and for his own profit. More often than not, the results of the surveys languished, unpublished, in the dark denizens of government offices. As maritime trade expanded, however, and as more and more ships were lost in uncharted waters, the British government established the Royal Hydrographic Office in 1795.

The Hydrographic Office became a clearinghouse for incoming information from expeditions, including charts, maps, and soundings, and by the 1830s it had become the research and development wing of the Admiralty. This was partly due to the dissolution of the Board of Longitude in 1828, until then the acting research and development department within the Admiralty. But it was also due to the ascendancy of Francis Beaufort as the fourth hydrographer to the Admiralty. Beaufort became hydrographer in May 1829 and held this position for twenty-five years.

During Beaufort's tenure, the tonnage carried by the British merchant fleet more than doubled, and Beaufort worked relentlessly in charting and surveying the world's oceans. In the process, he was the link between the scientific elite and the British government. He was responsible for acquiring funds for almost all of the British voyages of discovery of the mid-nineteenth century. He drafted instructions specific to each voyage and provided each ship with scientific instructions, and often the scientists themselves. Beaufort also worked with scientific societies, including the Royal Society, the Royal Astronomical Society, and especially the newly formed BA.

With the ascendancy of Beaufort to the position of hydrographer and the founding of the BA, the connections between government and the British scientific elite began to formalize. The creation of a scientific branch within the Admiralty in 1831 further solidified this process. The scientific branch was composed of the Hydrographic Office, the two astronomical observatories at Greenwich and the Cape of Good Hope, the Nautical Almanac Office, and the Chronometer Office. Largely owing to Beaufort's own scientific interests as well as his position within the scientific and maritime communities, the Admiralty placed him in charge of the budget for the entire scientific branch. His close relationship with men of science led to increased funding in the new, empirically based projects in the geophysical sciences.

Humboldtian Initiatives

A variety of global geophysical initiatives were begun in the early Victorian era, and all expertly demonstrate the increasingly productive relationship between the British Admiralty and the scientific elite in Britain. Research on tides, terrestrial magnetism, and meteorology illuminate the type of science practices within Britain's global imperialist network. All possess similar traits. First, eminent scientists themselves usually initiated the projects, which were then furthered by government. Second, all demonstrate the complex negotiation between nationalism and internationalism. Scientists often used nationalism as an argument to instigate international cooperation, while the resulting collaboration in turn advanced the sciences that were important for nationalistic, often militaristic ends. Third, all demonstrate Humboldt's own interest in the interconnectedness of natural forces and the need to take measurements from a large network of observers placed geographically around the globe. The initiatives became massive, global research projects that utilized the resources of a multitude of European maritime nations and their possessions.

Tides

The military rivalry between England and France shifted after the end of the Napoleonic Wars to include a strong element of scientific competition, and for the British scientific elite, study of the tides was a subject full of national prestige. The Institute of France had taken a commanding lead in the study of terrestrial magnetism, and the initial efforts in that subject in Britain would not prove fruitful until the late 1830s. The study of the tides, however, represented a field in which the British could take a commanding lead. For the British Admiralty and scientific elite, therefore, the reward would not only be practical and profitable but also political and prestigious.

Isaac Newton had discovered the general law governing the tides, using tidal data from voyages of discovery as supporting evidence for his theory of universal gravitation. Newton's theory, however, had no practical consequences; one could not construct accurate tide tables from his general laws. As such, it was a theory barren of consequence that gave only a general guidance as to when the tides actually occurred on the coasts of Europe. A predictive mathematical law for the tides was still wanting.

Two options were open to scientists interested in creating correct tide tables for the major ports in Europe. Long-term observations could be made at every port to determine the correct mathematical variables for Newton's tidal theory. This was the path initially used by John William Lubbock and William Whewell, who resurrected the theoretical study of the tides in England in the 1830s from almost a century of relative neglect. Through this type of analysis, Lubbock and Whewell produced tide tables for London, Liverpool, and other British ports where long-term observations had been taken. The process was time-consuming, however, as nineteen years of observations were needed for accurate tide tables. Thus, Whewell adopted a new approach that entailed constructing a theory of the tides based on their progression as they moved across the oceans. If the course of the tides was known and if accurate tidal tables existed for one port, Whewell reasoned, a researcher could then extrapolate from one port to the next and eventually to all ports in Europe and beyond. To attain such a theory of the progression of the oceanic tides, short-term observations taken at a large number of places along the coast were needed rather than years of observations at each and every port.

A tidal chart of the world, ca. 1850, drawn and engraved by John Emslie and published as part of a set of forty-four educational charts by James Reynolds. Emslie used Whewell's co-tidal lines to demonstrate the progress of the tide through the ocean and onto the shores of the major ports and estuaries of the world. (Science Museum Pictorial)

But there was a significant obstacle: these observations had to be simultaneous, which meant that they had to be organized above and beyond what a single researcher could do.

Such a project proved a perfect subject for the young BA to foster. First, through the sales of tickets to its meetings, the BA had quickly become wealthy enough to cultivate such initiatives through grants of research. The first BA grant for research went to Whewell and Lubbock for their tidal studies. Second, the tides were accessible to even the most humble of researchers, and their study could be advanced by the participation of local and foreign observers living near the sea. Whewell solicited such help and received tidal data from all around Great Britain, Ireland, and France that he then tabulated and attempted to mold into a mathematical law to show how the tides traveled around the coast of Great Britain.

The project also proved a perfect subject for the British Admiralty. In practical terms, the Admiralty was the institution that would most profit from a correct theory of the tides. Whewell attained the participation of Beaufort and the scientific servicemen at the Hydrographic Office as well as Beaufort's connections with other hydrographers in other nations. Once Beaufort became involved, the project changed in scope. Beaufort suggested to Whewell that this type of short-term but simultaneous tidal analysis should be extended beyond the British Isles to include the entire globe. Beaufort wrote to all his surveyors and to captains of vessels on voyages of exploration to have tidal measurements recorded and sent back to the Hydrographic Office. He also wrote to all the foreign hydrographers to have them contact their captains, masters of ships, and naturalists sailing in foreign waters.

In June 1835, nine countries and many of their surveyors and explorers took measurements of the tides every fifteen minutes, day and night, for two weeks. More than 650 tide stations contributed data. For the first time in history, observations of the simultaneous tides were tabulated, graphed, and mapped. Observers reported the state of the weather at all of the stations, along with the barometric pressure and force of the wind. Whewell produced an isotidal map based on Humboldt's method of data analysis that demonstrated how the tides progressed through the Atlantic and Pacific and onto the shores of the major maritime nations and their possessions. Whewell and Beaufort utilized the resources of British and other nations' imperial possessions to create tide tables for the major ports in Britain, Europe, and the major European possessions overseas. However, this was just a beginning to this type of global science—a prototypical venture that was repeated in the study of terrestrial magnetism, the study that seemed most likely to make advances through a global, coordinated effort.

Terrestrial Magnetism

European mariners had long used the magnetic compass to find their way in the deep ocean based on the principle that a magnetized needle pointed roughly to the magnetic pole. The principle, however, demonstrated some peculiar inconsistencies. Earth's magnetism seemed to vary considerably as one sailed around the globe. Some of the perturbations were local in character due to iron in mountains or from magnetic storms, while others seemed to be general in nature, changing throughout the decades, the seasons, and even daily or hourly. Magnetic north was not a fixed point, a problem that had vexed mariners and led to an astonishing number of lost vessels. Indeed, the question was still open as to whether Earth had one set of magnetic poles or two. As the process of international collaboration in the collection of geophysical data intensified, maritime nations were hopeful that the wandering nature of the magnetic pole or poles could be reduced to a mathematical law. Humboldt himself initiated the investigations, though the French and British soon overshadowed the limited undertakings of the great explorer.

Upon his return to Europe, Humboldt began to organize scientific ventures based on the methods he had formulated in South America. He was particularly interested in magnetic storms and the cause of the variation in the magnetic compass. On a trip to Russia, he proposed that magnetic observations be instituted throughout the country's vast possessions. He suggested that observations be taken hourly and include not only compass readings but also barometric pressure, humidity, temperature, and rainfall. The result was the first system of simultaneuous observations in terrestrial magnetism, an effort that was advanced by the Göttingen Magnetic Union, founded in 1834 by Carl Friedrich Gauss.

Gauss was a well-known and highly distinguished mathematician. He had met Humboldt earlier in Prussia, and the two had become close friends. When Gauss traveled to a conference in Berlin, it was natural that he stayed as a guest of Humboldt, then living in the Prussian capital. During his stay, Humboldt convinced Gauss to turn his mathematical acumen to the study of terrestrial magnetism and to help extend the observation points established in Prussia and Russia. Gauss, in turn, marshaled the help of his assistant, Wilhelm Weber, and together they helped set up a network of stations equipped with standard instruments from St. Petersburg to Peking, including several in Siberia and Alaska. From the combined data that these stations provided, in 1838 Gauss published a paper on the problem of terrestrial magnetism that allowed any observer to compute the values of the intensity, declination, and inclination of the magnetic force at any of the stations and to calculate the position of the magnetic poles. By 1840, he had published three seminal papers on terrestrial magnetism, including a map based on Humboldt's graphical representations of geomagnetism.

The Institute of France and to a limited extent the French government fervently continued and extended Humboldt's, Gauss's, and Weber's studies of terrestrial magnetism, instituting a systematic and extensive study within France and its colonies. By the mid-1830s, magnetic observatories with standardized instruments taking coordinated and simultaneous observations covered much of the globe, or at least that much of the globe not under British control. While progress was being made on the continent of Europe, advances were at a standstill in Britain. To remedy this situation, the British scientific elite was not above playing on nationalist sentiment. British scientists were dismayed that Britain, the premier maritime nation in the world, was not in a position to collaborate with the Prussians, Russians, Germans, and French. The lack of British involvement was anathema to the premier military and maritime power in the world.

The BA proved the perfect venue for the study of terrestrial magnetism. Here was a study open to all researchers. They could help advance one of the most important theoretical studies in the geophysical sciences, thus helping with British economic expansion. Moreover, it was a perfect study to involve international collaboration, thus leading to international harmony and good-will. At the fourth meeting of the BA, a special committee of recommendations met to confer with François Arago, a distinguished French scientist and foreign member of the BA. Arago suggested that the BA focus on the study of terrestrial magnetism in Great Britain and its colonial possessions that conformed in plan, instrumentation, and principles to the ongoing French project.

Through the support of the scientific elite in Britain and ultimately the British government, the recommendations of Arago were followed in England with the fervor of a religious calling. Thus, it became known as the Magnetic Crusade. The most ardent evangelical magneticist was Edward Sabine, an Irishman educated at the Royal Military Academy in Woolwich, London. Sabine served in the Royal Artillery and had been the astronomer on an expedition to search for the Northwest Passage in 1818. He had also toured the Southern Hemisphere in 1821 and 1822, and there he became interested in terrestrial magnetism, taking measurements throughout his journeys. Once settled back in England, he joined John Augustus Lloyd and James Clark Ross in studying the direction and intensity of the magnetic force in Ireland. Together, they produced a magnetic chart of Ireland based on the methods outlined earlier by Humboldt and Gauss. Sabine also produced a chart of the magnitude, intensity, and direction of the magnetic force throughout the entire British Isles.

Like Whewell's earlier work on the tides, Sabine used his researches on terrestrial magnetism around Britain as a template for a magnetic chart of the entire globe. Also like Whewell's work, this entailed a substantial change in the funding and implementation of the research. Sabine suggested that the BA approach the British government with Arago's suggestion to join the establishment of fixed magnetic observatories, thus adding to the international cooperation in the field. But he also intimated that he wanted a traveling observatory in the form of an Antarctic voyage to determine the position of the southern magnetic pole. According to Sabine, only through a large number of fixed observatories, paid for by the government and working in collaboration with other nations' observatories, along with a roaming magnetic observatory traveling the southern seas, could a correct map be produced to help the mariner determine the changing magnetic field.

With the backing of the BA and help from John Herschel, Sabine became the instigator for lobbying the government. His first step was to urge Humboldt to become involved. In April 1836, Humboldt wrote a now famous letter to the Duke of Sussex, a colleague of Humboldt's at Göttingen and now president of the Royal Society. Humboldt asked that the British government establish magnetic observation in all of their vast possessions overseas to link the circumlocation of magnetic readings. Humboldt was thinking of a collaboration among all the foremost maritime nations, including Britain, France, and Russia.

Humboldt's name carried a lot of weight, and progress came quickly. Through the influence exerted by the BA and the Royal Society, the Admiralty consented to financing fixed colonial observatories. Permanent magnetic stations were established in Greenwich, Dublin, Toronto, St. Helena, Jamaica, the Cape of Good Hope, and Tasmania. The effort culminated in a mobile magnetic observatory in the form of an expedition to the South Pole under the command of James Clark Ross from 1839 to 1843. The East India Company added observatories in India and Singapore, while Russia continued its stations in its

---

Sir James Clark Ross (1800–1862)

Naval captain and magneticist, James Clark Ross was the premier polar explorer in the Victorian era, traveling to both the Arctic and Antarctic regions. He entered the Royal Navy at the age of twelve, first serving under his uncle, John Ross, also a famed Arctic explorer. Along with William Edward Parry, they attempted unsuccessfully to discover the Northwest Passage in 1818. Ross returned to the Arctic on three successive voyages with Parry and on a further privately funded voyage between 1829 and 1833 with his uncle. On this last voyage, he sledged to the north magnetic pole, the first person to stand with the magnetic needle pointing vertically. Throughout these voyages, he amassed considerable knowledge of the Arctic. He made constant magnetic observations in addition to soundings of the ocean floor, measurements of the tides and oceanic currents, temperature and pressure readings, and studies of marine life.

Ross's expertise in geomagnetism led to his command of HMS *Erebus*, which sailed with HMS *Terror* on a geomagnetic and geographic voyage of discovery to Antarctica, sponsored by the BA as part of the Magnetic Crusade gaining steam at that time. Ross amassed considerable information on magnetism, geography, natural history, geology, and meteorology. His dream to be the first person to stand on the southern magnetic pole was thwarted by heavy ice, and he had to be content with the discovery of a large sea in Antarctica, now named the Ross Sea, along with Mount Erebus, and Mount Terror. Upon his return to England, he published a two-volume account of his journey, *A Voyage of Discovery and Research in the Southern and Antarctic Regions during the Years 1839 to 1843* (1847). This publication, along with his many voyages of discovery, established him as the most experienced Arctic explorer in the nineteenth century.

own territory, including eleven in Siberia and one in Beijing. In the end, a total of more than fifty permanent observatories registered terrestrial magnetism around the globe, truly a worldwide survey.

Meteorology

With some success in both tidology and terrestrial magnetism, scientists in the early Victorian era were hopeful that a similar global effort could prove fruitful in meteorology. The science of meteorology required sustained cooperation and seemed especially suited to gain from such a coordinated initiative. Individuals throughout Great Britain had taken measurements of rainfall and barometric pressure for some time, while government-sponsored voyages often included similar instructions for not only the naturalists on board but the captains of British vessels as well. The rooms of the Royal Society, the BA, and the Hydrographic Office overflowed with meteorological registers, as did the desk drawers of many local gentry. Prior to the nineteenth century, however, the observations were unconnected and proved fruitless. According to the natural philosophers in Britain, the science of meteorology could advance only from numerous and corresponding observations made uniformly throughout the globe. James Forbes, in his address on meteorology to the first meeting of the BA, stressed the unconnected state of British observations and the lack of any unifying theory. He suggested that manned stations be established around the British Isles and continental Europe where exact, numerous, and corresponding observations could be made on a uniform plan using standardized instruments.

Initially, the British took the lead. Forbes outlined detailed instructions for acquiring meteorological observations by traveling explorers. Beaufort had meteorological logs kept by all his surveying vessels, at home and abroad, and the BA offered several grants for meteorological projects, mostly to reduce the data coming in from around the world. Meteorology received its greatest boost, however, when Herschel returned to England from the Cape of Good Hope in South Africa.

At the apex of his career, Herschel was perhaps the greatest scientist in midcentury England. He was the son of a famous astronomer, Sir William Herschel, the discoverer of Uranus and mapper of the southern skies. In the early 1830s, the younger Herschel spent more than four years in the Cape of Good Hope completing the mapping of the Southern Hemisphere begun by his father. He undertook not only astronomical investigations but also meteorology, tidology, terrestrial magnetism—any subject in which data from South Africa could be helpful. Like Whewell, Herschel was a polymath, studying almost every subject under the purview of the physical sciences. He returned home to England the don of British science and became a strong supporter of and highly respected voice for the governmental involvement in geophysical global initiatives. He was, for instance, the person responsible for bringing the Magnetic Crusade to the halls of government.

Once back in England, Herschel helped set up a scheme for worldwide meteorological observations. "There is no branch of physical science which can be advanced more materially by observations made during sea voyages than meteorology," he exclaimed. Owing to the homogenous nature of its surface as well as its relatively more stable temperature, observations made at sea had far less disturbing influences than those made on land. The area of the sea also far exceeded that of the land, and thus "a much wider field of observation is laid open, calculated thereby to offer a far more extensive basis for the deduction of general conclusions."[3] It was the sea's accessibility, geographic extent, and natural uniformity that made it so propitious for global investigations.

Meteorology, like magnetism and the tides, had direct and often dire consequences for navigation. The weather over the oceans, was responsible for a vast number of shipwrecks and loss of life and needed to be brought to rule. Moreover, knowledge of the winds and other atmospheric phenomena could help predict the length of time needed for an oceanic passage, the formation of ice in the Arctic and Antarctic regions, and thus the correct season for transoceanic travel. Herschel found an enthusiastic collaborator in Beaufort and other hydrographers in Europe and America. A grant from the BA enabled him to work closely with William Radcliff Birt to reduce, arrange, and analyze the barometric and weather observations coming in from around the world. They studied more than 330 sets of observations made at sixty-nine places in Europe, South Africa, and America. They traced the magnitude and direction of what they termed "atmospheric waves" over Europe, creating maps in the form of isothermal lines. Further advances, however, were made not in Britain but in the large landmass of the United States.

In the study of meteorology, American researchers finally advanced beyond their British counterparts. In contrast to the islands making up Great Britain, the large landmass of North America was ideally suited to chart storms and other meteorological phenomena. Joseph Henry at the Smithsonian Institution, Alexander Dallas Bache of the U.S. Coast Survey, and Matthew Fontaine Maury of both the Naval Observatory and the U.S. Hydrographic Office all viewed meteorology as a science in which the United States could excel. Between 1840 and 1870, American meteorology took the lead.

Both Henry and Bache set up large networks of meteorological observations, coordinating hundreds of volunteers to observe meteorological phenomena around the continental United States from the Atlantic Ocean to the Mississippi River. But it was Maury who became America's international spokesman for meteorological research. Appointed superintendent of the Depot of Charts and Instruments for the Navy Department in Washington in 1842, he assembled meteorological reports that he found submitted in the depot's archives. From these logs, he produced global charts of wind and weather currents throughout the oceans, which he then augmented as new logs and new observations arrived. His surface temperature and wind charts attracted international attention, and by the time he had published his influential *The Physical Geography of the Sea*, a textbook that synthesized his global work, his system had been adopted worldwide.

Maury viewed meteorology as a global phenomenon, and he suggested that an international conference be held to establish a worldwide system of meteorological observations covering both the land and the sea. He also proposed that blank forms be sent to the ships of all nations, a plan soon adopted by most European maritime nations. The first such conference was held in Brussels in 1853, and Maury was appointed the U.S. representative to discuss his international system of data collection and reduction.

One direct result of the Brussels conference was the founding in 1854 of the British Meteorological Department, whose sole purpose was coordinating meteorological data, especially at sea. Robert Fitzroy was elected its first director. After entering the navy at the age of fourteen, Fitzroy sailed on the first HMS *Beagle* voyage at age twenty-three. He became famous for the *Beagle*'s second voyage, which circumnavigated the globe and carried with it the young and aspiring naturalist Charles Darwin. Fitzroy, however, was a respected natural philosopher in his own right, becoming a fellow of numerous scientific societies including the Royal Society. As head of the British Meteorological Department, he outlined a comprehensive plan of data collection that focused on the global nature of meteorological phenomena and the importance of standardized instruments. He allocated instruments to the navy and merchant marine along with detailed instructions on how to measure the temperature, barometric pressure, wind, rain, humidity, and other meteorological phenomena, including aurora and magnetic storms. Although he had an especially troubled career as Britain's first forecaster, Fitzroy laid the foundation for European global meteorology.

The Search for the Northwest Passage

One consequence of these geophysical initiatives in the tides, terrestrial magnetism, and meteorology was the incorporation of the Arctic into the purview of global analysis. A route north of the American continent had long been a goal of European maritime ambitions. Such a route would cut thousands of miles off the voyage from Europe to the Spice Islands in the East, and whalers, explorers, scientists, and soldiers, one after another met starvation, death, and despair attempting to open a passage in the Arctic.

Especially important to European nations was the discovery of the Greenland whaling grounds. William Scoresby, the son of a famous Arctic whaler, had an extensive interest in both whaling and science, and during multiple trips to the Arctic he surveyed the coasts of Greenland and studied the meteorology, terrestrial magnetism, and aquatic life in the Arctic regions. On a voyage to the Arctic in 1817, he noted that the ice that usually halted further progress north on the coast of Greenland was melting. Perhaps, he mused, the same forces could have opened up a passage all the way to the Pacific. Beaufort took this as added proof that a passage existed, arguing "that it would be an intolerable disgrace to this country were the flag of any other nation to be borne through it before our own."[4] British polar expeditions were about to begin with renewed vigor, a geographical, scientific, and national objective that excited the public imagination more than any other geographic assault in the nineteenth century.

The next year, the British government fit out four ships to search for the Northwest Passage from two different directions. John Ross and William Parry commanded the *Isabel* and the *Alexander* from the Pacific, while David Buchan and John Franklin sailed the *Dorothea* and the *Trent* from the Atlantic. The instructions for polar voyages gave primacy to both scientific advance and geographical exploration. Ross, along with his nephew James Clark Ross, made constant magnetic observations along with soundings of the ocean floor, measurements of the tides and oceanic currents, temperature and pressure readings, and studies of marine life. Because the voyage failed to find the Northwest Passage, Ross's *A Voyage of Discovery*, published in 1819, stressed the voyage's scientific accomplishments, focusing on Humboldt's program of the interconnectedness of the physics and biology of Earth.

Franklin's first overland expedition was horrific; eleven members of the crew died of starvation or scurvy, while the survivors were forced to eat lichen off of rocks. But even this profited science. John Richardson, the naturalist on the voyage, described up to forty-two lichens, two brand new to science, and described the medical effects of starvation and hypothermia. Starvation and death seemed always to be close at hand on Arctic explorations, and one lesson became clear. Exploration in the Arctic regions required an understanding of the science of the polar regions: the types of animals and plants to eat, the fluctuating magnetic readings to find one's place in the ocean and on land, and meteorological phenomena, including wind and ocean currents, to retreat once the Arctic had won. These early voyages acquired a considerable amount of scientific information on the Arctic to at least garner a hope that further explorations could find the Northwest Passage.

Parry returned to the Arctic the next year, the first of four expeditions he commanded. Franklin also returned, on another overland expedition on the north coast of America, his second expedition out of three. Franklin's second voyage added significantly to both the geography and natural history of the Arctic region. He and his officers attended lectures given by the president of the Royal Geographical Society before their departure and had their vessels stocked with the latest scientific equipment: transit instruments, barometers, thermometers, compasses, and chronometers. Parry's expedition likewise focused on the science of the Arctic, partly owing to the growing uncertainty of the existence of the Northwest Passage. Upon his return, he published his account of the voyage that outlined his magnetic, tidal, and lunar observations, for which he was elected a fellow of the Royal Society in 1821. He also had gone farther north than any European, demonstrating that Lancaster Sound opened a passage to the West.

The British continued to send vessels to the Arctic region even though it was becoming increasingly clear that a Northwest Passage, if indeed it existed, would be too difficult to navigate. National pride, scientific inquiry, and territorial, economic, and military gains all combined to form the basis of one of the more successful Arctic voyages of exploration, that of James Clark Ross. Ross had entered the navy at the age of twelve, had served with his uncle John Ross in the Arctic, and in 1829 led HMS *Victory* into the Arctic seas. After spending two winters in the ice (with two more to follow), Ross and several members of his crew reached a spot on 1 June 1831 where the magnetic needle was all but vertical. With his entire crew alive and in good health, Ross sailed back to England and dined with King William IV, informing him that the king's name had been added to the magnetic pole.

After Ross's return in 1833, the pace of surveying ships heading to the Arctic with scientific orders began to quicken rapidly. Both the Hydrographic Office and the Royal Geographical Society played instrumental roles in all of the major scientific voyages of discovery in the mid-nineteenth century, and both increasingly turned to the science of the Arctic. In a sense, they were forced to do so. Charts, sailing directions, wind and ocean currents, edible flora and fauna—everything needed to set sail and return a vessel from the Arctic—became a priority owing to one voyage of discovery in particular. This was Franklin's third and last expedition. Two ships, the *Erebus* and the *Terror*, were refitted against the ice, fitted with steam engines, and given enough provisions to spend two winters locked in the ice.

They set sail from London in May 1845, were spotted by two whaling vessels near Lancaster Sound, and were never heard from again. The crew may have disappeared, but the fate of the Franklin expedition gripped the world. After the first year, members of the Admiralty began to wonder; after two years, they began to worry; and after three years, they began sending out rescue expeditions. The loss of Franklin and his crew did more for Arctic science, especially Arctic geography, than any other voyage of discovery; by the time Franklin's fate was realized, numerous search expeditions from a host of different nations had crisscrossed through the Arctic Archipelago. In searching for the lost vessels and their crew, Britain and other nations mapped, sounded, and surveyed the entire Northwest Passage, from the Atlantic to the Pacific. Each voyage of exploration, whether scientific or search and rescue, added to the science of the Arctic. The explorers contributed to a wide array of sciences, advancing all the major disciplines studied earlier by Humboldt. And the Northwest Passage had finally been completed.

In 1853, John Rae, a tenured Arctic explorer, finally discovered Franklin's fate. He received reports from Inuits living in the region of white men who had been seen wandering through the desolate Arctic region. A trail of bodies, they said, marked their path. Rae purchased articles of the Franklin expedition from the Inuits as proof. These tantalizing clues caused Rae to redouble his search, and what he found was dreadful. The crew had all but perished from starvation and disease before wandering over land for hundreds of miles, only to have their lives end, in Rae's words, after they "had been driven to the last resource—cannibalism—as a means of prolonging existence."[5]

The Study of the Ocean

Up to the mid-nineteenth century, the ocean was used primarily as a sea-lane to transport ships, cargo, and crews to distant destinations. The Northwest Passage, for instance, became important not for the Arctic itself but as a passageway to the East. Indeed, after the near-dreadful outcome of Edmund Halley's stewardship of the *Paramore*, natural philosophers were no longer allowed to captain ships. The scientific study of the sea itself, therefore, remained episodic, unconnected, and often useless. Yet by midcentury, there was a growing interest in the world's oceans, not as a lonely abyss that one was forced to cross but rather for the sea itself. Several factors combined to make the ocean an especially ripe area of analysis, leading to large-scale government-sponsored voyages with the sole objective of studying the science of the ocean.

An interest in the sea followed directly from the growing popular fascination with the natural history of the seashore; collecting oceanic flora and fauna washed up on the beach became a middle-class obsession. Sojourns to the sea as a panacea from crowded cities and as a cure for numerous ailments also led many middle-class professionals to the seashore and its environs. Following Humboldt's creed of the interconnectedness of nature, natural philosophers themselves became increasingly hopeful that oceanic depth, temperature, and salinity could help answer age-old questions concerning the changes in the atmosphere and the movement of the oceans. Moreover, much of Earth's landmasses had been traversed, mapped, charted, and largely explored. The ocean, which covers 70 percent of the planet, was relatively unknown. Marine scientists realized that they were on the precipice of a massive and largely unexplored waterscape.

Several immediate causes also led to a new interest in the ocean itself. The first was an ambitious project in the 1850s and 1860s to connect Europe and North America with a transatlantic telegraph cable, which required soundings of the ocean floor. The telegraph was a preferred instrument of Britain's commercial and political empire, allowing instantaneous communication among merchants and armies. The telegraph allowed Britain to control its expanding possessions overseas and scientists increasingly turned to the study of the ocean to assess areas on the ocean floor that could support the laying of a telegraph cable. The second impetus to the renewed interest in the oceans came, surprisingly, from the radical theory of the transmutation of species. Darwin published his theory of natural selection in 1859, and its reception aroused both condemnation and converts. For both parties, the ocean floor offered the most promising area to support or disprove his hypothesis that species evolved according to changes in their natural environment. The deep ocean was considered changeless, a constant four degrees Celsius, with the same light, salinity, and chemical makeup throughout its long history. If the environment remained unchanged, life forms should remain unchanged as well. Darwin's theory awakened the study of the ocean depths.

Strategic military and commercial interests also combined to focus scientists' attention on the study of the ocean. In the 1860s, the United States, buoyed by the success of the Charles Wilkes expedition, planned a voyage in the Atlantic and Pacific, while Germany and Sweden had scheduled exploratory voyages to the Arctic seas. The British government, therefore, agreed to reequip the warship HMS *Challenger* into an itinerant laboratory for a four-year voyage of scientific exploration of the ocean. All but two of the seventeen guns were removed to make room for the naturalists and their wares. Charles Wyville Thomson, regius chair of Natural History at Edinburgh, was appointed scientific director of the voyage, and he gathered specialists from all areas of science to undertake as comprehensive a study of the ocean as possible. Their aim was to study the entirety of the sea, including its biology, chemistry, and physics; its depths and salinity; the movements of its tides and currents; and the interrelationship among latitude, longitude, barometric pressure, and temperature. It also aimed to catalog the varied animal and plants at all geographic locations and throughout the different depths of the ocean.

Between 1872 and 1876, the *Challenger* Expedition traversed more than 69,000 miles, crisscrossing the Pacific, Atlantic, Indian, and Antarctic Oceans. It crossed the equator six times and was the first steam vessel to cross the Antarctic Circle, where the ice played havoc with the vessel's unprotected hull. The naturalists on board studied the ocean off the coasts of Europe, North and South America, Australia, New Zealand, the Antarctic, and Japan, setting up more than 360 scientific stations. At each station, they collected water and biological samples with the use of tow nets at intermediate stages of the ocean. The comprehensiveness of the geographical distribution of its data over such a large area of Earth's surface was overwhelming. Scientists amassed hundreds of crates filled with thousands of animal and plant specimens, more than 1,400 water samples from different depths and geographic regions, and hundreds of seafloor samples containing fascinating animals such as sea cucumbers and sponges that spanned the colors of the rainbow. They filled notebooks with information on sea elephants, whales, and penguins as well as important anthropological information on the habits of native peoples. They also amassed artistic and technological artifacts from native cultures. It is no wonder that, similar to Humboldt's work, the actual publication of the results took ten times longer than the voyage itself.

Upon the *Challenger*'s return to Britain, close to a hundred scientists from several different countries converged on Edinburgh to help analyze the scientific booty. From America, Louis Agassiz and Theodore Lyman, both well-respected zoologists, began work on the sea urchins and brittle stars. After meticulous and painstaking work, Agassiz, exhausted by the investigation, barely finished his volume. The Germans Oscar Schmidt and Ernst Haeckel studied sponges and microscopic animals, while the Norwegian naturalist Georg Sars tackled the crustaceans. In the end, fifty volumes of published results appeared between 1877 and 1895. Taken together, they contained a detailed analysis of the currents, salinity, depths, and temperature of the oceans; the topography of the ocean floor; and the geology, zoology, and biology of its inhabitants. More than 750 sets of the published volumes were given away free to almost every major scientific institution in the world.

The scientific results of the expedition were as diverse as they were surprising. Although they had searched intensely for Darwin's missing links, none were found, and the gaps in the evolutionary tree remained, as did the debates on evolution. The results also failed to determine the circulation of the oceans, a complex interaction of forces that would have to await the computer age. But other investigations provided important scientific, and oftentimes economic, successes. The existence of life at all levels of the ocean, including its deepest parts, was firmly established. Temperature readings also led to the curious results that the temperature of the eastern and western sides of the Altantic differed appreciably. George Strong Nares, the *Challenger*'s captain, suggested that a mountain range dividing the ocean floor was responsible, a ridge that John Murray attempted to map in 1882. Murray also undertook the study of plankton, demonstrating their diurnal migration and profusion throughout the oceans.

The *Challenger* Expedition opened the ocean for direct scientific study, outlining the objectives and research methods of all future oceanographic voyages. Both the voyage itself and the publication of its results brought together an international community of scholars all interested in the study of the sea, all using a common technical language, and all sharing their research methods among an increasingly coherent community of scholars. For these reasons, the *Challenger* Expedition is often hailed as the beginning of oceanography. It helped transform the ocean from a mysterious and unknown space to a legitimate and exciting new place for scientific research.

The study of oceanography was from the beginning expensive. The financial and practical difficulties of studying the deep sea made it entirely dependent on the financial support of national governments. Although most of the major nation-states in Europe and North America sent voyages of discovery to study the sea itself, only Britain, could afford to advance such a momentous undertaking. It was a maritime nation, willing to pay handsomely for scientific exploration of the world's oceans and coastlines. From its roots in Humboldtian initiatives that required large amounts of data that covered large areas of the world, the grand-scale nature of these developments created a powerful stimulus to further study of the oceans. In the decades following the *Challenger* voyage, numerous other voyages were carried out under the auspices of national governments, including those in North America and the Scandinavian countries, where the understanding of ocean circulation, plankton studies, and habitats of marine animals grew by leaps and bounds. From depleted fisheries to the role of the ocean in accounting for Earth's climate, the scientific study of the ocean has become one of the most fertile areas of scientific investigation.

Conclusion

The Victorian era was a time of unprecedented specialization of the sciences; scientific journals and academic chairs blossomed as the amateur slowly gave way to the professional. Advances in almost every corner of science has earned the period the honor of being called the second scientific revolution. But the most profound change to occur in science during this period was the explosion of interest in the sciences of Earth and the environment, with its reliance on accumulating and subsequent reducing, tabulating, and graphing large amounts of data from all over the globe.

Humboldtian initiatives in midcentury Britain, despite the large amount of money invested, were only modestly successful. In the study of the tides, the Hydrographic Office produced tide tables for more than one hundred ports throughout the world by midcentury, compared to none in 1830. Whewell made other minor additions to tidal theory, including the empirical laws for the effects of the parallax and declination of the sun and moon as well as the importance of the diurnal inequality, the difference between the time and height of the two high tides each day, for any future theory. In terrestrial magnetism, Gauss discovered an analytic law allowing mariners to determine the magnetic intensity and declination at stations around the world. Sabine, moreover, determined from data collected from the Magnetic Crusade, and especially from the Antarctic voyage of Ross, that the variations of the magnetic needle were due to two distinct causes, one from within Earth and the other from without. He also established the regularity of magnetic storms in relation to sun spot activity, thus finally determining the origins of magnetic storms, first researched by Humboldt. In meteorology, the first system of storm warnings was instituted, first in the United States, then in Britain, with limited success. The *Challenger* Expedition was more successful, forming the foundation of the modern study of oceanography.

As with Humboldt's own research, however, it is important to judge these monumental efforts by more than simply the number of correct scientific theories they produced. More important was the foundation and precedence set by such grand undertakings. With observational stations around the globe equipped with the latest instruments to measure and ultimately compare observations on a worldwide scale, the standardization of measurements and apparatuses directly followed. It is no coincidence that at this very time, governments pursued the standardization of weights and measures, Greenwich Mean Time, and other scientific constants. Moreover, the standardization of instruments across borders speaks to a more far-reaching result of global, geophysical initiatives: the internationalization of the practice of science itself. The new science was on such a grand scale that not even an extremely rich maritime nation such as Britain could afford to single-handedly fund its development.

The transnational complexion of tidal studies, magnetism, meteorology, and oceanography matched a world that was increasingly linked together through European imperial and colonial trade. What occurred was a new type of international scientific collaboration among France, Germany, Russia, the United States, and Britain. By the mid-nineteenth century, when much of the landmasses and coastlines had been explored and charted, arguments for continued explorations were increasingly scientific. Science evolved into a large-scale, international practice that was increasingly funded by government and used the world's oceans and landmasses as its laboratory.

Bibliographic Essay

The main themes of this chapter were sparked by the seminal chapter "Humboldtian Science" in Susan Faye Cannon's *Science in Culture: The Early Victorian Period* (1978). A spurt of informative essays has recently followed, including R. W. Home, "Humboldtian Science Revisited: An Australian Case Study" (1995), and two especially interesting accounts in Nicholas Jardine, Emma Spary, and J. A. Secord, eds., *Cultures of Natural History* (1996): Michael Dettelbach's "Humboldtian Science" and Janet Browne's "Biogeography and Empire." Malcolm Nicolson in his "Alexander Von Humboldt, Humboldtian Science and the Origins of the Study of Vegetation" (1987) and "Humboldtian Plant Geography after Humboldt: The Link to Ecology" (1996) has investigated the central position of botanical studies and plant geography within Humboldt's system of inquiry, scholarship that has resituated the Humboldtian enterprise within German romanticism.

For analysis of specific Humboldtian initiatives in Britain, the best place to start is the section on Humboldtian science in chapter 8 of Jack Morrell and Arnold Thackray, *Gentlemen of Science: Early Years of the British Association for the Advancement of Science* (1981). It underscored the political aspects of Humboldtian science and the central role of the BA, as does W. H. Brock, "Humboldt and the British: A Note on the Character of British Science" (1993). For tidal studies, see Margaret Deacon, *Scientists and the Sea, 1650–1900* (1971); Michael S. Reidy, *Tides of History: Organizing the Ocean and Creating the Scientist* (forthcoming); and Isaac Todhunter, *William Whewell, D.D., Master of Trinity College, Cambridge: An Account of His Writings with Selections from His Literary and Scientific Correspondence* (1970). In the past two decades, scholarship on William Whewell has increased. See Joan Richards,

"Observing Science in Early Victorian England: Recent Scholarship on William Whewell" (1996), for an overview of this scholarship. The most engaging treatment of Whewell is given by Richard Yeo, *Defining Science: William Whewell, Natural Knowledge, and Public Debate in Early Victorian Britain* (1993).

For terrestrial magnetism and, in particular, the Magnetic Crusade, see the two seminal works by John Cawood, "Terrestrial Magnetism and the Development of International Collaboration in the Early Nineteenth Century" (1977) and "The Magnetic Crusade" (1979). Alison Winter gives a more detailed account of the problems with oceanic navigation owing to the changing magnetic field in "'Compasses All Awry': The Iron Ship and the Ambiguities of Cultural Authority in Victorian Britain" (1994). For the earlier period, Patricia Fara's *Sympathetic Attractions: Magnetic Practices, Beliefs and Symbolism in 18th Century England* (1994) takes a cultural approach that places terrestrial magnetism within the debates concerning the value of science in British culture.

Meteorology has been perhaps the most studied geophysical initiative in the past few years. Katharine Anderson's *Predicting the Weather: Victorians and the Science of Meteorology* (2005) and "The Weather Prophets: Science and Reputation in Victorian Meteorology" (1999) focus heavily on the life and work of Robert Fitzroy. For the earlier period, see Vladimir Jankovic, *Reading the Skies: A Cultural History of English Weather, 1650–1820* (2000), a heady and sophisticated cultural account. His "Ideological Crests versus Empirical Troughs: John Herschel's and William Radcliffe Birt's Research on Atmospheric Waves, 1843–50" (1998) offers a shorter and more readable account of the work of John Herschel and the calculator William Birt.

For American meteorology, see James Rodger Fleming, *Meteorology in America, 1800–1870* (1990). For American Humboldtian initiatives more generally, see Hugh Slotten, *Patronage and Practice, and the Culture of American Science: Alexander Dallas Bache and the U.S. Coast Survey* (1994), and Thomas G Manning, *U.S. Coast Survey vs Naval Hydrographic Office: A 19th-Century Rivalry in Science and Politics* (1988). Helen Rozwadowski's *Fathoming the Ocean: The Discovery and Exploration of the Deep Sea* (2005) is an exciting narrative that covers the Anglo-American interest in the deep sea in the nineteenth century. For internationalism in science during this period, see Elisabeth Crawford, *Nationalism and Internationalism in Science, 1880–1939* (2002), and the accounts in W. R. Home and Sally Gregory Kohlstedt, eds., *International Science and National Scientific Identity* (1991). Material concerning the Northwest Passage is available through accounts of the actual voyages of discovery themselves, including those of Parry, Ross, and Franklin, or through accounts of Arctic exploration more generally. Two of the more comprehensive accounts include Leslie H. Neatby, *The Search for Franklin* (1970), and Ann Savours, *The Search for the North West Passage* (1999).

For material covering the history of the Royal Navy and the British Admiralty, J. R. Hill, ed., *The Oxford Illustrated History of the Royal Navy* (1995), offers many readable accounts and is a good beginning. See also Paul M. Kennedy, *The Rise and Fall of British Naval Mastery* (1976), and Michael Lewis, *The Navy in Transition, 1814–1864: A Social History* (1965). An important shorter article is Andrew D. Lambert, "Preparing for the Long Peace: The Reconstruction of the Royal Navy, 1815–1830" (1996). For the science of the Royal Navy in particular, see Trevor H. Levere, *Science and the Canadian Arctic: A Century of Exploration, 1818–1918* (1993), especially the opening chapters that, like Lambert's article, cover the reorganization of the Royal Navy.

For the role of Sir Francis Beaufort, see Alfred Friendly, *Beaufort of the Admiralty: The Life of Sir Francis Beaufort 1774–1857* (1977), and Archibald Day, *The Admiralty Hydrographic Service, 1795–1919* (1976). A more recent and popular account is Nicholas Courtney, *Gale Force 10: The Life and Legacy of Admiral Beaufort* (2003). An important shorter article is Roger Morris's "200 Years of Admiralty Charts and Surveys" (1996). Beau Riffendburgh, *The Myth of the Explorer: The Press, Sensationalism, and Geographical Discovery* (1994), explores the sensationalism of the British press in popularizing the voyages of exploration.

For the role of institutions, see especially Morrell and Thackray, *Gentlemen of Science*, discussed above, and the collected entries in Roy MacLeod and Peter Collins, eds., *Parliament of Science: The British Association for the Advancement of Science, 1831–1981* (1981). Although a bit dated, Marie Boas Hall's *All Scientists Now: The Royal Society in the Nineteenth Century* (1984) covers the supporting role of the Royal Society in the exploration process. David Philip Miller, "The Revival of the Physical Sciences in Britain, 1815–1840" (1986), explores how the mathematical scientists during this period co-opted the leadership roles of both the Royal Society and the BA for their own geophysical initiatives.

Humboldt's own writings have been extremely difficult to acquire, but recently several of his works have been republished in English. This includes his *Personal Narrative of a Journey to the Equinoctial Regions of the New Continent*, translated with an introduction by Jason Wilson and a "Historical Introduction" by Malcolm Nicolson (1995), and *Cosmos: A Sketch of a Physical Description of the Universe*, translated by E. C. Otté with an introduction by Nicolas A. Rupke (1997). Both introductions, especially the one by Nicolson, place Humboldt within the history of scientific ideas. Secondary sources on Humboldt are now numerous, as more and more humanists are reinventing him. Helmut de Terra, *Humboldt: The Life and Times of Alexander von Humboldt, 1769–1859* (1955), and L. Kellner, *Alexander von Humboldt* (1963), are classics, as is the chapter on Humboldt in Victor Wolfgang von Hagen's *South America Called Them* (1945). Anthony Smith's *Explorers of the Amazon* (1990) also examines Humboldt within the context of South American exploration but is less historical in content. Humboldt takes center stage in David N. Livingstone's *The Geographical Tradition* (1992), which treats geography as "the science of imperialism par excellence." Mary Louise Pratt's *Imperial Eyes: Travel Writing and Transculturation* (1992) is an exceptional account focusing on the many different uses of Humboldt by Europeans and Creoles but is perhaps most useful in graduate-level courses.

For material on the *Challenger* expedition, see Margaret Deacon, *Scientists and the Sea, 1650–1900* (1977); Margaret Deacon, Tony Rice, and Colin Summerhayes (eds.), *Understanding the Ocean* (2001); Muriel L. Guberlet, *Explorers of the Sea: Famous Oceanographic Expeditions* (1964); and Susan Schlee, *Edge of an Unfamiliar World: A History of Oceanography* (1973).

Natural History in the Nineteenth Century

Britain launched expensive voyages of overseas scientific exploration in the nineteenth century to survey and map the world, collect flora and fauna, and study native populations. In the process, they built an empire. As the politics of imperialism intensified and the economics of a world-wide trade expanded, Britain emerged as a technological and scientific leader. Voyages of exploration, moreover, led Europeans to question their own place within the natural world. The findings of the naturalists dispersed throughout the globe led to doubts concerning the trustworthiness of Scripture to unlock the secrets of the origin of Earth and man, thus undermining deeply entrenched views of the age of the planet, the fixity of species, and man's overarching relationship to nature.

Eighteenth-century voyages of exploration helped bring to the European reading public a vision of native populations. Through the publication of travel narratives or through the satirical accounts of these voyages in the new genre of the novel, voyages of exploration challenged Enlightenment thinkers to incorporate non-Europeans into their philosophical and ethical systems. Most of these earlier travelers wrote on the supposed backwardness or lethargy of native populations as stemming from their centuries of despotic rule; unorganized social, political, and religious systems; or sometimes simply the tropical climate. These travelers perceived culture as the correct gauge of civilization, with the lack of science and technology merely a corollary of other cultures' backwardness.

As voyages of exploration with naturalists on board reached far-flung areas of the globe throughout the nineteenth century, Europeans' perceptions of their own scientific and technological superiority began to influence their interactions with native populations overseas. Two concomitant changes occurred. First, Europeans began to view their achievements in science and technology as the determining factor of their supposed superiority and the most meaningful way to evaluate the differences between Western and non-Western cultures. Second, the naturalists who accompanied voyages of exploration in the second half of the nineteenth century helped formulate the modern discipline of biology, based on a highly explanatory theory of evolution through natural selection. Naturalists extended the eighteenth-century taxonomies to include not only animal and plants but also humans.

These two simultaneous developments combined to help subjugate native societies throughout the world as European governments moved increasingly toward imperial rule in the last half of the nineteenth century. Europeans transformed their realization of superiority over other societies into a right to rule, a duty to rule, and, ultimately, a justification of that rule. It was their moral obligation to overthrow corrupt despotism, improve the plight of natives, and bring peace and Christianity to areas through a "civilizing mission." An unabashed faith in progress, the power of rationality, and advances in science and technology all combined throughout the second half of the nineteenth century to place humanity within a hierarchy of beasts and to rank other cultures within a hierarchy of humans.

Charles Darwin

Charles Darwin holds a place in the history of the biological sciences similar to the place that Isaac Newton holds in the physical sciences. Like Newton, Darwin unified an inchoate science under one unifying law, a foundation from which all else would have to draw. Darwin's theory likewise reached far beyond the sciences to include theological, philosophical, political, and economic significance. Both Newton and Darwin were influenced by the cultures in which they lived, and both needed their followers, the Newtonians and the Darwinians, to in turn transform those cultures. Neither, moreover, got all the answers right; they are revolutionary figures partly because they asked the correct questions. They differed, however, in the manner in which they came to ask these questions. Newton was an experimental natural philosopher who never traveled beyond the confines of England. Darwin, by contrast, emerged because of his five-year circumnavigation of the globe on board HMS *Beagle*. As Darwin acknowledged, "the voyage of the *Beagle* has been by far the most important event in my life and has determined my whole career."[1]

Darwin suffered through most of his early education, content with running through the outdoors near Shrewsbury, where his father practiced medicine. As the second male child, his prospects for a career, though not entirely limited, were not that numerous. He could join the military, go into the church,

Charles Robert Darwin (1809–1882)

Co-discoverer of the theory of evolution through natural selection, Charles Darwin was also a geologist, botanist, popular writer, and the epitome of the middle-class naturalist in mid-Victorian England. The grandson of the irascible Erasmus Darwin and the son of a highly respected physician, Darwin was born on 12 February 1809. At age sixteen, he followed his older brother to the University of Edinburgh to study medicine, where he attended lectures in chemistry and anatomy. Unable to stomach gruesome surgeries, he moved to Cambridge in 1827 to prepare for a life as a clergyman in the Church of England. At Christ's College, Cambridge, Darwin received his only formal training in natural history, largely from Adam Sedgwick, one of the premier geologists in England, and John Stevens Henslow, an exciting lecturer with an exceptionally broad base of knowledge and interests in natural history and geology. Darwin often accompanied Henslow on long jaunts through the forests and hills around Cambridge with geological hammer and insect net in tow. It was Henslow who suggested that Darwin accompany Fitzroy on a five-year circumnavigation of the globe aboard HMS *Beagle.*

The voyage of the *Beagle* furnished Darwin with his first in-depth scientific training and convinced him that he would pursue science as a vocation. The *Beagle* voyage also offered Darwin the opportunity to trek inland on largely unexplored islands and continents to study the geographical distribution of the world's flora and fauna, their relation to changes in the environment, and their resemblance to now extinct species, subjects that Darwin would pursue for the rest of his life. Upon his return to England, he published his account of the voyage, *Journal of Researches into the Geology and Natural History of the Various Countries Visited by H.M.S. Beagle* (1839); his theory of the formation of coral reefs, *Coral Reefs* (1842); and other geological research works, including *Volcanic Islands* (1844) and *Geological Observations on South America* (1846). These early publications made Darwin well known in the scientific community. He also began to formulate his views on the mutability of species, opening his private "Notebook on Transmutation" in 1837, two years after the return of the *Beagle.* He undertook lengthy studies on barnacles and plant and animal husbandry (domestic breeding) before publishing his monumental *On the Origin of Species* in 1859, and then only after Alfred Russel Wallace had sent Darwin his own identical theory. Darwin wrote three subsequent books that expanded on his theory of natural selection, including *The Descent of Man* (1871) that extended his theory to man. Darwin died on 19 April 1882 and is buried in Westminster Abbey not far from Sir Isaac Newton.

or follow his father into medicine. He chose the last option and accompanied his older brother to Edinburgh to begin his medical education. It was a bad decision. Unable to stomach the horrors of surgery in an era before anesthetics, he moved to Cambridge and entered Christ's College in the fall of 1827 to pursue a career in the Anglican Church. He spent the next three years of his life like most other college students, sharing his time between bouts of intense study and equally intense philandering.

The University of Cambridge produced ministers, lawyers, and statesmen, the type of "gentlemen" that the government needed to run an expanding empire. Cambridge was not intended to produce scientists. Darwin read closely the works of William Paley on natural theology, Thomas Malthus on population growth, and Adam Smith on political economy as well as the writings of contemporary scientists, including John Herschel and William Whewell. But one book in particular that Darwin read during his last year at Cambridge forever changed his life. He discovered and then devoured Alexander von Humboldt's *Personal Narrative* and was animated by the discussion of the excitement of scientific research and the need to travel to the interior of distant and exotic areas of the globe. While reading Humboldt's tantalizing portrait of South America, little did Darwin know that he, like so many other British merchants, collectors, and missionaries, would follow in the great traveler's footsteps.

Within twenty years of Humboldt's return to Europe, Paraguay, Uruguay, Brazil, Argentina and most other former colonies had attained their independence, opening their trade to other European nations. By the time Darwin received his degree from Cambridge, British economic interests in South America had blossomed. The growth of the whaling industry in the North Pacific made the western coasts of the Americas especially appealing to British traders. Before the construction of the Panama Canal, however, the trip to the western coasts of the Americas was fraught with difficulties. Vessels had to round the southern coast of South America at Tierra del Fuego, a treacherous passage filled with dangerous channels, rough currents, small islands, and uncharted coastlines. The Admiralty, therefore, sanctioned HMS *Beagle* to finish the survey of the coasts off the southern tip of South America. The *Beagle*'s former captain had committed suicide, and the position was offered to the young Robert Fitzroy, a gentleman who could trace his lineage to English nobility. Life on board ship in the early nineteenth century, however, was as hierarchical and stringent as British society itself, and the life of a gentlemanly captain, who spent most of his time attending to his own duties, could grow exceedingly lonely. Fitzroy was less worried about the previous captain's suicide than he was of his own uncle's; depression seemed to run deeply through Fitzroy's royal blood.

Fitzroy's solution was to acquire a companion on the voyage, someone of his own rank and station. He made this request to Francis Beaufort, who approached John Henslow, who in turn suggested his former student, the young Darwin. Darwin was elated, but his father was not. Darwin had gone from medicine to the ministry and now to what? There was no career in traveling as a naturalist; Darwin was merely a passenger on a journey around the world, paid for by his father, with no career prospects upon his return.

After Darwin's father acquiesced to the voyage, the young Darwin was a bustle of energy and enthusiasm. He needed to acquire instruments, buy books and paper, and procure the materials that he would need to preserve his botanical and zoological specimens. Henslow agreed to take charge of the material Darwin sent back during the long journey, and he gave the aspiring naturalist a parting gift, a book by Charles Lyell published only a few months earlier, that Darwin was to read on his long voyage. Even the title of the book, *Principles of Geology*, was exciting, a direct parallel to Newton's *Principles of Mathematics*. Lyell outlined not only a new theory in geology but also the principles behind scientific reasoning in general. His object was to free the science from theology, to base the study of geology on natural, not supernatural, processes. Although the text itself helped usher in the specialization of the discipline, at the time of its writing geology was still connected with the organic as well as the inorganic past. Thus, it included discussions of the origins of animal and plant life, including man. Lyell argued against the transmutation of species, what we would term evolution, but he did not object on religious grounds; he limited the discussion to natural phenomena, arguing that perceived changes should be studied according to causes, similar in kind and degree, to those now seen operating in nature. Whewell would later term this "uniformitarianism," where the slow, steady processes of change required an age of Earth far surpassing that laid down in Genesis. Published in 1830, Lyell's text gave scientists, theologians, and the educated public almost three decades to deal with the seemingly divergent views of science and religion before the shocking revelation of Darwin's *On the Origin of Species* (1859). With these ideas waiting for him on his long journey across the Atlantic, he and Fitzroy, two adventurous gentlemen in their early twenties, departed from the shore of the tiny island of Britain with the entire globe to explore.

The excitement turned quickly to seasickness as the landlubbing Darwin was tossed to and fro in his cramped quarters aboard the *Beagle*. How wonderful must have been the vision of the tropical forests of Brazil from the all-too-familiar deck of the *Beagle*. "I formerly admired Humboldt," Darwin gleamed, "I now almost adore him; he alone gives any notion, of the feelings which are raised in the mind on first entering the Tropics."[2] The *Beagle* sojourned for three months in the tropics of Brazil, offering Darwin his first experience of the abundant forests teeming with uncataloged plants and insects. From there, the *Beagle* disembarked to the coasts of Tierra del Fuego, the seat of its main surveying duties. Fitzroy and the crew of the *Beagle* spent almost two years traversing the route between Tierra del Fuego and Montevideo (in present-day

An 1890 painting of the HMS Beagle *in the Straits of Magellan. It was during the journey of the HMS* Beagle *that Darwin collected much of his scientific data that he later used to formulate the theory of evolution through natural selection. (Bettmann/Corbis)*

Uruguay) along the southeastern tip of South America. Darwin did not always stay with the ship but, following Humboldt, took extended trips to the interior of the continent as often as he could, including extended treks into the heart of the Andes and hundreds of miles into the interior of Patagonia.

Darwin's field trips around Shrewsbury and Cambridge, his enthusiasm for Lyell, and the popularity of geology itself made him travel with his gaze toward Earth and its geological strata. Darwin became an avid fossil hunter. Many of the fossils he found were similar to living species; others were different but closely related. Even more peculiar, in the Andes he found fossilized seashells thousands of feet above sea level. As he was pondering the relationship between fossils and living animals, a destructive earthquake rocked the coastline and interior of Chile while the *Beagle* was anchored off its shore. Hundreds of miles of the Chilean coast moved like the waves of the ocean, drastically changing the configuration of the land and opening up a geological laboratory for Darwin. He noted that near the shore, the land had been raised up, stranding above the ocean shellfish that then dried in the hot South American sun. Here Darwin saw firsthand Lyell's theory in action, the slow and gradual change of the land, additive through eons. This could explain the fossils in the mountain regions of the Andes, which Darwin reasoned had once been an ocean bed. Lyell was correct; geological time was expansive, even unfathomable. But more questions than answers followed. The fossils that Darwin found suggested that there were numerous extinct species, perhaps as many as were alive at the time, but they differed in size and variety according to their geographical relation to other fossils and existing species. What could explain the close relationship between fossils and existing species, and why were fossils similar but not entirely the same in different regions?

The expansiveness of geological time was not the only lessons that Darwin gleaned in his years of journeying through the inland of South America. Geographical space was equally important. Following Humboldt, Darwin studied the relationships among vegetation, animal life, and the geography of South America. Some species lived only in particular areas, while others seemed to thrive almost anywhere. Darwin found that closely related species occupied adjacent areas, a distribution that no past theory could explain. He was especially struck by the similarities of species that seemed to coexist in the same geographical areas. While searching for dinner, one of Darwin's traveling companions shot what appeared to them all as a premature rhea, a flightless bird resembling an ostrich. They prepared it for dinner and had nearly finished eating it when Darwin realized its significance. It was not a small rhea but rather a different kind of rhea entirely. He gathered what was uneaten and sent it back to England for analysis. Named *Rhea darwini*, it was the first new species that Darwin discovered. But why were there two different species of rhea in South America, and why did they differ from ostriches on other continents? Similar questions but few answers followed as the *Beagle* left the shores of South America for the Galápagos Islands.

South America offered Darwin firsthand experience of a large and largely unexplored continent, and he sharpened his skills as a biogeographer. He would put those skills to work as the *Beagle* voyaged some 600 miles off the coast of Ecuador to the Galápagos Islands, an island chain made up of many large islands and innumerable smaller ones. Formed of volcanic lava, these islands were populated by flora and fauna native only to the Galápagos. Of special concern to Darwin were the many different kinds of tortoises and finches. The finches on separate islands seemed to resemble each other closely but differed in significant ways, especially in the size and shape of their beaks. Again, more questions than answers followed. Why such abundance on such small and seemingly insignificant islands, and why the slight differences? As the questions mounted, Darwin became increasingly uneasy as to how the Scriptures and natural theologians attempted an answer. His observations of the massive number of species that populate the globe began to undermine his faith that God had created each species individually. But Darwin's own answers would have to wait for his return to England.

After a five-year circumnavigation of the globe, Darwin was indeed anxious to return home. Once the *Beagle* landed, Darwin surprised his family at their home, interrupting the middle of their morning breakfast. His father wryly noted that the shape of his head seemed different, a jocular way of expressing what everyone already knew. Darwin had been forever changed. He had left England a young, aimless student and returned home a mature, able-minded natural philosopher. Some of the letters that Darwin had sent back during his voyage had been published, and the scientific elite in London welcomed him into their circle.

In *On the Origin of Species*, Darwin wrote, "When on board H.M.S. *Beagle*, as naturalist, I was much struck with certain facts in the distribution of the inhabitants of South America, and in the geological relations of the present to the past inhabitants of that continent."[3] These opening remarks testify to the importance of the voyage for Darwin's theory. But it is well to remember that Darwin had not come to his theory on the voyage of the *Beagle*. He became increasingly skeptical of the accepted accounts of creation and was forced to ask what turned out to be the correct questions, but he had no answers upon his return. Even the significance of his research in the Galápagos would have to wait. It was only after returning to London that the ornithologist of the Zoological Society, John Gould, told Darwin that his collection of finches were separate species. Although Darwin's fieldwork throughout South America and the Galápagos certainly turned his attention to the transmutation question, he did not start formally thinking about an evolutionary theory until a couple of years after his return to London. Contemplating his research and preparing it for publication proved as important as the research itself.

Upon their return to England, Fitzroy immediately began to prepare his survey for publication and offered to have Darwin's account of the voyage attached as the third volume. As Darwin put his journal together, the answers to his many questions of the origin of species also slowly emerged—no flash of intellect, no eureka moment, simply the accumulation of clues and hints that began to disabuse the accepted theory and lead, little by little, to the astonishing conclusion that species were perhaps mutable. By the summer of 1837, Darwin began his first notebook dedicated to transmutation. He became more and more convinced that transmutation did indeed occur, but he had no mechanism for that change until he read the Reverend Thomas Malthus's *Essay on the Principle of Population*. Malthus had argued that human populations always outstripped their food supply, leading to disastrous consequences of famine, sickness, and death. It struck Darwin that if he applied Malthus's theory to the animal and vegetable kingdoms, the pressure of overpopulation could give Darwin the mechanism that he needed. He realized that all offspring varied from their parents. Those variations in a species that helped in the struggle for existence were passed to the next generation, while those variations that hampered survival naturally disappeared. Over many successive generations, nature thereby produced new species better adapted to their environment. All of this, however, would have to wait, as Darwin was thrown into the spotlight of success for entirely different reasons.

Both Fitzroy and Darwin published their accounts of the *Beagle* voyage in 1839, and though both accounts were received with enthusiasm, Darwin's was by far the more popular. It quickly went through many editions, propelling Darwin to the unexpected position of a publishing celebrity. He also worked on his geological research including his theory of coral reefs, which he had formulated on the return journey of the *Beagle*, outlining what is still our basic understanding of their formation. The *Beagle* voyage around the world had catapulted Darwin to the position of a rising star, not as an evolutionist but as a popular writer of an extremely successful travel narrative and one of the premier geologists in England. No one yet knew of Darwin's other leanings, what Darwin had worked on secretly from his country home in Down, 20 miles from the fashionable streets of London. Indeed, the secrecy tormented Darwin, and he began looking for someone in whom he could confide. He wrote a telling letter to a fellow naturalist and voyager in 1844: "At last gleams of light have come, and I am almost convinced (quite contrary to the opinion I started with) that the species are not (it is like confessing to a murder) immutable."[4] The murderous confession from the aspiring naturalist was his remarkable theory of evolution through the mechanism of natural selection, and the naturalist with whom Darwin confided was Joseph Dalton Hooker.

Joseph Dalton Hooker

Joseph Hooker was the perfect person for Darwin to invite tentatively, ever cautiously, into his confidence. Hooker was a botanist from a family of botanists and was much like Darwin in training and connections. Hooker's father, William Jackson Hooker, was a famous naturalist who had traveled throughout the Scottish Highlands, all over the Continent, and finally, on the recommendation of Sir Joseph Banks, to Iceland on a voyage of exploration. The elder Hooker accepted the chair of botany at the University of Glasgow and he added significantly to its botanical holdings with a herbarium that rivaled the top plant collections in the world. This led to his appointment in 1841 as director of the Royal Botanic Gardens at Kew, which had fallen into relative disarray since the death of Joseph Banks in 1820. Two generations of Hookers became directors of Kew Gardens. Together, over the next forty years, father and son transformed Kew into the premier botanical research institution in the world, the center for economic botany with connections to colonial botanical gardens overseas.

Joseph Hooker was by all accounts his father's equal. At the age of fifteen he enrolled in the University of Glasgow and, like Darwin, spent his college days botanizing and collecting throughout Scotland and England. Finished with his medical studies at the age of twenty-one, Hooker was ready to begin a career as a botanist. "From my earliest childhood," he confided to Darwin later in life, "I nourished and cherished the desire to make a creditable Journey in a new country . . . as should give me a niche amongst the scientific explorers of the globe I inhabit, and hand my name down as a useful contributor of original matter."[5] Hooker would go on two major voyages of discovery that, as with other aspiring naturalists, would help establish his career as a world-renowned botanist.

Sir James Clark Ross was set to sail to the Antarctic in search of the southern magnetic pole in 1839, and Sir William Hooker, who knew Ross personally, used his friendship and influence to have his son assigned as assistant surgeon on HMS *Erebus*. The objective of the voyage of discovery was decidedly scientific, and though Antarctica was hardly the bastion of botany, the four-year sojourn of the *Erebus* and *Terror*, with its stops in New Zealand, Tasmania, and the Faulkland Islands along with three separate jaunts into the Antarctic Circle, offered the young Hooker plenty of unknown species to plunder and ponder. He was a careful collector with a vast knowledge of botany, an exceptional researcher with an extraordinary stamina for work, and, like his father, a talented artist. Published in separate installments upon Hooker's return to England, *The Botany of the Antarctic Voyage of HMS Discovery Ships Erebus and Terror* established Hooker as one of the leading botanists of his day.

Hooker had met Darwin before this voyage, but upon his return they became correspondents tethered by an awesome secret. Within only four short months of Hooker's return, Darwin confided in him his theory of evolution through natural selection. This intimation and their continued friendship benefited both. Hooker was determined to go on yet another voyage of discovery, and with his father's support and a letter from the by then famous Humboldt, Hooker traveled to India and spent three years in the Himalayas, traversing the states of Sikkim and eastern Nepal. He was the first naturalist in history, including even Darwin himself, to explore a new world armed with Darwin's theory of evolution. Darwin, in turn, asked him questions in letter after letter concerning the

Himalayan flora and fauna—their distribution and habits, size and appearance. Hooker's choice of India was also propitious for both. As he acknowledged in his *Himalayan Journals*, he wanted to add a careful study of the natural history of the temperate zones to his previous research of Arctic flora and fauna. This was information that Darwin could also use, as Europeans had largely ignored the Himalayas owing to the extreme geographical conditions and unfriendly natives. "We were ignorant even of the geography of the central and eastern portions of these mountains," Hooker lamented, "while all to the north was involved in a mystery equally attractive to the traveler and the naturalist."[6]

Hooker traveled as a naturalist on a voyage that was supported by the Crown, and thus he had Crown duties to attend to. It fell to Hooker not only to gather the natural historical treasures of the Himalayas and send them back to Kew but also to further imperial ambitions. He was to report on the possibilities of cultivating cotton, tobacco, sugar, and other plants, a duty he performed as an economic botanist. Moreover, he had orders to explore, survey, and map the uncharted territories he visited. In Sikkim, "ground untrodden by traveller or naturalist," Hooker produced an exceptionally accurate and detailed map, published by the Indian Trigonometrical Survey and subsequently used for numerous military campaigns.[7] In Hooker, one finds the perfect blend of naturalist and imperialist.

To Hooker's accomplishments in botany and surveying we can also add mountaineering. On a three-month expedition to the interior, he explored and mapped several passes between Nepal and Tibet, banked by the sheer cliffs of the Himalayas. Resembling Humboldt both in style and content, Hooker's published accounts often invoked the sublime in describing the rugged terrain. "Any combination of science and art can no more recall the scene, than it can the feelings of awe that crept over me, during the hour I spent in solitude amongst these stupendous mountains."[8] He fought his way through ice and snow across the Donkia Pass between Sikkim and Tibet from which he attempted to scale the glaciers of Kinchinjunga, then thought to be the highest mountain in the world. Suffering from altitude sickness and little help from his guides, he climbed above 19,000 feet, even higher than Humboldt.

Hooker's adventures in mountaineering are important in exemplifying the personality of most Victorian scientific travelers, many of whom had a penchant for vigorous outdoor activity. Raised in polite society, Hooker was nevertheless at home sleeping on the ground, either huddled by himself in a homemade tent or simply beneath the stars. He hiked for miles each day, all the time botanizing, collecting, surveying, and thinking. He practiced his science outdoors, an experience that was part of the process of knowledge creation. Becoming one with nature is not merely a twentieth-century phenomenon but is evident in the

Sir Joseph Dalton Hooker spent three years traveling in the Himalayas, traversing the states of Sikkim and eastern Nepal, the first naturalist to explore the region armed with Darwin's theory of evolution. He later became director of Kew Botanical Gardens, a center of botanical research and an invaluable institution of British imperial expansion. (Hulton-Deutsch Collection/Corbis)

nineteenth century, from Ralph Waldo Emerson and Henry David Thoreau in America to Humboldt and Hooker in Europe. The sheer beauty of the wilderness and one's experience with it, oftentimes a rough and dangerous experience, was an active ingredient in the naturalists' tradition. Armed with the dangerous combination of imperialist mentality and moral sanctity, the Victorians spread though the uncharted wilderness like no other society before or since.

Hooker returned from the Himalayas to Kew in 1850 to arrange, name, and distribute his collection of more than 6,000 species of India flora to more than sixty public and private herbaria throughout Britain and Europe. Hooker described these new species in his *Himalayan Journals* (1854), which included maps, illustrations, and a powerful narrative filled with daring climbs and close encounters with the Sikkim authorities. He also discussed plant geography and offered veiled evidence that supported Darwin's theory. Hooker was not a convinced evolutionist on his journey to India, but his work in taxonomy and his research in the biogeography of plants helped sway his allegiance. He dedicated his *Himalayan Journals* to Darwin as a sign of friendship and respect, and by the time Darwin's *On the Origin of Species* appeared in 1859, Hooker was a staunch supporter of evolution, actively defending Darwin against vociferous critics.

Along with his role in the eventual acceptance of Darwin's theory, Hooker directed his genius to enhancing the position of Kew within both British society and its imperial possessions. He became assistant director of Kew in 1855 and director after his father's death a decade later and was intent on continuing his father's efforts to make Kew a center of economic botany. Again, his imperialist leanings shined through. He turned Kew into the center of an international exchange network that followed the Crown geographically and helped convey its authority politically. It consolidated materials discovered around the world and made them useful for Britain's industrial and imperial missions. Rubber seedlings were transported from Latin America to Kew and then from Kew to Ceylon, spawning a massive rubber industry in the East and West Indies. When telegraph cables instigated a need for insulation material, rubber was imported from Singapore. Sugar, tea, coffee, orchids, water lilies, and rhododendrons all traveled the trading routes, from their native shores to Kew and then from Kew to their nascent plantations in Britain's growing imperial possessions. Kew became a useful institution not only for the botanist and physician but also for the banker, the merchant, and the manufacturer.

Economic botany afforded the British Empire the possibility of new and economically viable crops while consolidating its power and reach. The case of cinchona is perhaps the most evident and familiar example. Cinchona is a genus of trees native to the Andes, and its bark had long been known to prevent fevers. The Andean natives had used it long before it was "discovered" by

Spanish conquerors and missionaries. Known by the seventeenth century as Jesuit's bark, Spain controlled its price and consumption, limiting its availability to wealthy Europeans. Both the British and the French attempted to secure cinchona seeds in the eighteenth and early nineteenth centuries. Hooker took up the effort in the 1860s, making Kew a center for the transportation of different varieties of cinchona seeds first to Kew for germination and then to plantations in the British colonies for harvesting. In midcentury, it was discovered that cinchona contains alkaloids such as quinine that could be used for fevers, especially malaria, thus opening a new door to the continent of Africa. Although the cinchona scheme was only mildly successful in Britain, its possibilities contributed to the partitioning and ultimate colonization of Africa in the last quarter of the nineteenth century. It also confirmed the central importance of Kew as a center of botanical research, giving legitimacy to the new science and the new scientist and thus adding to the prospect of professionalization so apparent in the mid- to late Victorian period. Kew Gardens became a symbol of power, legitimizing both science and the imperial process.

Hooker continued to be Darwin's confidant after returning from the Himalayas. Darwin's intimation of his thoughts on evolution to Hooker in January of 1844 suggests that Darwin was becoming more secure in his views and bolder in letting those views be known to the wider community of naturalists. Indeed, he was on the verge of publishing his theory when another book on evolution appeared that caused Darwin to pause. The *Vestiges of the Natural History of Creation*, published anonymously in 1844, was a work of obvious amateur status written in a rhetorical style that often verged on blasphemy. It included significant errors in biology, geology, and the physical sciences. Adam Sedgwick, Charles Lyell, and T. H. Huxley all denounced it in print, causing scientists to look upon evolutionary theories with added discretion. Darwin realized that the time was not ripe for another theory of evolution. Yet, *Vestiges* actually helped Darwin and the eventual acceptance of his theory in several ways. First, it was extremely popular. This moved the subject of evolution to the forefront of the scientific community and the general reading public. Moreover, not all of the book was unreasonable; it connected geological time with transmutation and argued that the biological sciences, like the physical sciences, were run according to fixed laws of nature. In many respects, therefore, it paved the way for Darwin's own argument. Second, the negative reviews by the leading naturalists in Britain prepared Darwin for what he could expect. He now knew the main lines of argument that would be waged against his theory, arguments he could pick apart in his own publication. A further surprise would hit Darwin fifteen years later, in 1858, when yet another evolutionary theory arrived at his doorstep. This one, however, had the opposite effect as

Vestiges; it caused Darwin to rush his own theory into print, as it contained a theory of evolution that was almost identical to Darwin's own.

Alfred Russel Wallace

The package at Darwin's doorstep in 1858, which included a cover letter and a succinctly written paper on natural selection, arrived from Alfred Russel Wallace, a little-known naturalist of an entirely different fold from Darwin. Whereas Darwin moved easily within polite society, unfettered by worries of career prospects or mounting debts, Wallace was from the lower end of the middle classes in British society. His pay as a teacher in a school in Leicester was modest as were his duties, which left him ample time to botanize and read. Two books especially stand out. The first, Humboldt's *Personal Narrative*, directed Wallace's attention to the tropics, while the second, Malthus's *Essay on the Principle of Population*, introduced him to what he termed "philosophical botany." Both were defining texts in Wallace's life, just as they were in Darwin's.

Wallace's growing interests in botany endeared him to another aspiring naturalist, Henry Walter Bates, two years his junior and as interested in beetles and butterflies as Wallace was in plants. The two spent their leisure hours together collecting as field naturalists. But Wallace was becoming restless in the confines of Britain. "I begin to feel rather dissatisfied with a mere local collection; little is to be learned by it," he lamented to Bates in 1847. "I should like to take some one family [of plants] to study thoroughly, principally with a view to the theory of the origin of species."[9] With no direct government sponsorship and no official positions in the Royal Navy, Wallace and Bates headed for South America into the jungles of the Amazon.

Wallace traversed the rain forests of the Amazon for four years, collecting birds, beetles, butterflies, and whatever else of nature's bounty that would catch a good price. Wallace's agent back in London, Samuel Stevens, helped find buyers by advertising Wallace's specimens at meetings of the Entomological Society. He also made certain that Wallace's letters were published in natural history magazines, thus ensuring that even though Wallace was tramping through the jungles thousands of miles from London, his name was becoming known in scientific circles. Wallace's travels through the Amazon helped him mature as a naturalist and allowed him to gather material on the geographical distribution of species, what he believed to be the key to the origin of different forms of life. After parting with Bates, Wallace returned to England in the summer of 1852 intent on preparing his momentous experiences for publication. Yet even before the publication of *A Narrative of Travels on the Amazon and Rio Negro* (1853),

Alfred Russel Wallace (1823–1913)

Born the eighth of nine children, Alfred Russel Wallace received little formal schooling, although he read voraciously, including travel tales such as *Gulliver's Travels* and *Robinson Crusoe*. In 1837, the year that Darwin began his secret notebook on transmutation, Wallace was fourteen and apprenticed to his brother to learn the art of surveying, a career that allowed him to spend the next six years outdoors. He also picked up a lively interest in botany, a hobby he would take from the rolling hills of England and Wales to the deep forests of the Amazon. After his brother's death Wallace moved to Leicester, where he began a second career as a teacher at the Collegiate School and continued his interests in botany. He also met Henry Walter Bates, an aspiring naturalist, and together they planned a trip to the Amazon to work as collectors for museums and natural history collections in Europe. They botanized and collected throughout the Amazon River basin, and Wallace intensely studied the geographical distribution of the continent's diversity of flora and fauna with an eye toward solving the origin of species.

Collecting in South America proved a dangerous occupation, with threat from disease, hostile natives, and even more hostile animals and insects. After a shipwreck on his return journey, Wallace sold what little was saved of his specimens and prepared his momentous adventures for publication. *A Narrative of Travels on the Amazon and Rio Negro* (1853) helped him make a name for himself as a scientific traveler. The trip to the Amazonian jungle had added to Wallace's thoughts concerning species, and he particularly noted the role of barriers, in this case the Amazon River, that geographically separated different species of flora and fauna. But he had not solved the question of the origin of species. He departed the following year for the Malay Archipelago, where he spent eight years collecting prize specimens, including a unique type of bird of paradise, now Wallace's Standard-wing, and undertook fundamental research in geology, geography, and ethnography. Wallace spent January to July 1857 on the Aru Islands, off the coast of New Guinea, and it was there, while fighting a bout of malaria, that he developed his thoughts on the mechanism of natural selection.

Wallace returned to England in 1862 with enough material to publish numerous books and articles on evolutionary theory and the distribution of plants and animals, including his classic travel narrative *The Malay Archipelago* (1869) and the two-volume *Geographical Distribution of Animals* (1876). He outlined the distribution of animals in the Malay Archipelago, dividing two very different kinds of fauna, a break that is now know as Wallace's Line. He continued to publish throughout his long life, extending his work on the geographical distribution of plants and animals to small islands in *Island Life* (1880). Yet increasingly after his return to England, Wallace separated man from the evolutionary process, a reflection perhaps of his growing interest in spiritualism. His later life was filled with fruitful controversy defending his evolutionary views from critics and his views on man from the evolutionists.

Wallace was planning another trip abroad, this time to the Malay Archipelago (what is now Indonesia and Malaysia) for an excursion that would last the better part of a decade. His adventures in the Malay Archipelago transformed him into a trained evolutionist. While in England, he read Chambers's *Vestiges* with a slightly different eye than the British scientific elite, focusing on its possibilities rather than its problems, on its overarching theory rather than its specific shortcomings. From then on, Wallace was convinced that species were indeed mutable. He therefore explored the Malay Archipelago not merely as a collector but as an aspiring naturalist keen on determining that mystery of mysteries, the mutability of species.

Wallace found an abundance of fauna in the Malay, species of insects and birds that Europeans had yet to catalog. He forwarded his prize possession, a unique type of bird of paradise, to the British Museum, where it was named "Wallace's Standard-wing," the first of its kind seen in London. He also came across the orangutan, an especially interesting creature for Victorians because it resembled so closely human habits and actions. Wallace anthropomorphized its description after shooting a mature female and keeping her infant as a pet. By far the most important period of his journey to the Malay was spent on the Aru Islands, off the coast of New Guinea, from January to July 1857. Not only were these islands teeming with new and exceedingly valuable specimens, but it was here that Wallace had his own Malthusian moment. As Wallace recounted some years later, he was hit by a horrible fever that confined him to his bed for days while he suffered through bouts of delirium. Between dream and reality, Wallace's thoughts turned to Malthus's *Essay*, a book he had read many years earlier. All seemed to become clear. "I waited anxiously for the termination of my fit so that I might at once make notes for a paper on the subject," recounted Wallace. "The same evening I did this pretty fully, and on the two succeeding evening wrote it out carefully in order to send it to Darwin by the next post."[10] Did Wallace know that his revelation would change our conceptions not only of animals but also of humans? He had produced a concise and masterful theory of evolution through natural selection, the exact theory that Darwin had been contemplating for more than twenty years.

Although historians have portrayed Wallace's letter to Darwin as a "bombshell," it was not necessarily a surprising or unexpected blast. Darwin and Wallace had been corresponding for two years, certainly though perhaps tentatively broaching the subject of evolution. Indeed, Wallace had published a paper titled "On the Law Which Has Regulated the Introduction of New Species" in 1855 in which he argued as an evolutionist. Darwin was impressed with the subtlety of Wallace's arguments and was advised more than once by Joseph Hooker and Charles Lyell to publish his own research before his ideas were scooped. With naturalists ensconced in a culture of competition, combing the seas, rivers, and mountains across British possessions, it was perhaps only a matter of time. The fact that not one but two British naturalists formed similar theories of evolution demonstrates most clearly the role of social and cultural forces in shaping scientific knowledge. Both had sailed on voyages of discovery and traveled inland in South America, both had read Humboldt and Malthus, and both could appreciate the struggle for existence among animals and plants as well as among imperial nations and native populations.

Darwin was caught in a dilemma, conscious of his own priority but torn by his Victorian sense of morality to be fair to the young naturalist. Unfortunately, he was spared from having to make any decisions on how to proceed. Soon after receiving Wallace's paper, Darwin's son caught scarlet fever and did not recover. Darwin was devastated by his son's death, and the quarrelsome problem of priority fell to others. Hooker and Lyell convened an impromptu meeting of the Linnean Society on 1 July 1858 at which they established Darwin's priority. Wallace's paper, a succinct account of the process of natural selection, was read only after Darwin's own account. With Wallace in the Malay Archipelago, and considering the differences between the two in terms of class and connections, it was probably the fairest of all outcomes. Darwin had indeed been working on natural selection for more than twenty years, and he combined his researches the following year in *On the Origin of Species*, the most significant text in the history of biology. As Wallace intimated after reading Darwin's *Origin*, he was struck by the comprehensiveness of Darwin's argument, and was content that he was not the one to have to bring the theory, in all its force, to the Victorian public. Moreover, the most dreaded part, convincing others of its validity and extending the theory to man, was still to come, and there the codiscoverers of evolution through natural selection increasingly parted ways.

Four years after their joint publication, with an abundant collection of thousands of specimens from the Malay Archipelago, Wallace returned to England with enough material to keep him busy for a lifetime. *The Malay Archipelago: The Land of the Orang-Utan, and the Bird of Paradise* (1869) became an instant success. It went through numerous editions and was widely read in Britain, leading to Wallace's reception among the British scientific elite. Other factors, however, prevented a warm welcome. Wallace entered a scientific world that was being reshaped; scientists were attempting to professionalize the discipline of biology. Although Wallace had successfully entered the world of science from the outside, the new profession's requirements were no less harsh and regimented than the old scientific establishment. The new army of professionals waged turf battles against spiritualism and other forms of alleged pseudoscience just as fiercely as they had attacked the old columns of religion and social hierarchy. Wallace's own spiritual leanings pushed him to the outskirts of the new community. In the mid-1840s, at the same time that he had read Humboldt and Malthus, he was also exposed to mesmerism and phrenology, studies increasingly excluded from the scientific establishment. His move to spiritualism upon his return to England, combined with his socialist ideas on land nationalization and forward-looking views on women's rights, added to his marginalization. His political and spiritual views form part of the reason that he has received less attention than Darwin as a codiscoverer of natural selection. More important, however, were his views on the application of natural selection to man.

Throughout his travels, Wallace had as keen an eye for anthropology as he had for looking down the barrel of a gun at exotic specimens. He undertook careful studies of the language and customs of the different peoples he encountered, especially the indigenous populations of the Indo-Malayan region. As his earlier anthropomorphized discussion of the orangutan confirms, he was always fascinated with man's close, perhaps filial, relationship to the other species. Yet once in England after 1862, his views began to change. Only humans had advanced reasoning faculties, a quality that Wallace believed natural selection could not explain. For Wallace, these higher faculties suggested a higher intellect at work, and he concluded that man was not formed through the process of evolution through natural selection. Although his views were far from the divine creator as propounded by the natural theologians earlier in the century, they smacked of his spiritualist ties. The codiscoverer of natural selection who had never seriously considered the existence of God as having a place in explaining natural phenomena had turned the advance of science upside down. To a scientific worldview increasingly antagonistic to theological causes, Wallace reverted to a Deity. "I hope," Darwin wrote to Wallace, "you have not murdered too completely your own and my child."[11] Darwin had no need to worry, for the implications of their theory were clear enough. Darwin unabashedly included man in the natural process of evolution only in his later publications, and it was left to Thomas Henry Huxley, Darwin's self-appointed "bulldog," a ferocious intellect yielding an equally dangerous pen, to keep their child from harm.

Thomas Henry Huxley

Thomas Huxley's decision to join HMS *Rattlesnake* as assistant surgeon on a four-year surveying mission was not made rashly. A voyage of exploration on a naval ship could be exceedingly dangerous, he realized, but it also could launch his career, as exemplified by Darwin and Hooker, who returned to England with established reputations as men of science. By the mid-nineteenth century, the Royal Navy had become one of the ways in which naturalists obtained training in the field. Huxley had finished his medical degree at the age of twenty-one and, too young to become a practicing surgeon and too in debt to forgo a paying position, followed the well-trodden path of aspiring naturalists and joined the Royal Navy. His first assignment was to a military hospital, where he worked under the naturalist Sir John Richardson, a famed Arctic explorer. There Huxley honed his skills in dissection and comparative anatomy, and it was Richardson who recommended Huxley as assistant surgeon on the survey expedition of HMS *Rattlesnake*. With debts mounting, Huxley jumped at the chance. He would get the chance to philosophize and botanize in far-flung regions of the globe. The captain of the *Rattlesnake*, Owen Stanley, introduced Huxley to imminent naturalists such as Edward Forbes, an expert at dredging sea creatures from the ocean floor, and Richard Owen, perhaps the greatest of British anatomists. When Huxley finally received his official appointment on the *Rattlesnake*, he was elated, certain that he was destined, like Darwin and other explorers before him, to become a practicing scientist.

Armed with an observing eye, an equally fiery sense of optimism, and the whirl of adventure gleaned from reading Darwin's *Voyage of the Beagle* and Humboldt's *Personal Narrative*, Huxley joined HMS *Rattlesnake* as it pushed off the coast of England for a four-year surveying expedition. Huxley was always troubled with his place in the hierarchy of British society, and his future career was at the forefront of his mind throughout the voyage. He fretted constantly about his career prospects upon returning to England, but the voyage itself helped solidify his choice of professions. He pinned his hopes on science, a dangerous and precarious move in mid-Victorian England. Indeed, Huxley was betting his future on a career path that he himself would have to invent.

The objectives of the *Rattlesnake* voyage followed the routine of surveying vessels intent on mapping the shores where British merchants and the British military were sure to sail. The *Rattlesnake* was to survey the northern coasts of Australia and New Guinea, including the Torres Strait. This was an area of particular interest to the Royal Navy, as it was a passage frequented by British vessels returning to England. Thus, like the *Beagle* voyage, the surveying duties of the *Rattlesnake* had a distinctly commercial motive: to make the seas safe for British vessels and to reconnaissance new areas for British colonies.

While the crew members attended to their surveying duties, Huxley created his niche as a naturalist. The southern Atlantic provided Huxley with his

Thomas Henry Huxley (1825–1895)

If Darwin represented the Cambridge-educated, upper-middle-class naturalist with the best breeding and the right connections and Wallace represented the self-educated, lower-middle-class collector with neither breeding nor connections, Thomas Henry Huxley falls somewhere in between. Born above a butcher's shop, his parents' seventh and youngest surviving child, he was a focused child, fascinated with the living world around him, immersing himself hodgepodge in the natural sciences by way of his father's library. At fifteen, Huxley moved to the bustling streets of North London to work as a medical apprentice to his brother-in-law. A scholarship at Charing Cross Hospital helped toward the completion of Huxley's medical degree at the age of twenty-one. He then joined the Royal Navy, where he eventually procured a position on the survey expedition of HMS *Rattlesnake* as an assistant surgeon.

Huxley's decision to join the crew of the *Rattlesnake* helped solidify his choice of a career and moved him even more toward development and invertebrate anatomy, helping him to form his own niche where Darwin and Wallace had yet to wander. Huxley returned to England in 1850 and published more than twenty scientific papers from the research he undertook during the voyage. In 1854 he was appointed lecturer in natural history at the Government School of Mines, where he taught for more than thirty years, innovatively incorporating laboratory work into his teaching. When he was appointed naturalist with the Geological Survey, his interests and research increasingly turned toward paleontology and geology. He became an ardent student of mammalian and reptilian fossils. This background first in invertebrate zoology, then in vertebrate zoology, and finally in paleontology and physical anthropology prepared him to relentlessly defend Darwin's and Wallace's theory of evolution through natural selection. Through learned journals, public debates, and working-class lectures, Huxley combined a strong wit and an equally subtle pen to defend evolution from its critics. He synthesized this work in his *Evidence As to Man's Place in Nature* (1863), unabashedly linking man to the process of evolution.

Huxley's interests and expertise spanned the entirety of the life sciences: biology, paleontology, geology, zoology, botany, anthropology, physiology, and developmental morphology. Although he is remembered as "Darwin's bulldog," the most relentless defender of evolution in the mid-nineteenth century—including Darwin himself—this was just one part of a larger goal of freeing science from the shackles of religious orthodoxy. Huxley worked assiduously in revamping science education and is one of the scientists responsible for introducing biology into the university curriculum. Both his war on theology and his interest in university education were part of his broader move to professionalize the discipline of biology. He used evolutionary theory as a double-edged sword to both attack theology and to replace the authority of religion with the temple of science. In this sense, he played a valuable role in the authority given to science and the scientists to the present day.

first experience with exotic oceanic flora and fauna, specifically crustacea, sponges, and the like. After minutely studying his most exotic find, a Portuguese man-of-war, he became all too conscious of the limitations of earlier taxonomic systems, especially that of Georges Cuvier. Huxley's attention began to coalesce around the classification of organisms. Trained in medicine with a special interest in physiology, he differed from the earlier voyaging naturalists such as Darwin and Wallace. Huxley specialized in dissection and the minute examination of the exotic organisms he caught in his nets or dredged from the ocean floor. While dredging on board the *Rattlesnake* on the long journey to Australia, he noted that the body of a jellyfish resembled those of polyps and similar sea creatures. He wrote up his research, and Stanley helped him to get it published. Stanley sent the paper to his father, the president of the Linnean Society, who forwarded it to the Royal Society of London, where, unbeknownst to Huxley, it was read and subsequently published in the *Philosophical Transactions* in 1849. This paper, along with several others sent while Huxley was aboard the *Rattlesnake*, helped him become known as a first-rate comparative morphologist and secured his election as fellow of the Royal Society while he was still abroad.

Sydney, Australia, served as the *Rattlesnake*'s home base between its four surveying missions that formed the bulk of its duties. It was very much like home, with an active medical community and an even more active social life. Parties, receptions, and other engagements continuously called Huxley away from his scientific work, but one development in particular caught his attention. Her name was Henrietta Heathorn. As rash and confident as ever, Huxley asked her to marry him after only a few meetings. Their actual marriage, however, would take seven years longer, because Huxley's uncertain future and family council all corroborated against any rash decisions. Huxley first had to prove himself in his chosen career.

After three months in Sydney, the *Rattlesnake* embarked on its four surveying missions in the waters north of Australia. These missions took Huxley to some of the most remote regions of the world, including sojourns to the interior of the relatively unknown Australian continent. With a fiancée added to his already difficult future prospects, however, Huxley fell into depression and complacency. The third surveying tour took him to the Great Barrier Reef, which overflowed with natural wonders, yet Huxley could do no more than wish he could work. It was only during the fourth surveying tour, to New Guinea, that he once again became fascinated with the living creatures around him. This time, it was not the jellyfish and crustacea as before but rather the aboriginal culture of the jungle island that captured his attention. On New Guinea, Huxley found an island on the edge of civilization. It was, he wrote, "perhaps, the very last remaining habitable portion of the globe into which European cruisers and European manufactures had not penetrated."[12] He studied everything he could about the aborigines' culture—their mode of fishing, eating, sleeping and how they dressed. He took notes, made drawings, and attempted to communicate with the natives. He also met Teoma, a young woman of twenty who first appeared to Huxley as a native. She turned out to be an Englishwoman from Aberdeen, shipwrecked on the Torres Straits, rescued by natives, and now a part of the culture that Huxley was so intent on studying. As a passenger on the *Rattlesnake* for the rest of the voyage, Teoma served as an informative guide for Huxley as he studied aboriginal culture.

Surveying was a delicate, time-consuming, and exceedingly repetitive process, especially hard on the captains of ships, as witnessed by the suicides of the *Beagle*'s captains. The same held true for Captain Stanley. After giving his heart and soul to his duties, leaving no reef or sandbar uncharted, he suffered an epileptic fit and died while on board the *Rattlesnake.* The ship and its crew were duly ordered to return to England, with Huxley intent on making his way in British scientific circles.

Huxley returned to England in 1850 a new man. Brimming with self-confidence, he now had four years of experience in the field. He had dredged, dissected, studied, and sketched the intricate structures of different marine life more closely than any other naturalist, and his publications preceded his arrival in England and assisted in his reputation as a scientist. He was now an experienced field biologist who had traveled to exotic regions of the globe and had seen nature, with all its diversity, firsthand. More importantly still, he had gained extensive knowledge of aboriginal culture. His experiences, like those of Darwin and Wallace, had widened his view of nature as well as man's place within it. Just as his careful examinations of the Portuguese man-of-war had led him to question the outmoded classification system of marine life, his studies of primitive cultures led him to consider the problems of the human species in zoological classification. His four-year voyage had prepared him to view species as transmutable, and for Huxley, man was one such species.

Darwin's publication of *On the Origin of Species* in 1859 was a revelation for Huxley. He became an early advocate of the theory; indeed, he thought himself quite blockheaded for not thinking of it himself. Darwin's and Wallace's theories unified Huxley's own research, giving it added direction. His work in comparative anatomy, physiology, and the geographical distribution of animals became part of a common goal to resurrect the lineage of Earth's species and place them within a proper taxonomy. Moreover, the theory of evolution through natural selection detatched the investigation concerning the origin of species from religion and placed it squarely in the purview of the new profession of science that Huxley was so incessantly fostering.

Darwin had shied away from including man in his theory of evolution, but Huxley well appreciated Darwin's cryptic intimation in *On the Origin of Species* that "light would be thrown on the origin of man and his history."[13] Indeed, Huxley became the luminous force casting shadows across the philosophical halls of London, the manufacturing districts of northern England, and the industrialized nations of the world. Four years after the publication of Darwin's *On the Origin of Species*, Huxley published *Man's Place in Nature*, the title of which summarizes the important subject of the book. Huxley brought Darwin's and Wallace's theory to its logical conclusion, demonstrating the close connection between man and the apes, cousins on the evolutionary bush. He also discussed the recent discoveries of the fossils of Neanderthal that in turn held implications for man's relationship with primitive man, the Feugians and the Tasmanians. His text brought up all sorts of intriguing and scary questions. Were we unique? Are Feugians our brothers? Do apes have morals? Huxley brought the origin of the different races of humans to the forefront of the scientific community.

Although Huxley became the main champion of Darwin's and Wallace's theory, he did not follow them blindly. He defended their theory in print, in his own research, and in public debates, the most famous being the debate with

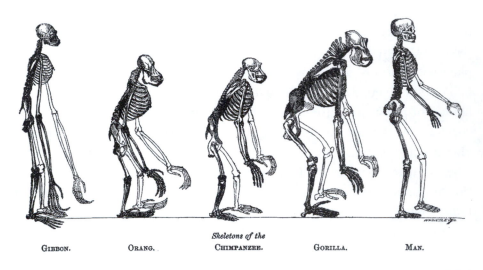

| GIBBON. | ORANG. | *Skeletons of the* CHIMPANZEE. | GORILLA. | MAN. |

The frontispiece from T. H. Huxley's Evidence As to Man's Place in Nature, *1863. As Darwin's most ardent and outspoken supporter, Huxley incorporated humans into Darwin's theory of evolution. (The British Library/HIP/The Image Works)*

Archbishop Samuel Wilberforce at the 1860 British Association for the Advancement of Science meeting. Yet as "Darwin's bulldog," Huxley often pulled his leash in brash and uncompromising directions. The origin of man was the last great bastion in science where, by midcentury, theologians still had a say in the discussion. Huxley used evolutionary theory to help professionalize biology, fostering the prospect of social advancement through paid positions in the new discipline. He demanded that those entering the field use naturalistic rather than theological arguments. He became a staunch defender of not only evolutionary theory but also of science more generally, including its administration, education, and role in society. Huxley was an evolutionist intent on spreading the gospel of evolution, but more importantly, he was a scientist equally intent on advancing the authority of science within broader culture.

Evolution through natural selection, therefore, was not accepted by the scientific community solely on the basis of the strength of its supporting evidence. Rather, Huxley and other middle-class scientists intent on professionalizing their discipline used evolution to justify their own views on the progressive developments occurring in mid-Victorian culture. Natural selection mitigated any sense of progress—a directed evolution toward something. However, the idea of progressive advancement toward higher levels of development (evolution but not natural selection) was widespread well before Darwin and Wallace published their biological theory. The German philosopher Georg Hegel advocated a progressive view of history, while the French philosopher Auguste Comte viewed human society as advancing in similar ways. In England, Herbert Spencer had reached an exceedingly popular audience with his views of the progressive development of civilizations. Spencer was a political economist who rose to become the spokesman for the rising middle class, the portion of society that was attaining an ever greater role in British society by midcentury. He advocated a free-enterprise, laissez-faire approach to governing the multitude whereby the natural laws of competition would lead to the advance of civilization. In his *Social Statics* (1851), published eight years before *On the Origin of Species*, he used evolutionary principles to justify a capitalist society free from government interference, arguing that the "survival of the fittest" would ensure progress.

The followers of Darwin, including Spencer and Huxley, incorporated Darwin's theory of evolution through natural selection into their already entrenched views of the progress of civilization. Biological evolution was seen as working through the same mechanisms as social evolution, where struggle and competition proved paramount. All too often, Darwin's "struggle for existence" was equated with Spencer's "survival of the fittest," where the "fittest" meant those of the industrial, rising middle class bent on thrift, industry, progress, and development. They extended the evidence of Darwin and Wallace from the realm of animals and plants to the organization of societies in line with what they experienced in the commercial and professional classes in which they were playing an active part. These advances, when combined with the imperialist acquisition of lands and the subjugation of native populations, had dramatic consequences. It often led to the justification of colonial and imperial ambitions and gave scientific credence to racist theories. To find a direct link between the advance of evolutionary principles in biology and racist principles in the advance of imperialism is extremely difficult, many scientists and most of the general public increasingly viewed the relationships among different races in hierarchical terms. This then justified the manner in which white Europeans subjugated native populations through the process of imperialism. With its roots firmly in the natural history of classification, which had gained so much ground in the eighteenth century, the direct study of native populations provided the means to produce hierarchies of races. Taxonomists placed white Europeans on top of such hierarchies, descending downward to other less "advanced" races.

Huxley found himself in a paradox. How can a law based on the ruthless competition in society lead to human morality? Huxley acknowledged that humans would never completely escape the law of evolution based on competition and self-interest; we would never escape our base instincts, the animal within us. This is what worried the Victorians the most and why many despised actions that lowered humans to the status of beasts, acts as different as slavery and vivisection, where the act itself, though harmful to the slaves and animals, was even more harmful to the dignity of man.

Huxley's answer in his later years was in direct opposition to Spencer and other staunch Social Darwinists. Huxley viewed the process of evolution itself as nonmoral, free from any inherent correctness or moral righteousness. Man's standards of right and wrong could not be justified, therefore, by reference to the "natural" process of evolution, as Spencer and other Social Darwinists of the nineteenth century increasingly referenced. For Huxley, man's intellectual capacity, so highly evolved from that of other beasts, had made the struggle for existence defunct; an ethical process that grew out of the evolutionary struggle for existence had taken its place. And, according to Huxley, our ethical process must revolt against our natural instincts. Ethics must combat nature. We should direct our attention, Huxley believed, not to the survival of the fittest but rather to fitting as many people as possible to survive.

Huxley believed that only civilized society could advance to this next step, a belief that links Huxley directly to the imperialism of his age and aligns him not only with Darwin, Wallace, Hooker, and even Spencer but also with the ideological underpinnings of Victorian society in general. Huxley was informed by at the same time that he informed Victorian imperialism. All of society was seemingly evolving, moving forward, from the more primitive to the most civilized. The advance of science played a major part in this progression, a belief with strong roots extending back to the Enlightenment. Science itself was also seen to be evolving—progressing forward and taking Western European countries with it. The societies with the most advanced science always appeared to be the most advanced culturally. In fact, the two were often viewed as the same in the popular literature of the time. Just as Huxley and Darwin were attempting the conquest of the hidden laws of nature, so too were imperialists attempting a conquest of nations. Mastery over nature and mastery over other nations were viewed as similar, both working though competition and struggle and both ending in progressive development.

For Huxley, Spencer, and others, moral progress could only come with social progress, the move from primitive man to civilized man. The moral responsibility, therefore, lay with the imperializing nations, a burden—the White Man's Burden, according to Rudyard Kipling—that they had to endure to help raise the more primitive cultures toward advanced civilization. This was, in the end, the essence of Social Darwinism, whereby survival of the fittest was the outcome of the conflict between races. For Spencer, this was a direct result of the laws of nature. For Huxley, it demanded a revolt against the laws of nature; we should use our morality to ensure progress within an otherwise unprogressive system of natural laws. Unfortunately, this did not always work according to Huxley's designs. The Tasmanians, of whom Huxley had firsthand knowledge, were on the verge of extinction, and other aboriginal cultures such as the American Indian were faring little better. As Wallace wrote, "It is the same great law of 'the preservation of favoured races in the struggle for life,' which leads to the inevitable extinction of all those low and mentally undeveloped populations with which Europeans come in contact."[14] Too often, progress and civilization equated into violent removal rather than peaceful progress forward.

Thus, like Darwin and Wallace, Huxley belongs within the tradition of the Enlightenment with its unabashed confidence in the power of rationality combined with a moral certainty in the imperial process. These ideological factors must be brought into any discussion of the link between evolutionary theory and racism. They demonstrate the extremely powerful and widely-held imperialist assumptions behind the scientist's work and how that work in turn was used to justify the process of European expansion and control. Assumptions about the power of evolution and the differences between human groups, therefore, developed concomitantly with both the advance of science and the progress of imperialism.

Conclusion

Voyages of exploration served many purposes in the nineteenth century. They surveyed uncharted waters, opened new territories to commerce, and carried missionaries, traders, and the military to the farthest ends of the globe. But increasingly in the nineteenth century, they also carried with them naturalists who mapped and taxonomized the flora and fauna of the world. The consequences were far-reaching for science and for European and indigenous cultures. The world served as a laboratory. Large, uncharted landmasses became home to some of the most important science practiced in the mid-nineteenth century. These voyages provided the naturalists with scientific training that they could not have received from a university education, private reading, or immersion in the scientific culture of the metropolis. Their descriptions of the world's astonishing diversity and the collections they amassed and brought to the European botanical gardens and museums transformed European consciousness, undermining eighteenth-century conceptions of the fixity of species and forever transforming Europeans' views of their own place in nature.

Voyages of discovery in the mid-nineteenth century increasingly became an accepted mode of entry into science for aspiring naturalists. Extending the reconnaissance of the great age of discovery in the late eighteenth century, where voyages charted the world's oceans and coastlines, the aspiring naturalists of the nineteenth century moved inland, following Humboldt's insistence on the necessity of exploring large landmasses. Darwin traveled hundreds of miles into Patagonia, Hooker scaled the valleys of the Himalayas, Huxley explored the interior of New Guinea, and Wallace boldly followed the course of the Amazon. All returned to England as acknowledged experts in natural history.

The material amassed by Darwin and Wallace during their exploratory voyages served as the empirical foundation for their work on the theory of evolution through natural selection. Hooker's published accounts of his Antarctic voyage on HMS *Erebus* and HMS *Terror* established his reputation as a botanist, and his research in the Himalayas, especially the distribution of plants, led him to tentatively accept Darwin's theory of evolution through natural selection. Likewise, Huxley's decision to join the crew of the *Rattlesnake* helped solidify the choice of his career, and his own work on comparative anatomy and geographical distribution of species led him to become the most outspoken champion of Darwin's and Wallace's theory. Perhaps owing to his own humble beginnings, Huxley also used evolutionary theory to help solidify biology as a viable career for future naturalists. And perhaps most importantly, Huxley's research into human ancestry was viewed as having profound importance concerning the relationship between different races.

These scientific advances occurred simultaneously with Britain's move from the most industrialized country in the world to the largest imperial power in the nineteenth century. Struggle and competition were at the root of Britain's industrial and imperial might, concepts that were everywhere and everyday experienced. Indeed, the naturalists themselves did not come to their views on evolution irrespective of their experiences with other cultures. While Darwin and Hooker were fascinated with other societies, Wallace and Huxley studied these cultures intensely. As Wallace wrote to his brother-in-law George Silk, "I am convinced no man can be a good ethnologist who does not travel, and not *travel* merely, but reside, as I do, months and years with each race."[15] The disciplines of ethnography and anthropology were reformulated in the late nineteenth century in an attempt to describe different human societies within a hierarchy that resembled the hierarchy of plants and animals on the evolutionary bush. The powerful new science of biology tended to give these racist views increasing authority as the metaphors of competition, survival of the fittest, and extinction easily flowed between the two. This, in turn, was often used to justify imperial expansion and the ruthless subjugation of native populations.

These views paralleled the simultaneous interaction between native populations and the European public at large. By the mid-nineteenth century, Europeans were emigrating from Europe in unprecedented numbers. Between 1815 and 1914, 30 million people emigrated from Europe. Twenty million of these were British citizens, moving mostly to North America, Australia, and India. But they also followed the trading routes to the jungles and mountain passes of East Asia, the deserts of South Africa, and the river valleys of South America. Everywhere they went, they brought with them the elements of "informal" empire—the exchange of ideas and institutions, whether religious, financial, or cultural. Science was used to help taxonomize Britain's place in relation to its colonies and imperial possessions. But what these developments in natural history in the nineteenth century demonstrate perhaps most clearly is the manner in which science and the scientist brought these ideas back home to the European consciousness. Naturalists increasingly published their accounts of exploration, and travel literature became extremely popular. Science was used as a means toward imperial rule while simultaneously being a result of the imperial process itself.

The four traveling naturalists came together one last time, on 19 April 1882. Wallace, Hooker, and Huxley served as the pallbearers at Darwin's funeral. That Darwin was the greatest scientist in England, second only perhaps to Newton himself, is exemplified by this gathering. He was buried in Westminster Abbey, a site reserved for nobility. His theories transformed not merely the scientific but also the social and cultural fabric of his age. He had placed man within a hierarchy of beasts and helped place different civilizations within a hierarchy of men. Perhaps recognizing the complexity and ramifications of Darwin's views and harkening the debates to come, the funeral announcement read, "No Person will be admitted except in mourning."

Bibliographic Essay

The framework of this chapter was inspired by the two volumes edited by Roy MacLeod and Philip Rehbock, *Nature in Its Greatest Extent: Western Science in the Pacific* (1988) and *Darwin's Laboratory: Evolutionary Theory and Natural History in the Pacific* (1994). The introductions to both volumes establish the Pacific as a "scientific laboratory" for natural philosophers sailing from the Atlantic and place the advance of Pacific science within its imperial context. The essays included in these volumes are especially strong on Darwin and Wallace and the enduring legacy of their science in the Pacific world.

Older but still relevant studies include the classic by Robert S. Hopkins, *Darwin's South America* (1969), and the classic by Victor Wolfgang von Hagen, *South America Called Them* (1945), which includes a general background on explorers in South America, including Humboldt and Darwin. For the social background of the popularity of natural history in Victorian culture, see David Elliston Allen's *The Naturalist in Britain: A Social History* (1976). For more recent literature, Peter Raby's *Bright Paradise: Victorian Scientific Travellers* (1997) offers a highly readable narrative of explorations into both South America and Africa. Mary Louise Pratt's *Imperial Eyes: Travel Writing and Transculturation* (1992) offers a nuanced approach that is perhaps most useful in graduate-level courses.

This chapter draws on vast and ever-increasing literature in evolutionary studies, and the best place to begin is with Darwin himself. Along with Darwin's published works, his posthumously published writings include Francis Darwin, ed., *The Autobiography of Charles Darwin* (2000); Francis Darwin, ed., *Life and Letters of Charles Darwin* (1887); and Frederick Burckhardt and Syndey Smith, eds., *The Correspondence of Charles Darwin* (1985–2003), which has now reached thirteen volumes covering the letters to and from Darwin from 1821, when Darwin was twelve years old, to 1865, when Darwin wrote *The Variation of Animals and Plants under Domestication*. The first volume covers the period from 1821 to 1836 and contains the letters to and from Darwin during his five-year voyage on HMS *Beagle*.

Exceptional biographies on Darwin keep appearing. Adrian Desmond and James Moore's *Darwin: The Life of a Tormented Evolutionist* (1991) offers an in-depth social history of Darwin and mid-Victorian England, while Janet Browne's definitive two-volume biography *Charles Darwin: Voyaging* (1996) and her *Charles Darwin: The Power of Place* (2002) will not be matched in breadth, depth, insight, or joy in reading. David Quammen's *The Reluctant Mr. Darwin: An Intimate Portrait of Charles Darwin and the Making of His Theory of Evolution* (2006) is a wonderful read that looks at Darwin's life after his return from the *Beagle* voyage. Charles Darwin, *On the Origin of Species* (1982), offers a quick and informative introductory essay by the editor J. W. Burrow. Michael C. Mix et al., eds., *Biology: The Network of Life* (1992), also offer short and readable chapters on Darwin's travels and influence in the context of the advance of the biological sciences more generally. On the voyage of the *Beagle* in particular, along with the texts by Robert S. Hopkins and Victor Wolfgang von Hagen discussed above, see the reprinted edition of Charles Darwin's *Voyage of the Beagle* (1989), especially the introduction by Janet Browne and Michael Neve. The lavishly illustrated *Darwin and the Beagle* (1969) by Alan Moorehead and Stephen J. Gould's *Ever Since Darwin* (1981) also offer informative accounts.

Studies on the impact of Darwin's ideas include the classics by John C. Greene, *The Death of Adam: Evolution and Its Impact on Western Thought* (1996), and William Irvine, *Apes, Angels, and Victorians: Darwin, Huxley, and Evolution* (1955), both of which are a bit dated. More recent scholarship includes the many seminal works by Peter J. Bowler, the most important being *Evolution: The History of an Idea* (2003). John Hedley Brooke, *Science and Religion: Some Historical Perspectives* (1991), covers the move from natural theology to evolutionary theory. Mark A. Largent, ed., *Sourcebook on the History of Evolution* (2002), offers a wonderful selection of primary materials on evolution, from Lamarck to the Scopes Trial, and is meant for a history of science course at the undergraduate level. On the impact of science and technology more generally as it applies to the way in which the West viewed non-Western cultures, see Michal Adas, *Machines As the Measure of Men: Science, Technology, and Ideologies of Western Dominance* (1990). The material on racism, science, and imperialism is extensive, with Simon Dentith's chapter "Ethnicity, Race, and Empire" in *Society and Cultural Forms in Nineteenth Century England* (1999) especially helpful.

On J. D. Hooker, his own *Himalayan Journals: Notes of a Naturalist in Bengal, the Sikkim and Nepal Himalayas* (1854), is fascinating reading. The exchanges with Darwin are in Burckhardt and Smith, *The Correspondence of Charles Darwin*, Vol. 4, 1847–1850. See also Leonard Huxley, ed., *Life and Letters of Sir Joseph Dalton Hooker* (1918). W. B. Turrill's *Joseph Dalton Hooker: Botanist, Explorer, and Administrator* (1964) and Mea Allan's *The Hookers of*

Kew (1967) both provide background, while more recent scholarship includes Lucile Brockway, *Science and Colonial Expansion: The Role of the British Royal Botanic Gardens* (1979), and Richard Drayton, *Nature's Government: Science, Imperial Britain and the 'Improvement' of the World* (2000), which covers Kew from Joseph Banks to the Hookers and adds to recent scholarship by focusing on economic botany and the role of Kew in British imperial expansion.

Alfred Russel Wallace's travel accounts, which include *Travels on the Amazon and Rio Negro* (1853) and *The Malay Archipelago: The Land of the Orang-Utan and the Bird of Paradise; A Narrative of Travel with Studies of Man and Nature* (1869), are perhaps the most enjoyable reading of all the Victorian scientific explorers. See also James Marchant, *Alfred Russel Wallace: Letters and Reminiscences* (1916), and Jane R. Camerini, *The Alfred Russel Wallace Reader: A Selection of Writings from the Field* (2002). For Wallace's life, see his two-volume autobiography *My Life: A Record of Events and Opinions* (1905) and the important works by Raby, including *Alfred Russel Wallace: A Life* (2001) and *Bright Paradise.* Also important are shorter pieces by Jane Camerini, including "Wallace in the Field" (1996) and "Evolution, Biogeography, and Maps: An Early History of Wallace's Line" (1994). Arnold C. Brackman in *A Delicate Arrangement* (1980) argues for the priority of Wallace's achievements. Exchanges between Wallace and Darwin are in Burckhardt and Smith, *The Correspondence of Charles Darwin*, Vol. 7, 1858–1859.

The work on T. H. Huxley inevitably extends to the fascinating debates in Victorian society brought on by Wallace's and Darwin's theory. The standard jumping-off point is Leonard Huxley, ed., *Life and Letters of Thomas Huxley* (1900), although it does not cover Huxley's voyage on the *Rattlesnake* in any detail. For this, see Julian Huxley, ed., *Thomas Henry Huxley's Diary of the Voyage of the H.M.S. Rattlesnake* (1972). A recent biography is Adrian Desmond's *Huxley: From Devil's Disciple to Evolution's High Priest* (1997). For Huxley's thoughts on biology and society, see Alan P. Barr, ed., *Thomas Henry Huxley's Place in Science and Letters* (1997). Along with including an excellent paper on Huxley's voyage by Joel S. Schwartz, it also covers Huxley's role in science education as well as discussions of his later writings on evolution and ethics. For the latter, see James G. Paradis, *T. H. Huxley: Man's Place in Nature* (1978); James Paradis and George Christopher Williams, eds., *Evolution and Ethics: T. H. Huxley's Evolution and Ethics with New Essays on Its Victorian and Sociobiological Context* (1989); and Paul Farber, *The Temptations of Evolutionary Ethics* (1998).

5

Scientific Exploration of a Manifest America

In the annals of exploration history, the presidency of Thomas Jefferson immediately calls to mind Meriwether Lewis and George Clark's Corps of Discovery, a daring and intrepid expedition designed to explore and map a sliver of the North American interior. During the same administration, the U.S. Coast Survey also received its mandate to create a detailed map of the eastern U.S. coastline. These were important years for the nascent republic, and the twin ventures mark two distinct styles of exploration that continue to define the parameters of exploration discourse. On the one hand, the Corps of Discovery represented an exploration style in the tradition of Christopher Columbus: an attempt to go where "no white man" had ever been before. On the other hand, the U.S. Coast Survey represented the desire to marshal the tools and techniques of a European science to advance trade and commerce. Both styles characterized the history of scientific exploration in the nineteenth-century United States, and both elements factored largely into the narrative of territorial conquest that lay at the heart of the U.S. spirit of Manifest Destiny.

The ideology that science should be of some practical use has a long history and continues to resonate today. This belief was nowhere more true than in the case of the exploration of the American landscape in the nineteenth century. But a counterideology holds that explorers do the work of exploration out of a spirit of unbridled curiosity. The goal was to discover what was around the next bend—for example, to uncover a fossilized skeleton of an extinct animal from an exposed riverbed. In truth, both ideologies underscore significant aspects of the history of American exploration. This history demonstrates how explorers contributed to expanding imperial interests and how the nature of exploration itself was given form and movement by American culture. In other words, exploration was not just a tool for American interests; its peculiar character was a product of the social environment of an expanding nation.

U.S. Coast Survey

With the advice and recommendations of Jefferson and a group of elite intellectuals from the American Philosophical Society, the U.S. Congress created the Coast Survey in February 1807. The initial charter was decisively practical: the Coast Survey was to map "the islands and shoals, with the roads or places of anchorage, within twenty leagues of any part of the shores of the United States."[1] Moreover, the Coast Survey was charged with conducting a hydrographic and oceanographic study of the entire eastern seaboard. The United States was following the well-trodden path of European states that, throughout the seventeenth and eighteenth centuries, enlisted a vast array of sciences to survey the environments of both core and periphery for the purposes of navigation, commerce, and defense. It was therefore fitting that Ferdinand Hassler, a Swiss geographer and surveyor trained at the University of Bern, was called on to head the early Coast Survey.

Surveying began somewhat belatedly in 1816, and Hassler immediately took a section of New Jersey's coastline as a basis for a geodetic survey that enlisted the labor of Coast Survey staff as well as the military and the Corps of Engineers. After only two years of work, Hassler temporarily lost control of the Coast Survey to the military but then regained the position in 1832 to oversee an organization that required the careful management of both civilian and military laborers. But Hassler was better at sighting a marker through a theodolite than negotiating with the politicians and bureaucrats of antebellum America on whose patronage his work depended. He bristled at inquiries into his work, especially regarding its expense. "Mathematics and nature cannot be commanded," he proclaimed.[2] Science, he thought, was not a subject that had to answer to congressional inquiries. Upon his death in 1843, the torch was passed to Alexander Dallas Bache, a man who combined Hassler's predilection for science by European standards with a certain political savvy that made the Coast Survey one of the premier scientific institutions of antebellum America.

Bache had already embraced the mantle of Humboldtian science by the time the Coast Survey came under his direction. He kept abreast of Europe's most advanced techniques for conducting geodetic surveys, especially those pertaining to the Magnetic Crusade and Britain's triangulation survey of India in the 1830s and 1840s. He brought these methods to bear on the Coast Survey's quest of compiling data on the United States and its environs. Within four years, Bache had extended Hassler's work to incorporate the territories of eighteen states, a task that continued as a result of U.S. expansion. Responding to a congressional inquiry as to when the work of the Coast Survey would be

Alexander Dallas Bache (1806–1867)

A great-grandson of Benjamin Franklin, Alexander Dallas Bache was born into an elite Philadelphia family that maintained a strong commitment to public service and self-improvement through higher learning. After a patrician childhood, Bache attended the U.S. Military Academy at West Point in 1821 at the age of fifteen. He studied under Sylvanus Thayer and began to appreciate the social and economic importance of scientific fields such as chemistry, mineralogy, dynamics, and astronomy. Bache graduated first in his class, spent a year teaching mathematics and natural philosophy, and then earned a position as lieutenant with the Corps of Engineers. While many of his fellow students continued in the corps to become prominent explorers of the western frontier, Bache decided to take a position as professor of natural philosophy and chemistry at the University of Pennsylvania in 1828. Philadelphia provided him with the perfect environment to develop his thoughts on the role of science in the cultural affairs of the nation, and he quickly became a rising star among Philadelphia's scientific elite. Moreover, amid conducting research into meteorology and terrestrial magnetism, he began to develop various strategies for pedagogical reform that he implemented while serving as superintendent of Philadelphia's Central High School from 1839 to 1842.

When Hassler, the director of the U.S. Coast Survey, died in 1843, Bache took the post and began to transform the Coast Survey into the preeminent federally-funded scientific institution in the nation. Bache had previously visited Great Britain and learned about the most scientifically advanced geodetic techniques of the day. His mission therefore was to bring the Coast Survey in line with the mathematically rigorous surveys of European standards.

Bache's legacy to the history of exploration was twofold. First, by embracing the European standards of surveying, he helped to make the triangulation survey the standard for all exploring expeditions; the cumulative work of the Coast Survey brought American science into line with a truly Humboldtian tradition. Second, he was able to ingrain the importance of civilian science for the American people, especially in terms of commerce and navigation. This was a common objective among the Scientific Lazzaroni, a group of scientists who argued for federal support of the sciences—a practice that had the potential for appearing a bit too Whiggish for the new democratic republic. While scientific explorers throughout the nineteenth century found it a daunting task to lobby the federal government for patronage, Bache had made the relationship far more acceptable.

completed, Bache was reported to have asked, "When will you cease annexing territory?"[3] As the work of the Coast Survey continued apace, Bache made it a habit to quickly publish maps and charts for distribution to both scientific institutions and agencies involved with navigation and commerce.

Despite congressional attacks on a federally-funded scientific agency, the work of the Coast Survey expanded between 1848 and 1852. Largely through

Bache's political adroitness, the Coast Survey's operating budget during those years increased from about $150,000 to more than $460,000, and Bache extended the work of surveyors to recent acquisitions along the coasts of Texas, California, and Oregon. By 1855 the Coast Survey employed 776 surveyors, and their united work elicited a comment from the American Association for the Advancement of Science that the Coast Survey was saving the country nearly $3 million a year by helping to prevent shipwrecks. Indeed, the success of the Coast Survey is largely attributable to Bache's acumen in promoting the exacting standards of Humboldtian science to the American public. He trained his fieldworkers to develop collaborative relationships with property owners on whose lands the Coast Survey relied for conducting triangulation readings. Bache also put the Coast Survey into the consulting business by utilizing scientific knowledge for harbor and river management. At the New York Exhibition of 1853–1854, the Coast Survey displayed the tools of its trade: theodolites, tide gauges, standards of weights and measures, electrotype plates for printing maps, thermometers for measuring deep-sea temperature, and maps and charts produced by the Coast Survey. The American public began to realize the value of science to national affairs. "This purely scientific work," noted a Philadelphia newspaper, "so important to the commerce of the country, is pushed forward year by year, by the indefatigable Superintendent, with a precision and certainty in its methods, that satisfy the requirements of the most exact of the sciences, and at the same time with a promptness and businesslike punctuality in its results that rival the annual balance sheets of the merchant's ledger."[4]

By the end of the Civil War, Bache had transformed the Coast Survey into the preeminent government-funded science institution of the day. Not only was a small army of surveyors under his command, but he also transformed the exploratory work of the Coast Survey into the American paragon of Humboldtian science. Through both the work of the surveyors and Bache's extended network of nongovernment correspondents, the Coast Survey helped to choreograph American research in astronomy, geodesy (the shape of the Earth), terrestrial magnetism, oceanography, hydrography, tidology, and natural history. In Humboldtian terms, these were not separate scientific disciplines; rather, they were related fields of inquiry that employed similar standards of exactness to investigate the interrelated principles of natural and physical phenomena. Even after Bache's death in 1867, the mission of the Coast Survey moved forward, and its budget continued to grow through the late 1870s. By 1878 the Coast Survey, then under the capable leadership of Charles Peirce, had effectively mapped and charted U.S. coastal territories and had begun the work of connecting the Atlantic and Pacific surveys with a triangulation of the American continent. It was at precisely the same time that the U.S. Geological Survey would take the lead in enjoying government largess for exploration. But the rise of the Geo-logical Survey was preceded by more than seventy years of government-funded exploration of North America's interior.

Corps of Discovery

While Bache's Coast Survey emblemized the importance of civilian science for the American people, Lewis and Clark highlighted science and explora-tion in its military mode. From a strictly practical perspective, the Corps of Discovery's meandering exploration of the Missouri and Columbia Rivers were abysmal failures. No easy trade route joined the eastern United States to the Pacific Ocean. Nor was the Corps of Discovery the first to complete the cross-continental mission. The fur trader Alexander Mackenzie had already under-taken a similar mission in Canada and even published an account of the expedi-tion two years before Lewis and Clark entered the field. Indeed, it is sometimes difficult to understand how the Corps of Discovery has taken on such mythic importance. But if the expedition fell somewhat short in practical significance, it made up for the dearth with an enduring cultural importance.

The Corps of Discovery's twenty-eight-month exploration (1804–1806) brought together two visions of American nationalism. The first belonged to Jefferson, its most stalwart architect. Jefferson and his Democratic Republi-cans envisioned an "empire for liberty" based on the nobility of the yeoman farmer. While it was certain that trade and commerce would continue to grease the wheels of America's growth, Jefferson hoped that a country of virtuous farmers would prevent the class divisions and vice of urbanity. Key to his vi-sion was land. With a few financial transactions, numerous strokes of pens, and some particularly good luck, the Louisiana Territory, a massive swath of land that stretched from the Mississippi River to the Rocky Mountains, fell into American sovereignty. Part of the Corps of Discovery's mandate was to investi-gate how this new territory would figure into the new nation.

The second national vision hailed from Europe and dates back to the Renaissance tradition of collecting either natural specimens or knowledge of plants, animals, rocks, minerals, and even humans. Such natural curiosities were signs of power back in Europe. Collections were amassed in menageries, cabi-nets of curiosities, botanical gardens, and eventually museums. Such archives of knowledge and objects connoted power and prestige and were often practical and symbolic signs of imperial prowess. This tradition enjoyed abundant growth in America's nascent scientific institutions and held particular sway with Ameri-can philosophes of Jefferson's ilk. In 1803 Jefferson commissioned his personal secretary, Captain Meriwether Lewis, to conduct a grand reconnaissance of the American West from St. Louis up the Missouri River, over the Rocky Mountains, and down the Columbia River to the Pacific Ocean. The general hope was that he would find the holy grail of three centuries of exploration: an easy path to the trading ports of Asia. In short, it was a mission that was both a commercial venture and a scientific expedition. Jefferson appears to have had no problem securing the unprecedented allotment of $2,500 from the War Department.

It would be a stretch to call Lewis "scientific," even by early nineteenth-century standards. He was a military figure of uneven temperament, born into southern plantation wealth, modestly educated by several Virginia tutors, and destined by family right to become a profitable farmer. A youthful wanderlust tugged at the planter, and in 1794 he answered George Washington's call for volunteers to quash the Whiskey Rebellion. Fittingly enough, Lewis's first civic involvement was to establish order in the Ohio River Valley, then part of the expanding West. For a time, Lewis had fought alongside an Ohio Territory rifle company commanded by Lieutenant William Clark. The two became friends, and ten years later Lewis wrote to Clark and requested that the latter co-lead the Corps of Discovery.

The mission was carefully planned at the core of American cultural and political power. Scientists such as Caspar Wistar, Benjamin Rush, Andrew Ellicott, and Robert Patterson advised Lewis on particular phenomena that were of scientific interest, including how to collect and preserve specimens. Lewis honed his surveying skills and became adept at making astronomical observations with the sextant and theodolite. Jefferson himself gave Lewis and the entire team the mandate to carefully journal all of their observations and contacts with Native Americans. They were to create a general map of the region and look for productive resources. They were to classify metals, limestone, coal, saltpeter, and other valuable minerals. They were to report on soil conditions and locate sites suitable for development. They were to study Indian culture, including their relations with other tribes, their economic and military endeavors, and their laws, customs, and commerce.

The Corps of Discovery was not headed into a land that was completely unfamiliar. Obviously, this frontier wilderness was filled with Native Americans, some of whom had grown accustomed to white people of various empires attempting to establish the valuable fur trade. Explorers too, namely James Cook and George Vancouver, had tackled the Pacific Northwest via the Pacific Ocean, although their reports of the Columbia River were still rudimentary. The tale of the Lewis and Clark expedition—its Indian encounters, the members' hunger and privation, the maintenance of martial order, and the leadership of their Indian guide—has taken on epic and mythic importance. In the sense that the

Frederic Remington's 1906 painting of Lewis and Clark and the Corps of Discovery at the mouth of the Columbia River during their exploration of the Louisiana Territory. (MPI/Getty Images)

crew returned having lost only one member on a long and arduous journey, the expedition must be characterized as a success. They brought back reports of Native Americans eager for trade; some 60 reasonable maps of the explored region; more than 150 botanical specimens; reports of fertile lands suitable for cultivation; and volumes of journals that Lewis and Clark edited and reedited for both government and public consumption. The West was, indeed, a vast territory of almost unlimited land and resources. Many Americans had already realized this fact; Lewis and Clark gave it the imprimatur of a federally-funded expedition. Needless to say, their return to the East was greeted with much fanfare.

But if the Corps of Discovery was sent out to gather information and bring this data back to "civilization" for digestion by American expansionists, it also brought much of American culture into the wilderness. Lewis and Clark entered the West as representatives of a people who prided themselves on their racial and moral superiority. They were to be a beacon of light in a howling wilderness, and while it was never their mission to be diplomats, they chartered a course of empire that would transform all that was wild into an emblem of civility. Nevertheless, Clark brought his slave York along as a servant. Both Lewis and Clark may have regretted such a decision when they realized that Native American peoples were more captivated by a mysterious black man than the white bearers of higher civilization. Lewis also secured the services of a young Shoshone woman who had been captured and enslaved by a Hidatsa tribe and sold to a French fur trader. Although she was instrumental in the expedition's success, Lewis and Clark went to some effort to minimize her contribution, another testament to the Corps of Discovery leaders' thoughts on racial superiority. The team members were also avid hunters. Foreshadowing the effects of the Americans soon to come, they seemed to wage an indiscriminate war on snakes, and hunting big game for either sport or food was a daily activity. In short, the Corps of Discovery was a symbolic and sometimes literary conquest of the American landscape, a planned and federally-funded expedition that was to chart the course of a nation. Perhaps this is the explanation for the enduring power of the Lewis and Clark story. But its real significance lay not so much in the maps or the collections or the journals but rather in its legacy of America's first effort to join the scientific work of exploration with the political and economic work of settling an expanding nation.

Army Corps of Topographical Engineers

Funded by the War Department and led by two prominent officers, the Corps of Discovery established American exploration as a military venture. Even before the party had returned to St. Louis, General James Wilkinson, governor of the Louisiana Territory, dispatched Lieutenant Zebulon Pike, twenty-three soldiers, and fifty-one Indian porters to cross the Great Plains and explore the eastern range of the Rocky Mountains. Ten years later, Major Stephen H. Long left the East with a small army of explorers and soldiers to establish military posts in the Yellowstone and upper Missouri basins to protect American interests in the fur trade. In short, the U.S. Congress had given the mandate for regional and territorial exploration to the U.S. Army. In 1824 Congress appropriated monies for establishing the Army Engineers to plan and implement road and canal construction to facilitate military and commercial objectives. After several name changes and alterations in personnel, the Army Corps of Topographical Engineers came into its most enduring form in 1838. It played an important role as diplomat in an age of continental expansion and helped to chart the West for railroad routes. Many of the corps members were geographers and engineers, often educated at West Point where they received training in analytical mathematics and engineering for the purpose of producing military officers who were enlightened "men of science." A select number were bent on the task of mapping the contours of the continent, but East Coast scientists often accompanied survey crews.

On the most general level, the work of the Corps of Engineers was simply the task of settlement. Upon its inception, Oregon, Washington, and much of the far West and Southwest territories had not yet been incorporated into the sovereignty of the United States. The corps was largely in charge of determining borders and investigating the topography of the interior for military, commercial, and civilian movement. Although their techniques would become more sophisticated in the late 1840s, early on the Corps of Engineers used a very simple meandering method that required little more than an odometer and a compass. The motivation to survey these territories was spearheaded by Senator Thomas Hart Benton of Missouri. Benton was a Manifest Destiny zealot who believed that he was personally commissioned by Jefferson to conclude the continental design of the United States. Benton passed that mandate on to his son-in-law, John Frémont.

Dubbed a "pathfinder" in the popular press, Frémont's expeditions into the Oregon Territory, the Rocky Mountains, and California during the period 1842–1845 were heralded for its leader's ingenuity and daring. Frémont was born into an Episcopal working-class family in Charleston, South Carolina, and by 1835 had earned a Masters of Arts degree from Charleston College. That same year, the U.S. Navy commissioned him a professor of mathematics, and thus began his long career of surveying the western frontier. Frémont cut a dashing figure, the paragon of the romantic Western explorer who combined bravado with science. The *New York Times* even published Frémont's phrenological characteristics, claiming that he had "great length from the ears to the forehead . . . , showing very large perceptive and prominent reflective organs. Thus all the organs necessary for the clear thinker, the civil engineer, and the scholar in natural science, are decidedly large. . . . Few men have as much heroism and ability to lead and control difficult and dangerous enterprises, and fewer still exhibit as much simplicity and modesty in general intercourse with society."[5] After Frémont contributed to expeditions in Georgia and the Dakotas, Benton gave him the task of blazing a northern trail to the Rocky Mountains to facilitate westward immigration into Oregon. More than any concrete scientific results, the successful expedition advertised the Rocky Mountains as one of America's most sublime possessions.

The following year, Frémont was to carefully map the Oregon Trail and provide a more detailed guide to the resources and dangers that migrants would encounter along the way. He was also to bring the party southward into California with the aim of linking the survey with previous Californian maps. Over the next two years, he led a party of thirty-nine men who mapped the northwestern territories and collected numerous biological and geological specimens. Frémont's reports of his expeditions to the Rocky Mountains, literature set within

John C. Frémont, the "pathfinder" of the U.S. Topographical Corps. (Library of Congress)

the genre of the heroic and romantic age of exploration, fueled the dreams of American citizens looking forward to new lives in western territories. However, Frémont was not above becoming a more direct instrument of Manifest Destiny. Given secret orders by his father-in-law, Frémont fomented tension with Mexico by leading the Bear Flag Revolt, imprisoned General Mariano Guadalupe Vallejo, and slipped a howitzer cannon into his list of scientific instruments.

With the establishment of its northern and southern borders and with general maps of the continent's interior completed, the Corps of Engineers next began the task of establishing a path for a transcontinental railroad. The task was both scientific and imminently political. Sectional strife led to fierce competition over the path of a transportation artery, with the location of the easternmost terminus an especially volatile issue. An ardent booster of the railroad, Benton, who was fond of saying that the destination of Western commerce was India, charged Frémont with finding a suitable path from St. Louis to San Francisco in 1848. While Frémont later declared the venture an unmitigated success, the team of thirty-five engineers encountered enormous difficulties. Native American resistance and inclement weather led to the deaths of several members and even claims of cannibalism. Frémont's ill-fated exploration was just the first of eighteen that entered the field between 1848 and 1850, though the location of what would be a federally-subsidized-project continued to be a controversial political issue.

The Army Corps of Engineers was the analog of the great navigators during the age of exploration stretching from the fifteenth through the eighteenth centuries. The latter had the job of charting the rough contours of the world's oceans and establishing trade routes that made the project of European colonization and imperialism possible. But in the eighteenth century, the ships were laden with naturalists, astronomers, and various scientific gear. The rise of the sciences during the Enlightenment gave tremendous impetus to assembling an archive of natural knowledge needed to better appreciate the ordered variety of the world. Knowledge of the world's natural objects, moreover, had as much practical economic significance as it did the ability to slake the scientist's natural curiosity. Just as eminent European scientists such as Joseph Banks accompanied European navigators such as Captain Cook, so too did American scientists accompany the Corps of Engineers on its voyages to the West. Their job was simply to conduct a grand reconnaissance of the plants, animals, geology, and people of the West or, rather, to create a detailed inventory of knowledge of the West so that eastern scientists, politicians, prospectors, and settlers could plan out the specifics of America's Manifest Destiny.

At times, the Corps of Engineers simply gathered specimens and sent them back to scientific institutions such as the American Philosophical Society or the recently developed Smithsonian Institution. But it was often the case that scientists such as Spencer Baird at the Smithsonian and John Torrey at Princeton advised explorers on what to look for and on methods of collection and recommended that specific scientific savants accompany the surveys. The latter was especially the case for the Pacific surveys on which more than a hundred scientists worked both in their museums and in the field. Perhaps the most enduring testament to the scientific side of these surveys was Jefferson Davis's thirteen-volume *Pacific Railroad Reports* published between 1855 and 1860.

The centerpiece of the *Pacific Railroad Reports* was G. K. Warren's map that combined all the efforts of the various surveys to compile a single map of the trans-Mississippi West. The geological findings did not measure up to the topographic success, but six geologists accompanied the Pacific surveys and provided basic trans-Mississippi geological knowledge, with some rudimentary historical speculations on the Great Basin and the Grand Canyon. Torrey compiled a systematic treatment of all of the botanical specimens, and Baird produced three volumes dedicated to describing the birds, reptiles, fish, and mammals of the American West. The *Pacific Railroad Reports* also included some ethnographical information on various Native American tribes. These too were classified into linguistic groupings in the vast compendia of natural knowledge about the West. The cumulative efforts of explorers and scientists who were either part of the Corps of Engineers or traveled along with it between the 1840s and 1860s was very much a continuation and elaboration of Lewis and Clark's Corps of Discovery. The work of nation-building required a systematic knowledge of all natural objects in the American West. This political motivation for exploration was perfectly matched by a corresponding scientific curiosity. Pure science and applied science were not as distinct as they seem today.

The Great Surveys

While the Corps of Engineers was responsible for considerable geological and natural historical work, its primary job was that of trail blazing and surveying. But under the corps' auspices, this order of business began to change after the Civil War. As the peppering of Western settlers became a smattering, further detailed explorations into various regions became a necessity. Moreover, the history of American expansion into the West was entering its bloodiest phase as settlers increasingly became embroiled with Native American resistance. The Corps of Engineers met this need, in part, by assigning semipermanent advisors to the six western military districts. But more than ever before, this new phase of exploration was mostly headed less by military heroes than by serious academic savants, a phenomenon that also signified the movement of the work of exploration from the War Department to the Department of the Interior.

Cumulatively, the four major surveys that entered the field in the late 1860s and early 1870s are known by historians as the Great Surveys. They included Clarence King's survey of the Great Basin along the 40th Parallel, John Wesley Powell's exploration of the Colorado River and the Grand Canyon, Ferdinand Hayden's work for the U.S. Geological Survey of the Territories, and George Montague Wheeler's geographical surveys of the territories west of the 100th Meridian. These surveys, more than most before the 1860s, were characterized by close scientific scrutiny of the western landscape. They made considerable

Clarence King (1842–1901)

Born into a Rhode Island family of merchant traders, Clarence King spent an adventurous childhood exploring the Green Mountains and his New England environment. From 1859 to 1862, he matriculated at Yale's Sheffield Scientific School, where he deeply imbibed in geology-intensive studies. He was also captivated by the many surveys that were being published about the exploits of explorers in the West. After graduation, King and his classmate, James Gardiner, set off for California. A lucky encounter with William Brewer, a prominent member of the California Geological Survey, led to both King and Gardiner becoming official members. King worked with the survey through 1866, charting out the High Sierra and Yosemite and gaining fame as perhaps the first person in America to view an active glacier.

With a desire to mount his own campaign to study the Great Basin's geological history, King lobbied Congress in 1867. When the bill passed, Henry Adams claimed that it was the first modern act of Congress, and the chief of the Corps of Engineers gave the highly sought-after commission to King. King spent the next twelve years orchestrating the 40th Parallel Survey—one of the four Great Surveys—and publishing the results. In comprehensiveness, efficiency, and scientific acumen, King's survey of the Great Basin was beyond comparison. Moreover, he was consistently able to demonstrate the commercial value of the expedition with its work on the mineralogy of the many prospecting lands of the West. Based on this success and some backstage politicking, in 1879 King was awarded the directorship of the U.S. Geological Survey, a new agency housed within the Interior Department, and was given charge of coordinating scientific investigations of the country's geology, a post he held for two years before it was taken over by John Wesley Powell.

King's geological work and his leadership ability were only matched by his skill as a writer. *Mountaineering in the Sierra Nevada* (1872) was a literary effort that ranks as one of the more important articulations of America's version of the sublime. A concept that hailed from romanticism, King's characterization of the Sierra Nevada as both terrible and awe-inspiring piqued the curiosity of American readers, who began to seek out wilderness experiences for the first time. This new characterization of the mountain wilderness helped to spur on American environmentalism at the end of the nineteenth century.

progress in addressing issues relating to western zoology, botany, meteorology, geography, and geology. But even with their scientific bent, and given the fact that they were chiefly civilian efforts, these surveys continued to work hand in hand with western settlement. King's 40th Parallel Survey and Powell's expedition through the Grand Canyon are representative of this new spirit.

King's survey was peculiar for its scientific pedigree. King himself was born into a Newport, Rhode Island, family of wealth. He attended Yale's

Sheffield Scientific School, one of the preeminent scientific schools then emerging in the Northeast, and studied under such luminaries as the chemist George Jarvis Brush and the geologist James Dwight Dana. In 1862 King advanced his studies by working at the California Geological Survey. Befriended by Baird, Josiah Dwight Whitney (the head of California's surveys), and Louis Agassiz, King had no trouble petitioning Congress for the $100,000 needed to conduct a three-year survey of the 40th Parallel that would roughly follow the projected path of the Central Pacific Railroad. His team included his Yale classmate, James Gardiner, who served as chief cartographer; Samuel Franklin Emmons and the brothers James and Arnold Hague went as the team's geologists; and the botanist William Whitman Bailey and the amateur ornithologist Robert Ridgeway rounded off the scientific contingent.

With a number of prominent photographers, a few camp men, and twenty men of the 8th Calvary as escorts, King's survey promised to be by far the most all-encompassing and widespread expedition to have ever entered the West. The territory included Utah, Colorado, northern and central Nevada, and southern Wyoming. The task was to map the region's geological structure with exacting detail and make a complete survey of the plants and animals. In short, the 40th Parallel Survey might be worthy of the Humboldtian tradition that sought to examine the interconnections of a region's living and nonliving material through exact instrumentation. Such a reconnaissance was not without its practical implications, as the region was increasingly looked to by miners, speculators, and town boosters affiliated with the Central Pacific Railroad.

The team rendezvoused just outside of present-day Reno, Nevada. About eleven horses, two freight wagons, an instrument cart, and a small army of mules accompanied the scientific staff into the Great Basin. The team's first job was to create an accurate topographic map of the region that would later be enriched with geologic and economic data. King conducted a triangulation survey of the area employing the same methods used by the Corps of Engineers on the Mexican Boundary Survey and Bache with the Coast Survey. The team split into several parties, and almost all succumbed to both malaria and desiccating heat. King himself was struck by lightning atop Job's Peak. But by the time the team wintered in Virginia City and Carson City, both towns that had developed within the silver mining district, King could boast of a successful survey of the area with accompanying geological and barometrical data.

The team's emphasis on the region's geology meant that it was perfectly equipped to examine and advise the network of mining operations scattered throughout Nevada. Part of a geologist's training was intensive study of chemistry, and the team found that more than $40 million had been wasted at the Comstock Lode because of faulty smelting techniques. James Hague went to work on new formulas to remedy the situation. King also thought that richer veins of silver would be found deeper in the mine, a fact confirmed in 1876 when miners tapped into one of the richest lodes in Western history. The team's geological work, including the location of harvestable coal deposits, made clear that the expedition had a decisively practical and economic purpose, and King made sure that the first compilation of the survey's results would emphasize its resourcefulness for American mining.

The 40th Parallel Survey continued through the summer of 1872, punctuated by King's spectacular discovery of the first active glacier found in the United States and the team's unmasking of a diamond speculation hoax. The team members then spent the better part of the 1870s sifting through their specimens, compiling results, and publishing a survey report that set the scientific standard for similar expeditions. King's own volume, *Systematic Geology* (1878), proved a masterpiece of geological synthesis. He even presented a paper at Yale that brushed against the grain of uniformitarian geologists, who had too easily believed that rate and scale of geological change had always been perfectly constant. The 40th Parallel Survey, in contrast, found evidence of violent volcanic

W. H. Jackson's photograph, "Pack Train of the U.S. Geological Survey" depicting the travels of one of the surveying teams along the Yellowstone River in 1871. (National Archives)

action and glaciers that sculpted the landscape of the West in fits and starts. The great success of King's survey, a model of efficiency, utility, and exacting science, was sufficient reason for him to be offered the position of director of the newly formed U.S. Geological Survey in 1879. The utility of civilian science to the nation had secured another foothold.

To the south of the region covered by the 40th Parallel Survey, at precisely the same time, John Wesley Powell's exploration of the American Southwest was making great strides in geology, natural history, and anthropology. The expedition was the stuff of legend, with all the daring bravado characteristic of the Corps of Discovery. But Powell brought exploration to a completely new level. More than a quest of discovery and a survey of mines, Powell gave specific recommendations for how the American Southwest should be settled, a troublesome task given the region's extreme aridity. Such recommendations competed with typical American settlement patterns based on intense agricultural use of the land. Powell marked the apex of the explorer as a national advisor, and he created much of the national bureaucracy that continues to manage the landscape today. But Powell's history also makes clear that there are certain limitations to the explorer serving as a national architect.

Powell's background was far more humble than many of the previous explorer-scientists trained at either West Point or one of New England's scientific schools. His father was an itinerant Methodist preacher and a staunch abolitionist. The family moved from upstate New York to Jackson, Ohio, where the young Powell studied natural history under George Crookham. The family then moved to Walworth County, Illinois, where Powell began teaching natural history in local schools while taking courses at Illinois Institute and Illinois College. His personal reconnaissance of the natural history of Illinois and the Ohio and Mississippi Valleys were thorough enough to merit election as secretary of the Illinois Natural History Society. In 1860 he enlisted in the Union Army and, despite losing his arm at the battle of Shiloh, quickly rose to the higher echelons of military leadership.

Powell returned to Illinois after the war and secured teaching positions at Illinois Wesleyan College and later the Illinois State Normal University. His heart, however, pulled him elsewhere. He began securing funding and support from Illinois education facilities, the Natural History Society, various railroad companies, and the U.S. Army to mount a reconnaissance of the Colorado River canyon, much of which was still a blank space on official maps. In 1867 and again in 1868 he led small parties to Colorado, variously examining the region's geology, natural history, and anthropology as a way to prepare for a more significant survey. Powell had the events widely chronicled in the *Rocky Mountain News.*

On 11 May 1869 in Green River City, Powell assembled a ten-man team. When compared to King's survey, Powell's was decidedly amateur, with little professional training in natural history beyond Powell. Moreover, in sheer size and support, the Colorado River exploring expedition must have appeared to be a modest undertaking. But the team was a hardy group willing to put their lives at risk to explore the unknown Colorado River. Powell had previously commissioned the construction of four boats designed to accommodate the torrents of the Colorado. The entire affair was funded by the members' own resources in addition to some funding from Illinois universities and several railroad companies. Although modest in outfit, the exploration's methods were the same as King's. They were to chart the river, fixing their position with sextant and barometer on the river's banks. Powell also took every opportunity to climb the canyon walls to piece together a geological history from the strata that the Colorado River had revealed. The river trip from Green River City to Salt Lake City took them sixty-nine days and continues to be an inspiring story full of dangerous rapids, long portages, fearsome hunger, and the anecdote of Powell clinging to a canyon wall with his one arm only to be saved by a survey member who thought quickly and lowered his long underwear to pull Powell to safety.

Powell returned from the journey a national hero. The expedition's tale circulated widely through newspapers and magazines, and while much of the astronomical data was lost during the course of the mission, Congress thought that Powell was successful enough to award him $10,000 for further geographical and topographical work in the region. By and large, the task of surveying the last unmapped geography in America was done. But to an extent, Powell's second expedition played a more pivotal role in defining his mind-set about western settlement. The second expedition was more about what was happening adjacent to the water's edge on the Colorado Plateau. It began in mid-July 1871 from the same Green River station. When the party reached the Uinta River, Powell left for an Indian agency and then returned to the river six weeks later. In October he left the party once again, passed through Kanab, and then traveled to Washington, D.C., leaving the rest of the team to triangulate the north rim of the Grand Canyon and the northern plateau into Utah. Powell returned to the team in November and spent much of his time visiting Mormon and Native American communities. The team returned to the river a year later but withdrew in September 1872 because the rapids proved too dangerous. Team members then worked on a proper topographical map of the Colorado Plateau, and thus the expedition ended in 1873.

While the second expedition possessed little of the climactic adventure of the first, Powell's work and conversations among both Native American and Mormon communities proved pivotal. The Mormon communities of Utah had been growing in the Great Basin since 1847, and through the charismatic leadership of Brigham Young they had displaced the resident Native Americans and developed a large network of self-sustaining communal settlements that often worked within the natural constraints of the desert environment. Powell had many conversations with Young and had visited many of these communities during his second expedition. Through this experience, Powell began to realize that settlement of the American Southwest required a different model than one that prized individualism and water-intensive agriculture. He brought these thoughts together in his classic *Report on the Lands of the Arid Region* (1875), a document similar in spirit to the many explorers' reports that had filtered through Washington, D.C., offices since Lewis and Clark that promised wealth and prosperity in the West. Powell claimed that the model of development that had been forged in the eastern "humid" regions could not be transplanted into the frontier of the American Southwest. In short, he was typical of the nineteenth-century explorers who tried to envision frontier landscapes into America's economic and political life. He stands out among the others, however, in proffering an alternative theory of development that even today is well-respected by many environmentalists and historians. But Powell was not alone. Perhaps the greatest nineteenth-century explorer who articulated a widely divergent plan for the West was John Muir. To appreciate his significance, we need to move back to a time before the Great Surveys when explorers were puzzling together California.

The California Geological Survey and Environmental Reform

Shortly after the Corps of Discovery's meandering survey, state governments and individual territorial governments began to call for more intensive surveys to better plan for the development of the nation. East Coast states such as Connecticut, Massachusetts, New York, and North Carolina took the lead in developing detailed maps relating to land use, especially in terms of potential sites for agriculture, mining, and transportation. More than anything else, the state surveys were a legacy of the Enlightenment vision of a polity and culture with natural and scientific understanding at its core, perhaps best exemplified by Thomas Jefferson's *Notes on the State of Virginia*. Throughout the antebellum period, most states continued to develop their own civilian surveys. The gold rush of 1848 set Josiah Whitney, a Yale-educated Michigan surveyor, to thinking about the scientific problems that the little-known region of California presented.

For another decade, exploration of California geology remained in the hands of a small number of pioneers such as John B. Trask, who outlined the rough geology of the Sierra Nevada, and the California railroad explorers. In 1859 Whitney marshaled the support of eastern scientists, and the state survey became a reality in 1860. Whitney's task, according to the formal congressional bill, was to make an accurate and complete geological survey of the state. There was no mention of the practical side of such a survey. Whitney noted that it was "not the business of a geological surveying corps to act . . . as a prospecting party," although he had previously published a study on the metallic wealth of the United States some six years earlier.[6]

Laying a model for the Great Surveys, Whitney first assembled previous information from the Pacific Railroad Survey and the U.S. Coast Survey to create a detailed topographical map of California that he then filled in with geological data. As previous maps were woefully incomplete, Whitney had his small team conduct a triangulation survey of the entire state, a technique that was becoming popular in the 1840s and 1850s with geographers across the globe. For the next two years (1861–1862), the small team traveled the north-south axis that lay to the west of the Sierra, charting its topography and estimating the area's potential mineral wealth. In 1863 the survey headed north to explore some of California's most sublime landscapes, including the Yosemite Valley, whose glacial and geological history the survey immediately began to piece together. But several members of the survey also realized that they were experiencing a rather unique and awesome spectacle. Whitney began lobbying President Abraham Lincoln to grant the area to the state of California so that it would remain undeveloped and used only for public enjoyment. Yellowstone would eventually become the first national park, but Lincoln's 1864 creation of Yosemite as a state "public pleasure ground" was the first legislative move, not without its commercial purposes, to preserve the natural beauty of the West's most spectacular landscapes.

The year 1864 presented a regional challenge to the existence of the survey because local interests began to criticize Whitney's work and question its value to California's economy. The survey survived, but with a new mandate to narrow its focus on the mineral wealth of the state. While Whitney busied himself with legislators, the survey's work continued. A year earlier, King had joined the survey and went with a team to the Mariposa region, where he found evidence for dating Jurassic gold-bearing slates and the auriferous gravels of the Pliocene. Another spectacular achievement of the year was a full triangulated survey of the High Sierras, which featured King's and Richard Cotter's daring ascents of Mount Tyndall and Mount Whitney. King later reported the ascents in his classic *Mountaineering in the Sierra Nevada*, a widely-read exploration narrative that invites comparison to Frémont's *Narrative* and Powell's *Colorado River*.

Despite successfully revealing an oil speculation hoax, which pitted Whitney against Yale chemist Benjamin Silliman, the survey continued to suffer legislatively from parties more interested in a prospecting map than a scientific survey. Also, King and several other prominent members of the crew had moved on to other endeavors. By 1868 the survey was truly moribund. One historian has noted that it was not necessarily Whitney's scientific stature that hampered public opinion but rather his conservationist and conservative ideas that urged caution, utility, efficiency, and sometimes preservation of the California environment. Whitney, like Powell, exercised an alternative vision of how the western landscape ought to be utilized. It remained for another explorer of the California landscape to solidify the links between scientific exploration and the new politics of environmentalism.

John Muir's story is one of an amateur naturalist ambling his way through the woods and is therefore very much different from the history of federal- and state-funded surveys discussed so far. But his exploration of California's landscape brought him into debates with such widely respected scientifics as Whitney and King, and his activities offered an enduring and hotly contested vision of America's use and development of the landscape. Auspiciously enough, he arrived in San Francisco in 1868 just as the legislature was bringing the California Geological Survey to its demise. Muir grew up in a humble environment. He had emigrated in 1838 from Scotland after his family was displaced from their herding grounds to make room for larger sheep ranches. Muir's father, filled with a religious passion, settled the family in Wisconsin to make a go of farming. Muir's childhood was preoccupied with chores, some self-education, and a keen desire to tinker with time-saving technologies. His early-rising machine led him to seek further education at the new University of Wisconsin, where he developed a desire to study geology. He vacillated for many years between becoming an inventor or a naturalist while working in broom handle factories and other seats of mid-nineteenth-century commerce. A slip of a file left him temporarily blind in 1867, and from that point on he devoted his life not to working in factories but rather in "Nature's workshop."

Giving up factory for field and prioritizing practical experience over book learning, Muir took a 1,000-mile walking tour from Indiana to Florida and then to California to see for himself many of the natural wonders in the West described in newspapers and magazines. He wandered south and found himself in Yosemite Valley, where he stayed for four years, tending sheep and working at a sawmill. Most of his time was spent examining the geology of the granite valley and piecing together the mystery of its history. Arguing against the prevailing wisdom of Whitney and King, who claimed that the valley must be a result of a geological catastrophe, Muir read evidence of a valley carved by the uniformitarian work of glaciers. He quickly became a local celebrity of Yosemite, and in 1871 and 1872 he was visited by Torrey, Asa Gray, John Tyndall, and Ralph Waldo Emerson. Muir preached a story of the interdependence of all natural phenomena whose sublime workings manifested themselves in mountain cathedrals. At the same time, he began criticizing both academic naturalists who let learning through books inhibit understanding through experience and the commercial interests, such as sheep farmers and loggers, who destroyed these natural temples. In short, Muir struck out against the anthropocentrism that lay at the very heart of the explorer's mission.

This was a pivotal development in the history of exploration. Muir's criticism joined Powell's alternative vision of the use of the landscape. Both echoed the sentiments of George Perkins Marsh's *Man and Nature* (1869) that criticized the heedless destruction of nature at the hands of fast and steady economic expansion. After Muir's four-year exploration of Yosemite, he returned to San Francisco and began the task of transcribing the Sierra's sublimity into readings for the appreciation of all Californians. His popularity steadily increased, and he led the campaign to transform Yosemite from a state park into a national park in 1890. In 1892 he rallied 182 California residents, some scientists but most just enthusiasts of the California landscape, to further lobby for the preservation of Yosemite and other wilderness spaces. The Sierra Club quickly became a reflection of Muir's wilderness ethic. But new visions of the landscape would continue to challenge Muir, the Sierra Club, and Yosemite Valley, most notably the lost battle in 1913 to save the Hetch Hetchy Valley from becoming a reservoir for San Francisco's mushrooming population. By the end of Muir's life in 1914, the American explorer was assuming new social roles as conservationists and preservationists.

A Long Conclusion: An End to Exploration?

In 1890, the U.S. Census Bureau declared the nonexistence of a line separating civilization from the wilderness. The announcement provoked the young historian Frederick Jackson Turner to write "The Significance of the Frontier in American History." Turner's essay reflected a dawning sentiment that a postfrontier America was a direct threat to American exceptionalism. Moreover, the end of the frontier wilderness threatened, at least symbolically, the resource base of cheap land, minerals, lumber, and water. Some explorers, very much like Powell, argued that America must learn to conserve its scarce resources.

Other explorers, such as Muir, fought vigilantly against the expansionist forces that threatened those few wild places that still existed. Newspapers and magazines began questioning whether there was anything left to explore. The answer was on display in Portland, Oregon, in 1905.

The Lewis and Clark Centennial Exposition served as a capstone for one hundred years of continental exploration. The event was a celebration that praised the heroic expeditions that materialized themselves in effective coast-to-coast colonization, a phenomenon similar to the Census Bureau's declaration of the closing of the frontier. But the exposition was as much a glimpse backwards as it was a look into the future. The Smithsonian Institution, the National Museum, and the Bureau of Fisheries created exhibits that displayed the natural and cultural worlds of Alaska and the Philippines. These extracontinental exhibits made explicit an imperial policy that was coming to fruition at the end of the nineteenth-century. After a century of continental expansion, the spirit of Manifest Destiny went oceanic, and American explorers played an important role here as well.

If America's destiny lay in the West, the Pacific Ocean did not bring an end to its steady course. William Seward's orchestration of the 1867 purchase of the Alaskan Territory from Russia added another immense swath of land to the Union. Initially, America sent in explorers with some hesitancy, as the task of settling the continental West was still very much in process. One incentive for exploration was a new interest in the North Polar Ocean, a seascape that perhaps offered a new northwest passage and an ocean full of whales for the butchering. In 1879 the American whaling ship *Jeannette* was stuck in the Arctic ice some 150 miles north of the New Siberian Islands. That same year the steamship *Corwin*, in the service of the U.S. Treasury Department for enforcing sealing regulations off the Fur Seal Islands, was given the task of making inquiries into the fate of the *Jeannette*. Accompanying the rescue ship was none other than Muir, an explorer who could not turn down an opportunity to lay his eyes on new wild areas and study evidence of glacial action in the Far North. Observing the whaling operations of the North Pacific, he drew an obvious analogy to the American West. "Newly discovered whaling grounds, like gold mines, are soon overcrowded and worked out, the whales being either killed or driven away."[7] The work of exploration and frontier economics had simply moved to another region. The *Corwin* continued to ship other explorers through the North Pacific, most notably Charles Haskins Townsend.

Townsend was the emblem of the old spirit of exploration working in new frontier areas. He grew up in a religious Pennsylvania family and enjoyed public education but did not attend university. By dint of his love for fishing and nature, he found himself working for the U.S. Fish Commission in 1883 in the area of salmon propagation in California. With the Pacific Northwest's fisheries in decline, Townsend headed north to Alaska and the North Pacific on a search for stocks of harvestable fish and to conduct a general natural history survey of the Arctic. In 1885, he was doing some routine surveying in the waters of St. Paul Island in the Aleutian Islands when he encountered the *Corwin* and requested passage to the Arctic for further study. On the basis of his Alaskan work, Townsend won the position of chief of the Division of Fisheries in the U.S. Fish Commission. Securing Townsend was not accidental, for the Bureau of Fisheries was then mounting a massive survey of Alaska's coastal waters. Townsend was to develop that potential, and from 1888 to 1892, the *Albatross* conducted a detailed hydrological and ichthyological survey that proved instrumental for development of the North Pacific fishery. In the process, Townsend played a pivotal role in international diplomacy between the United States and Russia over fishing rights in the Bering Strait in 1902. He also served on the Bering Fur Seal Commission, dedicated to preserving the sustainable resource of North Pacific furs.

The *Albatross* would eventually be called to Hawaii and the Philippines to conduct similar work in other territories won by the United States in the Spanish-American War. Townsend, however, went east, where he became director of the New York Aquarium and spent more than thirty years displaying the wonders of the deep to visitors in Battery Park and later Coney Island. The aquarium was under the administration of the New York Zoological Society, a civilian organization dedicated to the education of citizens about the need for conservation. Townsend's role changed accordingly, and he began lobbying on behalf of organisms such as the fur seals, the Alaskan reindeer, and the Galápagos tortoise in far-flung territories. Even in his educational and conservation roles, however, he exercised an imperial vision that truly situates him within the history of American exploration. In 1919 he went so far as to publicly call the United States to annex the arm of the Sonora Desert that connected the Gulf of California to Arizona so that "a seaport from which to exploit the fisheries of the gulf" could be developed.[8] The extension would also ease both commercial and scientific navigation between the United States and the Gulf of California and open up the area for wider mineral extraction and agricultural development.

Townsend was not alone. He was one of hundreds of government scientists who served in various bureaus beginning in the 1880s. Along with the federal government, civilian institutions such as museums and universities began mounting exploratory expeditions that were every bit as grand as the Great Surveys of the 1860s. They explored new territories, some of which became American possessions; others would be subject to various kinds of informal colonialism. Some of those explorers headed into the oceans; others would eventually go into space. Some looked for resources to control and conquer, others pieced together strategies for sound conservation of resources, and still others became outright critics of America's heedless rush to exploit every square inch of territory. But all are part of the history of nineteenth-century American exploration that joined science and nation to the frontier.

Bibliographic Essay

The role of exploration in the growth of the American empire has been splendidly chronicled by William Goetzmann. See both his *Exploration and Empire: The Explorer and the Scientist in the Winning of the American West* (1966) and his more recent reevaluation, *New Lands, New Men: America and the Second Great Age of Discovery* (1986), that situates the nineteenth-century story of American exploration back to the seventeenth century and, perhaps most helpfully, off the continental landscape into Japan, the Arctic, and the oceans. See also John W. Simpson, *Visions of Paradise: Glimpses of Our Landscape's Legacy* (1999), which is not focused on exploration but does an excellent job of describing the role of the explorer's cultural vision of landscape. Simpson was basing his work on the still-important books that discuss the role of wilderness and the West in American culture. See Henry Nash Smith, *Virgin Land: The American West As Symbol and Myth* (1950); Roderick Nash, *Wilderness in the American Mind* (1967); and Bernard Augustine DeVoto, *Course of Empire* (1952).

For treatments of American science in the nineteenth century that describe the purpose, role, and difficulties of establishing a scientific tradition within the American republic, see A. Hunter Dupree, *Science in the Federal Government: A History of Policies and Activities* (1986); Nathan Reingold, *Science in America since 1820* (1976); George H. Daniels, ed., *Nineteenth-Century American Science: A Reappraisal* (1972); and Robert V. Bruce, *The Launching of Modern American Science, 1846–1876* (1987). Many of these treatments tend to minimize the history of geography and natural history. An important exception is Charlotte Porter's *The Eagle's Nest: Natural History and American Ideas, 1812–1842* (1986), which does an excellent job of describing the history of exploration into the Ohio River Valley and beyond by focusing on the amateur and professional scientists in Philadelphia. See also John Rennie Short, "A New Mode of Thinking: Creating a National Geography in the Early Republic" (1999).

Historians of science have recently made some important progress in analyzing the culture of exploration in science. See, for instance, Mary Terrall, "Heroic Narratives of Quest and Discovery" (1998), as well as Jennifer Tucker,

"Voyages of Discovery on Oceans of Air: Scientific Observation and the Image of Science in an Age of 'Balloonacy'" (1996); Jane Camerini, "Wallace in the Field" (1996); and Bruce Hevley, "The Heroic Science of Glacier Motion" (1996). Historians of American exploration have not yet taken up this call, although the subject holds much promise.

A definitive history of Alexander Dallas Bache and the U.S. Coast Survey is found in Hugh Richard Slotten, *Patronage, Practice, and the Culture of American Science: Alexander Dallas Bache and the U.S. Coast Survey* (1994).

The literature on the Corps of Discovery is legion. The latest and perhaps most interesting take on the expedition is Thomas Slaughter, *Exploring Lewis and Clark: Reflections on Men and Wilderness* (2003), which not only helps to explain the cultural significance of the expedition but also historicizes the mythic stature of Lewis and Clark over the last two centuries. For recent topical treatments of Lewis and Clark, see James Ronda's *Lewis and Clark among the Indians* (1984); his more recent *Finding the West: Explorations with Lewis and Clark* (2001); and Albert Furtwanglers' *Acts of Discovery: Visions of America in the Lewis and Clark Journals* (1993). For more comprehensive treatments of the expedition in its more traditional narrative form, see Stephen Ambrose, *Undaunted Courage: Meriwether Lewis, Thomas Jefferson, and the Opening of the American West* (1996).

A historical treatment of the Army Corps of Topographical Engineers is William Goetzmann's *Army Exploration in the American West, 1803–1863* (1959). The literature on individual explorers is often beset with hagiographic portrayals of the great explorers, but there are some good exceptions. On John C. Frémont, see Ferol Egan, *Frémont: Explorer for a Restless Nation* (1977); John Rolle, *Charles Frémont: Character as Destiny* (1991); and "John Charles Fremont—Phrenological Character and Biography" (1856). A concise introduction to the Great Surveys is Richard A. Bartlett, *Great Surveys of the American West* (1962); see also Goetzmann's *Exploration and Empire* and Debora Rindge's "Science and Art Meet in the Parlor: The Role of Popular Magazine Illustration in the Pictorial Record of the 'Great Surveys'" (1999). On Clarence King, see Robert Berkelman, "Clarence King: Scientific Pioneer" (1953). On John Wesley Powell, see Donald Worster, *A River Running West: The Life of John Wesley Powell* (2002). Worster, both an environmental historian and a key architect of the New Western History, has done an exemplary job of retelling the life of Powell in a way that fully explores the many contesting issues over land development in mid-nineteenth century America. Also see Scott Kirsch, "Regions of Government Science: John Wesley Powell in Washington and the American West" (1999), and Wallace Stegner's *Beyond the Hundredth Meridian: John Wesley Powell and the Second Opening of the West* (1953).

On the Geological Survey of California, see Gerald D. Nash, "The Conflict between Pure and Applied Science in Nineteenth-Century Public Policy: The California State Geological Survey, 1860–1874" (1963), and Michael L. Smith, *Pacific Visions: California Scientists and the Environment, 1850–1915* (1987). Smith's book focuses on California scientists instead of just the Geological Survey of California and sets the standard in portraying the changing roles of exploration over a common landscape. For another important discussion of Muir and his environmentalist role, see Stephen Fox, *John Muir and His Legacy: The American Conservation Movement* (1981). For the classic treatment of conservation history that describes the people and economic forces that challenged Muir in the early twentieth century, see Samual P. Hays, *Conservation and the Gospel of Efficiency* (1959).

On the end of the frontier, see David Wrobel, *The End of American Exceptionalism: Frontier Anxiety from the Old West to the New Deal* (1993). Although Wrobel only hints at the changes to American exploration, his analysis is extremely valuable for understanding how and why explorers began moving outside of the continental United States. For the role of federal scientists in the late nineteenth century, see Philip Pauly, *Biologists and the Promise of American Life: From Meriwether Lewis to Alfred Kinsey* (2000). A helpful text that discusses conservation history in its extracontinental mode is Kurkpatrick Dorsey, *The Dawn of Conservation Diplomacy: U.S.-Canadian Wildlife Protection Treaties in the Progressive Era* (1998). On the history of exploration, science, conservation, and fisheries in the Pacific Northwest, see Richard White, *The Organic Machine* (1995), and Joseph Taylor, *Making Salmon: An Environmental History of the Northwest Fisheries Crisis* (1999).

6

The Exploratory Tradition in the Ocean Sciences

If the practice of exploration entails a certain movement from core to periphery, then by definition exploration was at the very heart of the development of the ocean sciences. Inquiry into oceanic geography and water life goes far back into ancient and medieval times. But it was within the last two centuries, as nationalism joined hands with a renewed emphasis on global oceanic travel and commerce, that ocean phenomena garnered unprecedented scientific scrutiny.

Oceanic exploration existed in different modes of practice. In the nineteenth century, three clearly discernible modes solidified into separate but tightly related practices: military, commercial, and imperial/expeditionary. National navies generally headed the military mode of ocean exploration. Naval personnel, often trained in military schools that prized the fields of mathematics and engineering, took charge of investigating the oceans to better enable the navigation of both commercial and military vessels. Naval ocean science was merely a predecessor of bureaucratic ocean science, those initiatives headed by governmental entities that sought increased knowledge of the oceans. Explorers who went to sea for profit also developed a commercial mode of exploration. Similar to the Victorian tradition of amateur scientists, many of these profiteers took the time to write up their observations of the oceanic world. Finally, what can be called an imperial or expeditionary mode of exploration was spearheaded by states that prized scientific knowledge of the ocean for the honor and power that it bestowed upon the nascent nation-state. These three related modes are most clearly visible in the nineteenth century, but their traditions continued to operate throughout the twentieth century.

While the history of the modern ocean sciences can be mapped along the axis of the various modes of exploration, it can also be traced along an axis of technological innovation. If anything distinguishes oceanic from other forms of exploration, it is the enormous inaccessibility of the ocean itself. It is for this reason that the secrets of the starry heavens, as the common adage tells us, have revealed themselves more expediently than the secrets of the abyssal depths. Throughout the modern period, naturalists and scientists have relied on technology to provide a murky window into oceanic natural properties. The stories of technologies such as coal-powered ocean vessels, deep-water dredges, seine nets, and sonar are very much at the center of the human drama of exploration in the ocean sciences. A more daunting technological problem has been the difficulties of human exploration beneath the ocean surface. This will be the subject of the next chapter.

Military Exploration and the Bureaucratic Imperative

Matthew Fontaine Maury's *The Physical Geography of the Sea* (1855) is one of the definitive texts that brought the field of oceanography into existence. While it is clear that military ocean navigators from Zheng He (1405–1433) to the present have routinely observed ocean winds, currents, tides, and storms, it was not until the middle of the nineteenth century that much of this data was brought together in a systematic treatise. Maury's contribution to the ocean sciences was the result of the U.S. government's wish to make ocean transport safer for both commerce and military vessels.

Maury was eager to learn the science of navigation, but he was frustrated with the state of navigational texts and by the failure of the navy to provide adequate means of education. He began the work of routine observation wherever his commission took him, noting the tides, winds, and currents in the Mediterranean, the South Pacific, and the coastline of South America. Simultaneously, he developed his skills in mathematics and the sciences. An 1839 horse carriage accident left Maury lame for three years, and in 1842 he was assigned to the Depot of Charts and Instruments in Washington, D.C. The Depot was essentially a clearinghouse for all navigational and astronomical instruments and data. Despite holding an unglamorous post, Maury was in the perfect position to explore the ocean not through his own direct observation but rather through the navigator logs and journals that routinely passed over his desk. From the masses of data that streamed through his office, he began to synthesize an understanding of the currents, winds, tides, and seasonal fluctuations of the oceans. From this portrait, he outlined sea-lanes that would most likely provide the safest and most reliable course for naval traffic. During an era that witnessed the slow transition from sail to steam, Maury mapped oceanic steamer lanes that served "like a double track railway—everything on the same track

*Matthew F. Maury, of the U.S. Depot of Charts and Instruments, and later the super-
intendent of the National Observatory, is remembered for carrying out the first
systematic survey of ocean currents and winds. His work served as a vital aid to
navigators in antebellum America. (Library of Congress)*

Matthew Fontaine Maury (1806–1873)

Matthew Fontaine Maury is best known as the author of one of the founding texts in physical oceanography. *The Physical Geography of the Sea* (1855) presents a form of oceanic exploration characterized by naval and federal scientists in the nineteenth century. During a time when scientists needed to convince bureaucrats that federal patronage would lead to beneficial results, Maury's primary goal was to help make ocean transport safer for both commerce and military vessels.

Maury hailed from a Virginian farming family of humble means. In an attempt to escape the drudgery of farm life, the modestly educated Maury went against his parents' wishes and followed his brother into military service. In 1825 he began his career in the U.S. Navy as a midshipman and began his naval education aboard the USS *Brandywine*. Maury was eager to learn the science of navigation, although he was frustrated with the current state of navigational texts and by the failure of the navy to provide adequate means of education. So he began the work of routine observation, noting the tides, winds, and currents in the Mediterranean and the South Pacific and along the coastline of South America. Simultaneously, he began developing his skills in mathematics and the sciences. He was assigned to the New York Navy Yard in 1830 to study for his officer's candidacy. After passing the examination, he was assigned to the sloop-of-war USS *Falmouth* as sailing master, and his experience led to his first scientific paper, "On the Navigation of Cape Horn." The humble publishing success boosted Maury's motivation, and in short order he wrote and published a new navigator's guide that he hoped would rival Nathaniel Bowditch's *The Practical Navigator*.

Maury was assigned to the Depot of Charts and Instruments in 1839. The Depot was essentially a clearinghouse for all navigational and astronomical instruments

and in the same lane, going one way."[1] His work became both practical and theoretical as he attempted to harness the vagaries of the ocean to provide navigators with a safe and speedy passage.

Maury's responsibilities widened, and in 1844 he was appointed the superintendent of the new National Observatory, where his staff grew to include departments of hydrography, astronomy, and meteorology. His popularity continued to mount, especially given the instrumental role of his navigation charts in the famous clipper ship races of the 1850s. He also became interested in the depths of the Atlantic Basin and the composition of the seafloor. He commissioned various sailors to employ rather simple surveillance technologies, such as a plummet made of a thirty-two-pound cannonball attached to a long line of twine. Working with his staff, he tinkered with this equipment to gather more information, an activity well suited for the superintendent of Charts and Instruments. Working with the midshipman John Brooke, Maury created a simple and data. From the masses of data that streamed through his office, he began to synthesize an understanding of the currents, winds, tides, and seasonal fluctuations of the oceans. From this portrait, he began to realize sea-lanes that would likely provide the safest and most reliable course for naval traffic. In 1843 he extended his knowledge of the oceans by requesting that merchant navigators fill out abstract logs and send them to the Depot in exchange for what would come to be known as Maury's *Sailing Directions*. He also petitioned navigators to periodically cast adrift bottles overboard that contained information on location and time.

Not only was such information crucial for the navigation aids that Maury was publishing at a steady clip, but it also provided the grist for some of his more theoretical work. For instance, he delivered a paper titled "The Gulf Stream and Currents of the Sea" at the 1844 conference for the National Institute for the Advancement of Science. There he scrutinized the speculations of three centuries' worth of navigators and commentators to arrive at a stunningly detailed explanation of the size, shape, and movement of the Gulf Stream.

Maury's responsibilities widened, and in 1844 he was appointed the superintendent of the new National Observatory. He also became more interested in the depths of the Atlantic Basin and the composition of the seafloor. He commissioned various sailors to employ rather simple surveillance technologies, such as a plummet made of a thirty-two-pound cannonball attached to a long line of twine. As all of this data came in, Maury and his staff patiently compiled the information and redistributed it in the form of navigation aids and charts. In 1853, Maury published the sixth edition of *Explanations and Sailing Directions to Accompany the Wind and Current Charts*. In it was a ninety-page chapter titled "The Physical Geography of the Sea." This chapter provided the foundation for the eponymous book that became an instant success when it was published in 1855.

cannonball plummet that collected a sample of the seabed in a hollowed-out section of pipe that was plunged onto the ocean floor. He also played a role as an advisor for Cyrus W. Field's attempts to lay a transatlantic telegraph cable. As all of this data arrived, Maury and his staff patiently compiled the information and redistributed it in the form of navigation aids and charts. In 1853, he published the sixth edition of *Explanations and Sailing Directions to Accompany the Wind and Current Charts*, which included a ninety-page chapter titled "The Physical Geography of the Sea." The chapter provided the framework for the same-named book that became an instant success when it was published in 1855.

At first glance, Maury may appear to be an odd character in the annals of exploration in the ocean sciences. A naval officer who spent more time behind a desk than on ships, he hardly stirs the imagination in recalling the intrepid work of other ocean explorers. But such is the work of military and bureaucratic surveillance, a practice that marshaled the energy and resources of countless quotidian observations to create a mosaic of scientific knowledge. The United States was by no means a pioneer in this practice; the French Dépôt des Chartes, Plans, Journaux et Mémoires Relatifs à la Navigation and the British Royal Hydrographic Office served as the chief models for the U.S. Depot of Charts and Instruments. Moreover, this form of exploratory practice continued to play an important role in the twentieth century.

Maury's significance reached out in other directions as well. He had the effect of turning navigators into amateur naturalists in their own right. One shipmaster returned his log abstract with a particularly telling comment. "I am happy to contribute my mite towards furnishing you with material to work out still farther towards perfection your great and glorious task, not only pointing out the most speedy route for ships to follow over the ocean, but also teaching us sailors to look about us and recognize the wonderful manifestations of the wisdom and goodness of the great God."[2] If Maury had a role in transforming sailors into naturalists, it took place within the wider context of an amateur natural history tradition that developed on board commercial vessels.

Commercial Exploration

There is perhaps no better example of the commercial mode of oceanic exploration than the body of work that emerged through contact with nineteenth-century whaling ventures in England and the United States. Cetaceans had long been a subject of ancient and medieval bestiaries, but it was not until the nineteenth century that their natural histories were more thoroughly investigated. The increasing sophistication of cetological studies developed alongside the New England whaling industry. Several liberally-educated ship captains and surgeons acted as both company agents and amateur naturalists, similar to the way that nineteenth-century natural historians were also town parsons, wealthy amateurs, museum workers, and university professors. Whaling personnel were no exception to this order. Their careers aboard whaling vessels gave them a unique opportunity to dispel the many myths about the size, structure, and behavior of whales and dolphins, and they sometimes conducted ingenious investigations of other oceanic phenomena. An early example of this mode of commercial exploration is William Scoresby, a British whaling captain who published an *Account of the Arctic Regions* (1820), a study of the Arctic Ocean and various whales that were being harvested by whalers.

Scoresby was born in 1789 into a whaling family in Yorkshire, England. His father, a whaling captain of solid reputation, began taking the young

Scoresby to the whale fisheries of the Arctic aboard the famous *Resolution*. The elder Scoresby had, by this time, befriended the famous British botanist Joseph Banks, who had traveled on James Cook's celebrated circumnavigations of the globe some years earlier. The correspondence resulted in the younger Scoresby's study of physics at the University of Edinburgh. He intermittently served for the Royal Navy and continued to travel the Greenland Sea. The aging Banks prodded him for accounts of the Arctic region, and Scoresby began delivering papers to learned societies on meteorology and terrestrial magnetism. His father turned over command of the *Resolution* to him in 1810, and Scoresby continued to ply Arctic waters for the next twelve years. As assuredly as the *Resolution* brought back profitable whale oil, so too did Scoresby spend these years exploring the ocean as an extraordinarily inventive naturalist.

Scoresby was one of those enviable people whose every observation became the subject of inquiry. Given the wide scope of his interests, he often improvised and turned his commercial ship, at least in part, into a floating laboratory. For instance, part of his "Survey of the Greenland Sea" was an investigation into the ocean's various colors. To accurately gauge the water's color, he looked down the trunk of the ship's rudder to shield his view from the contaminating lateral rays of the sun. He then looked for the "colouring substance" of the seas by procuring a sample of snow from a piece of floating ice. "A little of this snow, dissolved in a wine glass, appeared perfectly nebulous; the water being found to contain a great number of semi-transparent spherical substances."[3] Placing the sample under a compound microscope, Scoresby came to appreciate the extraordinary abundance of plant and animal life that made their home in the sea. He also investigated the seawater's salinity and specific gravity, and he even created some of the first vertical temperature profiles of the ocean by lowering devices that trapped samples of water from various depths. Such experiments are clear evidence of the influence of a culture of experimental physics that had defined the cutting edge of eighteenth-century science. But Scoresby was also something of a naturalist, and the second volume of his *Account of the Arctic Regions* reveals masses of information on the habits and structure of cetaceans as observed through the inquisitive gaze of a whaling captain who was reaping a harvest from the sea.

Scoresby was by no means alone in the practice of commercial exploration. The ship surgeon Thomas Beale detailed his experiences with the sperm whale in *The Natural History of the Sperm Whale* (1839). The American whaling captain Charles Melville Scammon compiled his natural historical knowledge in his widely cited *Marine Mammals of the Northwest Coast of North America* (1874). These treatises were amalgams of natural history and economic enterprise. Scoresby was attempting to provide a general survey of the Arctic as a potential geography of natural resources for British extraction. Similarly, Scammon envisioned his manuals as guides in aiding his fellow mariners to locate and harvest populations of marine mammals. These works also described the mechanics of whaling in as much detail as they depicted the cetaceans under investigation. All were written during the era of classical whaling, and they are filled with anecdotes of daring races and courageous battles. Nevertheless, these nineteenth-century whaling naturalists established the foundations for twentieth-century cetological science.

Imperial Science and the Expeditionary Tradition

The practices characteristic of both military and commercial exploration drew from the cumulative experiences of thousands of navigators whose work and lives brought them into an intimate relationship with the ocean. The mode of imperial science, in contrast, was characterized by large-scale expeditions, often funded by wealthy patrons or nation-states, whose primary purpose was to gather natural knowledge of the ocean. The motivations for oceanic expeditions were multiple: they provided natural and social information on distant colonies, aided in the practice of navigation and commerce, and sometimes were simple displays of personal or national wealth and power. In this latter instance, these activities assembled an imperial archive, a compendium of natural knowledge that symbolized, demonstrated, and stood in proxy for actual imperial power and control. Britain's nineteenth-century empire became a key motivator in the quest for oceanic knowledge. Such expeditions often hosted a small battery of scientists and naturalists who spent several years fending off seasickness to wrest the secrets of nature from the ocean. They would then return home and spend years poring over the preserved specimens of their haul before publishing their findings in massive volumes. Ballyhooed with much fanfare, these expeditions were also public ventures that demonstrated the power of modern nation-states, and it is just as certain that these expeditions were given military and commercial mandates.

Two such expeditions highlight many of the problems and potentials of this type of exploratory activity. A flurry of nationalistic exploring expeditions in the eighteenth and early nineteenth centuries brought seagoing science to a heightened level of activity. England had sent out Captain Cook, Sir John Ross, and Robert Fitzroy to map and catalog the contours of its far-flung empire. France, too, could boast of the expeditions of Dumont d'Urville and Louis de Bougainville. These expeditions—all of them grand and expensive—highlighted the value of extensive scientific activity at the peripheries of empire.

One of the most famous examples of the expeditionary tradition in the ocean sciences was the U.S. Exploring Expedition commanded by Captain Charles Wilkes. The expedition cannot be properly called an oceanographic undertaking because the objectives of the mission were far more general, reminiscent of the surveying and collecting expeditions of Cook. The U.S. Ex Ex, the popular shortened title of the expedition, was a venture that melded commerce with science. It set out to chart the Pacific for the New England whaling, sealing, and bêche-der-mer fleets; discover the predicted existence of Antarctica; and survey the Pacific Northwest with the partial aim of establishing a U.S.-Canadian border that would favor the interests of the United States.

Wilkes led three warships, one storeship, and two tenders. Delegation of scientific duties revealed Wilkes's desire to keep the expedition under military control, a dynamic that led to numerous personal problems while at sea. Studies in astronomy, surveying, hydrography, geography, geodesy, meteorology, and physics remained the province of naval officers. Nine civilian scientists, downsized from the original plan of thirty-two, were responsible for zoology, geology, mineralogy, botany, and conchology. In essence, naval personnel were to conduct scientific investigations at sea, and the civilian scientists would do their surveys on whatever island or coast Wilkes chose to deposit them.

For nearly four years the expedition circumnavigated the globe, concentrating its efforts in the South Pacific and along the Antarctic coast, which Wilkes found during one of the most heralded accounts of navigation. Early into the expedition, off the coast of Peru, one schooner was lost with all hands. Another schooner also went down, and although the crew found safety, a large cache of collections found their home on the floor of the Pacific. The expedition arrived home in 1842 to little fanfare. Examination and publication of the expedition's collections limped along for thirty-two years and resulted in twenty-four volumes. Many of the preserved specimens eventually found their way into the newly-erected Smithsonian Museum. There they took their place on display shelves next to and overshadowed by the specimens, steadily increasing in number, that were the fruits of exploring expeditions in the American West. When all was said and done, the true fruits of the U.S. Ex Ex were less important for their theoretical value than for the number of new species the ships' naturalists had discovered. Despite the successes of the expedition, the larger priority of American exploration shifted to the American West, and it remained for other empires to fully articulate an imperial form of exploration that would make the oceans its sole object of investigation.

Britain's *Challenger* Expedition is often the venture given the sacrosanct status of the first true oceanographic expedition, and the status is not without some justification. For reasons both economic and technological, deep-sea basins had resisted the exploratory practice of science. Indeed, most scientists conceived the ocean depths as a stagnant and lifeless region; it was known as the azoic zone. British naturalist Edward Forbes set out to investigate this hypothesis in the 1840s and initiated the technological development of a dredge for surveying deep-sea specimens. His limited investigations of the eastern Mediterranean led him to conclude that beneath 300 fathoms, there was little probability of finding life. But the azoic zone theory continued to be challenged, especially by those involved with laying and maintaining deep-sea telegraph cables that tended to accumulate encrusted evidence of deep-sea life. In the late 1860s and early 1870s, several British-funded scientific expeditions, including the *Lightning, Porcupine,* and *Shearwater,* punched a few more holes in the azoic theory. Darwin's theory of natural selection played a role as well. Indeed, several *Challenger* naturalists held high hopes for the discovery of "living fossils" that would still exist in the relatively unchanging environment on the ocean floor. Whatever the speculation, by the early 1870s there was a growing consensus that the ocean floor and its intervening layers contained a trove of beautiful and fantastic life forms worthy of exploration. The Royal Society of London and the British Association for the Advancement of Science successfully lobbied the British empire for a full-scale expedition to set matters straight.

The *Challenger* was a 2,300-ton naval corvette owned by the British Admiralty. Just as Banks had reconstructed the decks of Captain Cook's military ships, so too was the *Challenger* reconfigured with a dizzying array of relatively new exploratory technologies: sounding devices, new thermometers, forty dredges, and hundreds of miles of hempen rope. The corvette's 1,200-horsepower auxiliary engine and its 18-horsepower winch, which was used for raising the dredges, was just one sign that the science of oceanography was evolving in step with industrial and technological development. There were 269 crew members aboard, including a coterie of 6 scientifics, most of whom specialized in biology. C. Wyville Thomson, a naturalist with experience on previous oceanographic expeditions, was chosen as director of the scientific staff. The *Challenger* set sail in 1872 for what would be a three-and-a-half-year expedition that rounded Cape Horn, explored the southern reaches of the Indian Ocean, and then sailed on to Australia, New Zealand, the Philippines, Hong Kong, Yokohama, Hawaii, Tahiti, Valparaiso, St. Thomas, Bermuda, and Halifax. The empire's colonial network served as the chief ports of call; the oceans between and around them were the subject of inquiry.

In many ways, the *Challenger*'s survey of the world's oceans was a simple affair, akin to the tradition of Western explorers encountering a new landscape. Dredges routinely groped sections of the ocean floor, while conical trawl nets sampled various depths of the ocean for biological life. Specimens

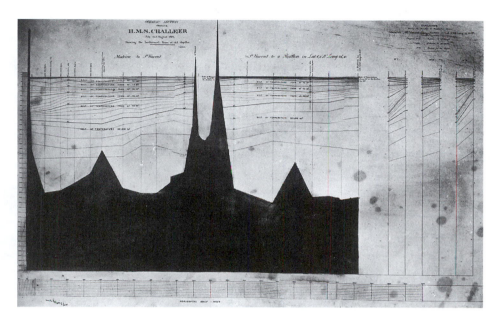

Chart of temperatures at various ocean depths taken between Madeira and St. Vincent during HMS Challenger's *scientific voyage around the world, 1873. (Time Life Pictures/Getty Images)*

were cataloged and preserved in jars for further examination at home. Numbers marking oceanic depth, temperature inversion, salinity, surface current, and subsurface current filled journals. Chemical and physical oceanography, however, was not the expedition's strength. More than anything else, the expedition was a search for ocean life.

The *Challenger* traveled almost 70,000 miles and brought back more than 13,000 kinds of animals and plants and 1,441 water samples. Thompson and the young Edinburgh-trained naturalist John Murray began the task of sorting through the collections and publishing the results. Specimens were distributed to seventy-six scientists throughout Europe and even the United States, and after nineteen years, the *Challenger* Expedition boasted a beautifully illustrated fifty-volume work titled *Report on the Scientific Results of the Voyage of H.M.S. Challeger.* Alexander Agassiz noted that it remained for other expeditions to "fill out the grand outlines laid down by the great English expedition."[4] In the next decade, the United States, Germany, Norway, Russia, Italy, and Monaco all commissioned similar expeditions, although they were more humble in time and scope.

By the end of the nineteenth century the three modes of ocean exploration—military/bureaucratic, commercial, and imperial—were well established. These modes would mix and mingle throughout the twentieth century as the science of oceanography became firmly established as a specialized scientific pursuit. The completion of an oceanographic museum in Monaco in 1910 and the establishment of the Institut Oceanographique de Paris in 1911, both lavishly funded by the extraordinary patron and naturalist Prince Albert I of Monaco and his Monte Carlo wealth, were two of the more visible signs that ocean exploration in the twentieth century would receive tremendous government, private, and scientific support.

Scandinavia, Oceanography, and Cod

Scandinavian scientists played a leading role during the heady days of establishing oceanography as a specialized science, but instead of merely surveying the biological contents of the ocean, they were often more interested in uncovering the secrets of a dynamic ocean filled with currents, upwellings, and drifts that moved both latterly and vertically. One of their prime motivations was to understand the fluctuating migrations of cod, herring, and other marine animals on which Scandinavia depended. Their investigations, however, were sometimes less about the cod itself than the chemical, physical, and biological elements of the cod's environment.

The development of northern European oceanography marked a new period in the history of ocean exploration that was made possible by technological and industrial developments in marine vessels and fishing fleets. The overwhelming shift from wind and human muscle to harnessing the power of fossil fuels increased the potential for stressing marine environments, but it also created the opportunity for frequent and widespread surveillance of ocean phenomenon, a prerequisite for understanding physical and biological oceanic characteristics. This period of oceanography also pointed to the value of international collaboration among scientists. The oceans are peculiar environments that resist the tidy geopolitical maneuverings of nation-states, and oceanographers around the North, Baltic, and Mediterranean Seas quickly realized the scientific payoff for joining forces, planning research programs, and sharing data. All three modes of exploration provided the key motivations and support for the scientific investigation of the oceans throughout the twentieth century. But more than ever before, the people doing the work of exploration were more specialized in research interests and were better trained in universities.

In the last quarter of the nineteenth century, several Scandinavian scientists discovered that populations of food fish seemed to occupy and enjoy regions of the Baltic and North Sea that had distinctive temperature and chemical characteristics. Fishermen would likely benefit not only from a knowledge of those characteristics but also from a map of oceanic layers and especially an understanding of the seasonal drifts and shifts of those layers. The Swedish chemist Otto Pettersson, who had a particular interest in heating and cooling dynamics, turned to the issue in 1890 and became one of the primary architects of physical oceanography. Working with fellow chemist Gustav Ekman, Pettersson began an intense investigation of Scandinavian seas with the use of five steamers that simultaneously took hydrographic readings from various established stations. The result was a static and dynamic profile of physical oceanic characteristics. This new method of exploration represented an important shift in that the practice broke down national boundaries and emphasized collaboration. By 1894 Sweden, Denmark, Germany, Scotland, and Norway had made four collaborative hydrographic surveys.

Pettersson's intensive historical and exploratory work on Scandinavian oceans in local environments was matched by the equally intriguing extensive exploratory work of Fridtjof Nansen in the North Polar Sea. Nansen's goal was to complement Pettersson's coastal studies with a more extensive understanding of the importance of ocean circulation. The Norwegian explorer planned to allow his ship to be captured in polar ice and then drift with the pack from the new Siberian Islands to the Greenland Sea while the crew performed meteorological and oceanographic research. The legendary *Fram*, a schooner with a hull that was rounded so as to withstand the pressure of the pack ice, left Norway in July 1893 and was trapped in the ice by September. The ship then resembled a research laboratory from which the crew conducted soundings and dredges and took water samples for temperature and salinity. Nansen hoped to discover the relationship between ocean currents and the drift of the ice, but this turned out to be a complicated affair. Ocean circulation proved to be a function of various layers of ocean water with distinct temperature, salinity, and density profiles.

These were heady days for Scandinavian oceanographers and meteorologists. Many were using the data obtained from Nansen's expedition, as well as numerous other expeditions, to determine the fluid dynamics of the ocean. Scientists combined their data with theoretical mathematics to create the dynamic method, a system of equations that captured a host of variables including the Earth's gravity, rotation, ocean temperature, salinity, and density. The Norwegian seas and the Arctic Ocean were thus filled with a host of standing waves and whirling vortices. The method proved extremely useful. The German *Meteor* Expedition (1925–1927), hailed as one of the most important surveys of the Atlantic Ocean, created fourteen vertical profiles of the entire Atlantic Ocean. These profiles not only demonstrated the complex fluid dynamics of the ocean but also showed that the dynamic method, which had previously been used only

Fridtjof Nansen (1861–1930)

Fridtjof Nansen was one of the cadre of Scandinavian oceanographers who founded the dynamic theory around the turn-of-the-century while examining problems in oceanic fisheries. But he was more widely known in the annals of exploration history as one of the more daring explorers of the Arctic Ocean and the Greenland Sea. He grew up in Oslo as an exceptionally athletic sportsman. His studies at the University of Christiana emphasized physics, mathematics, and zoology, and he went on to curate the natural history collections of the Bergen Museum for some six years and earn a doctorate for his work on the nervous system.

Before taking the job as curator, Nansen traveled aboard the *Viking* in 1882 to observe the Arctic environment. The trip was the inspiration for his 1888 east-to-west jaunt across the Greenland icecap. The success of the expedition, though of questionable scientific value, was received with great fanfare, as was the tradition of European Arctic exploration. The success brought in the necessary patronage to mount another expedition designed to test a theory he had about Arctic Ocean currents. The Norwegian explorer planned to allow his ship to be captured in polar ice and then drift with the pack from the new Siberian Islands to the Greenland Sea while the crew performed meteorological and oceanographic research.

The legendary *Fram*, a schooner with a hull that was rounded so as to withstand the pressure of the pack ice, left Norway in July 1893 and was trapped in the ice by September. The ship then resembled a research laboratory from which the crew conducted soundings and dredges and took water samples for temperature and salinity. Nansen hoped to discover the relationship between ocean currents and the drift of the ice, but this turned out to be a complicated affair. Ocean circulation proved to be a function of various layers of ocean water with distinct temperature, salinity, and density profiles. Many scientists were to use the data obtained from Nansen's expedition as well as numerous other expeditions to determine the fluid dynamics of the ocean. Especially helpful was Nansen's invention of bottles that collected ocean samples at specific depths for shipboard analysis.

Data met with theoretical mathematics to create the dynamic method, a system of equations that captured a host of variables including Earth's gravity, rotation, ocean temperature, salinity, and density. The Norwegian Sea and the Arctic Ocean were thus filled with a host of standing waves and whirling vortices. Before turning to the work of statecraft, politics, and humanitarianism for which he was awarded the 1922 Nobel Peace Prize, Nansen directed a laboratory in Christiania that was in the service of the newly established ICES.

in the Norwegian Sea and the Arctic Ocean, could be extrapolated for use on ocean basins.

These advances in physical oceanography were matched by exciting developments in biological oceanography, and here, too, Scandinavian scientists played an important role. They joined hands with other scientists in the European community in one of the most important organizations to push forward the science of oceanography in the early twentieth century, the International Council for the Exploration of the Sea (ICES). ICES was established as a body of people interested in examining some of the environmental pressures that the oceans faced due to the modernization of the fishing industry in the late nineteenth century with the introduction of steam-powered ships and the wide use of purse nets and trawl nets. In 1902, after three years of planning, hydrological and biological scientists from Germany, Denmark, Norway, Sweden, Finland, Britain, Holland, and Russia went to work devising a research program of ocean exploration. Physical oceanography and the dynamic method continued to shape the research program of ICES scientists who, on a quarterly basis, surveyed oceanic currents, layers, temperatures, salinity, and density. The thinking was that fishery fluctuations were likely a function of physical oceanic properties. But other scientists examined the fish themselves.

ICES members Johannes Petersen and Johan Hjort, for instance, worked on the connection between fish population and plankton fluctuations. The question had already received its most sustained attention at the celebrated University of Kiel School of Planktonologists in the Baltic Sea. Among many other scientists, Victor Hensen, a physiologist, began examining the extraordinary array of microorganisms from expeditions such as that undertaken by the *Challenger*. In 1887 he coined the term "plankton" to refer to all those minute organisms, and he quickly realized, along with many others, that there was an extraordinary diversity of these creatures. Plant plankton, or phytoplankton, Hensen surmised, may hold the key for physiological questions about oceanic productivity. He therefore devised various technologies for quantitatively calculating the production of phytoplankton. In order to discover this very important matter, Keil scientists crisscrossed the Atlantic Ocean in 1889 on the *National*, an expedition that became known as the Plankton Expedition. The expedition demonstrated that the abundance of plankton was far greater in colder northern waters than in warm tropical waters, a finding that challenged the logic of terrestrially-based humans.

Whether directly through ICES or by way of various fishery bureaus in Europe and America, exploration of ocean phenomena proceeded along a well-established tradition of monitoring variables and correlating them with ocean dynamics and oceanic life. This was by no means a humble undertaking; employment of the dynamic method required painstaking exploratory work, precise and systematic sampling techniques that ranged over time and place, and a careful process of analysis followed by synthesis of large quantities of data. Such exploration differentiated the extensive expeditions such as the *Challenger* and the U.S. Ex Ex from the more intensive investigations of local and discrete ocean spaces of Scandinavian scientists. The dynamic method was well in place, but in the second decade of the twentieth century, oceanography and ocean exploration began to become more tightly linked to governments and military forces. As the case of the United States amply demonstrates, an infusion of federal monies and nationalistic mandates dramatically transformed the patronage structures and the technologies of ocean exploration. War and a new kind of oceanic warfare provided the key impetus.

American Oceanography and the Military Imperative

Prior to World War I, the tools of oceanographic exploration adapted to the many desires of researchers interested in marine natural history and the chemical and physical properties of the ocean. Submarines drastically transformed the landscape of technological innovation. In their attempt to more fully understand the frontier of undersea geography, ocean scientists developed new tools that had military, economic, and scientific value. While seine nets, dredges, and sampling bottles continued to play important roles in oceanographic exploration, ocean scientists created new devices that extended their view of ocean phenomena, though mostly by proxy. As the techniques of oceanic exploration began to change, the scientific landscape gave rise to oceanography as a new academic discipline.

Since the early years of the 1900s, technological entrepreneurs in America had been interested in creating sounding and listening devices to prevent ships from running aground. The Submarine Signal Company, a Boston-based private firm, developed the first oscillator devices capable of sending signals through the ocean. The task was given greater importance, however, with the threat of German U-boats during the early years of the World War I. In 1915, the Naval Consulting Board, headed by Thomas Edison, set up the Committee on Submarine Detection by Sound. In 1916 the National Research Council (NRC) was established and set up its own committee for submarine detection research. Both of these committees fostered a strong relationship between marine scientists and the U.S. Navy centered on the development of acoustical equipment to monitor sounds under the surface of the ocean. The NRC constructed a laboratory composed of university physicists at the U.S. Naval Experiment Station in New London, Connecticut, in order to institutionalize the relationship. Early echo-ranging devices were developed in this context, and thus was invented a new wireless method of plumbing the depths of the ocean floor. The fathometer, as it was called, quickly proved its worth in the early 1920s as new maps of the continental shelf proved invaluable for both civilian and military navigation, especially those done by the American Coast and Geodetic Survey.

Physicist Harvey Cornelius Hays was responsible for developing a sonic depth finder (SDF) for the U.S. Navy. Despite limited federal funding for oceanography shortly after the war, the navy wanted to conduct further tests with Hays's SDF. In 1922, the destroyer *Stewart* made 900 soundings greater than 3,000 feet. The spectacular success of the *Stewart*'s ocean floor survey convinced the scientific community that a close relationship between the navy and ocean science could prove fruitful. For instance, Yale's Herbert Gregory had long been a proponent of surveying the Pacific region to guide economic, social, and political development. He therefore forwarded a proposal to investigate the rabbit damage in the Hawaiian Islands Bird Reservation. Furthermore, he wanted to investigate the hydrography of the region's islands. The expedition was widened to include the Wake, Midway, and Johnston Islands and took place in 1923 aboard the minesweeper *Tanager* over a five-month period. A permanent relationship between science and the navy was in the works.

Hays, who was heading the Naval Research Laboratory, suggested that the navy should sponsor basic research. His voice joined those of other prominent scientific administrators such as Thomas Wayland Vaughan, chair of the NRC's Committee on the Oceanography of the Pacific; Frank Lillie, director of the Marine Biological Laboratory (MBL) at Woods Hole; and John C. Merriam, president of the Carnegie Institute of Washington. They approached the National Academy of Science (NAS), which created the Committee on Oceanography. The committee nominated the MBL in Woods Hole, in conjunction with the home of the Bureau of Fisheries, as the perfect location to establish a community of ocean scientists. Before approving the grant, Henry Bigelow was asked to write a report on the state of American oceanography. Bigelow's report, titled *Oceanography: Its Scope, Problems, and Economic Importance* (1931), formed part of the foundation for American oceanography. Bigelow accurately noted that European countries had far outpaced America in the sciences of the sea. He therefore recommended that the government and American universities devote personnel, money, and facilities so that the United States could better harness the economic and military fruits of the ocean. One of his most enduring recommendations was for the establishment of an East Coast analog to the Scripps Institution of Oceanography (SIO). The NAS committee and the Rockefeller Foundation moved quickly to devote $2.5 million to provide the necessary support for facilities, scientists, and a research vessel, the *Atlantis*. Thus, the Woods Hole Oceanographic Institute (WHOI) was founded with Bigelow at the helm.

The SIO was for America's Pacific what WHOI was for the Atlantic. In line with the late nineteenth-century tradition of establishing biological laboratories in beautiful seaside areas, SIO was established by William Emerson Ritter in 1903 as a marine laboratory in La Jolla, California. In 1923 Ritter and his friend and patron E. W. Scripps decided to turn the laboratory away from fundamentally biological questions to embrace oceanographic research. Vaughan, a U.S. Geological Survey scientist, was selected to run the institute. Vaughan set a precedent for undertaking exploratory research at a small laboratory with a new mandate and relative modest funding. Through his links to the world of federal science, he drew from the U.S. Coast and Geodetic Survey and the Bureau of Lighthouses to supply sea temperature and salinity data. He also established a relationship with the U.S. Navy Hydrographic Office for data from the navy's new sonic depth finder.

In 1936 the navy worked with Henry Bigelow and Columbus Iselin of WHOI to investigate some problems with echo-sounding devices. Iselin found that temperature variations refracted echo sounds, thus vitiating the technology's usefulness. Because sonar effectiveness worsened as the sun heated the surface layer of water, this anomaly was called the "afternoon effect." As a result, new technologies were invented to more easily detect temperature variations. Most important was the bathythermograph, a device first developed by MIT's Carl Gustav Rossby that created a continuous temperature profile as it was lowered into the ocean. Previously, ocean scientists relied on either insulated water bottles or reversing thermometers to detect ocean temperature at various layers. Both techniques had inherent problems, not the least of which was the large amount of time and energy required in preparation and recording. The BT, as the bathythermograph became known, allowed scientists to create a continuous sketch of ocean temperature with incredible ease. The BT quickly joined the fathometer as one of the most important technologies that eased the burden of oceanographic work.

Advances such as these must have pleased Bigelow. Clearly, they were a sign that American oceanography was headed in the right direction. Indeed, oceanography as an academic discipline enjoyed considerable growth during the interwar period, especially at the sites of marine stations such as SIO and WHOI. Oceanography, like the sleepy little towns of La Jolla and Woods Hole, roared with energy during the explosive growing years of the early 1940s. More than anything else, it was World War II that catapulted oceanography as an academic discipline. While physicists and engineers around the country mounted a total effort to build devices that released the energy of the atom, oceanographers were enlisted to provide vital knowledge of the oceanic environment. The mutual benefit between oceanography and the military was arguably greater than any other relationship. In contrast to the creation of atomic bombs, where the place of warfare presents relatively slender logistical problems, submarine warfare put a premium on a comprehensive knowledge of oceanic temperature, salinity, and density gradients. Oceanographic translators were especially adept in demonstrating the value of oceanography to military uses.

The wider context was a new relationship between science and the federal government. In 1940 the National Defense Research Committee, and later the more comprehensive Office of Scientific Research and Development, were given life to forge a tight relationship between science and the military. Together they coordinated the mobilization of science to military ends throughout the 1940s and into the postwar period. The decisiveness of this influence was a result of the military's selection of oceanographic research problems. For instance, wave dynamics and submarine acoustics were two previously underresearched areas for which the military mobilized ocean scientists. Oceanography emerged from the war completely changed in proportion, aim, and allegiances. Physical studies of the ocean prevailed over the previous emphasis in marine biology. Economically, oceanography became tied to government support and government policy.

The navy's greatest need in 1941 was detection of German submarines. The National Defense Research Committee set up a project on the West Coast through the university at the Navy Radio and Sound Laboratory on Point Loma, soon named the University of California Division of War Research. The navy chartered the SIO's only research vessel, the *E. W. Scripps*, and enlisted Harald Sverdrup, Martin Johnson, Richard Fleming, Eugene LaFond, Walter Munk, Francis Shepard, and Roger Revelle as participants. SIO was to be the West Coast analog to WHOI, focusing on the knowledge of the Pacific Ocean. They researched the thermocline and found that the scattering layer, so named because sound waves scattered when hit, was really a mass of oceanic organisms. SIO scientists also studied sound reflectivity in this context. Different surfaces have unique sound signatures that indicate mud, coral, bedrock, or solid steel. Topographical charts of the ocean would thus describe the nature of the seafloor, crucial information when the navy wished to moor mines to the ocean floor.

In sum, the U.S. Navy mobilized American oceanography to get answers to specific military questions. Information was needed on echo-sounding technologies, ocean temperature gradients, ocean currents, wave dynamics, weather prediction, seafloor topography, seabed structure, and both the fouling and sound-producing activities of oceanic fauna. To answer these questions, military funding had transformed small-scale oceanographic laboratories into monster research facilities heavily dependent on federal support. The end of the war thus presented an important problem: How would the military continue its relationship with ocean science in the postwar period?

Roger Revelle (1909–1991)

Roger Revelle was one of the foremost translators who stood between the worlds of science and the military before and after World War II. His scientific and administrative work helped to bring oceanography to maturity. Revelle grew up in Sacramento, California, and received a bachelor's degree in geology from Pomona State in 1929. Throughout the early 1930s, he worked toward his Ph.D. in oceanography by doing his graduate studies at the SIO when it was still a young oceanographic institute. He studied under Vaughan and worked on chemical analysis of deep-sea sediments. In 1936 Revelle spent a year studying at the Geophysical Institute in Bergen, Norway, and upon his return he made great strides in dovetailing his geophysical studies with SIO director Harald Sverdrup's emphasis on the dynamic method.

In 1940 Revelle's status in the U.S. Naval Reserve was activated, and he began training radar operators at the Point Loma U.S. Navy Radio and Sound Laboratory. In 1942 he supervised BT data collection under the navy's hydrographic office and then worked as the head of the sonar section in the Navy Bureau of Ships. He therefore served as one of the most important links between oceanography and the navy. Other than creating sonar charts for naval vessels, he also coordinated research on ocean floor geology and wave forecasting. Continuing to serve in the Bureau of Ships, he was appointed director of ocean science planning at Crossroads and played a large role in convincing the navy that oceanographic work was necessary to complement the already wide range of research conducted at Bikini. He increasingly believed that the federal government would be the primary patron of basic, as opposed to applied, oceanographic research, and Crossroads effectively brought together the military and the SIO, as well as other institutions, into a tight dialogue.

Revelle left the navy in 1948 and returned to La Jolla, where he expedited naval support of SIO at just under $1 million annually. He was the mastermind of MIDPAC (1950) and the Capricorn Expedition (1952); he also played important roles in the IGY (1957) and Project Mohole. Revelle emphasized physical over biological oceanography, although after becoming director of the SIO in 1950 he saw to the establishment of the Institute of Marine Resources (IMR), an entity originally planned for research into California's fisheries. However, the IMR ended up doing most of its research on beach erosion and ocean engineering and searching for underwater minerals. His penchant for institution-building culminated in the establishment of the University of California, San Diego, where he played a large role in establishing graduate studies in the sciences.

Oceanography continued its uneasy alliance with the military throughout the Cold War era. The quintessential developments of nuclear-powered submarines and the launch-ready Polaris submarines ensured that a large part of the Cold War would be waged under the surface of the oceans. Oceanographers were poised to meet the demand for the reconnaissance of this military geography, especially in the extraordinarily vital period in the early 1950s that witnessed an abundance of expeditionary riches, mostly by way of scientific translators who had become adept science/military brokers during the war. By 1958, the Office of Naval Research (ONR), the Hydrological Office, and the Bureau of Ships combined to fund 80–90 percent of all U.S. oceanography.

The navy emerged from World War II in a precarious position. It seemed as if the nuclear arsenal would make the U.S. Air Force the main instrument of America's military. Submarines still remained a potent conventional weapon and ensured some future for the navy, but the success of submarines depended on advances in American ocean science. Such was the pressure that initiated the navy's funding of postwar oceanic science. The ONR was thus constructed in 1946 as the administrative arm of the navy for funding basic science, especially oceanic science. The ONR was quickly followed by the Atomic Energy Commission's patronage of oceanography. Perhaps the most conspicuous collaboration between oceanography and the military was during the Pacific atomic bomb tests.

Operation Crossroads, a joint military and scientific experiment concerned with the detonation of atomic bombs, was significant for laying the groundwork for postwar oceanographic exploration in two fundamental ways. First, the military continued to fund oceanographic exploration so long as civilian scientists continued to play a role as advisors and experts on ocean phenomena. Second, the style of oceanographic research at Bikini, a kind of before-and-after survey that resembled a natural field experiment, was one of the first studies to bring a baseline methodology into the province of the ocean. Such baseline data became increasingly central to oceanographic exploration as scientists monitored the overall health of the world's oceans just as they began to show evidence of anthropogenic degradation. The relationship between oceanographers and nuclear technology continued throughout much of the 1950s.

The Pacific expeditions had their analogs in the Atlantic, although in the Atlantic the emphasis was not on monitoring nuclear expositions but rather on a combination of deep-sea topography and continued work on sound transmission. The 1947 Mid-Atlantic Ridge expedition, a reconnaissance of the Mediterranean Sea's topography also in 1947, the 1948 *Caryn* expedition to Bermuda, and Operation Cabot's 1950 exploration of the Gulf Stream all involved WHOI personnel and equipment with various collaborations among Columbia University, the navy's Bureau of Ships, and the navy's Hydrographic Office. A steady stream of new monitoring equipment was put into action, including underwater cameras, seismic instrumentation, current motion instruments, and new tools to measure gravitational field variation necessary for internally guided missile technology.

Environmental and Remote Ocean Exploration in the Late Twentieth Century

Oceanography emerged from the war with an important national mandate and access to resources enjoyed by relatively few scientific fields. In the 1970s, this relationship developed into a new concern for the global environment as national planning of ocean policy joined hands with new remote-sensing technology. In the rush toward antisubmarine detection in the mid-1950s, oceanographers teamed with the military, Bell Labs, and AT&T Western Electric toward the deployment of SOSUS (Sound Surveillance System), a network of listening devices that fringed North and South America to monitor submarine traffic. The SOSUS array was able to listen for submarine traffic some 1,000–2,000 miles away. The Cold War thaw in the 1990s loosened the security on the navy's tightly guarded secret. For instance, in the late 1990s, the navy invited Cornell bioacoustic scientist Chris Clark to hear whale songs. The array is currently being used by scientists to monitor seismic events and even to study hydrothermal vents. As a technology that employs the natural potential of ocean physics, SOSUS is a highly idiosyncratic form of remote sensing. A much more common method of examining ocean phenomena has been from above the surface of the ocean. For example, in the early 1970s oceanographers began using plane-mounted multispectral line scanners and infrared radiometers to correlate sea-surface temperature with water color. Other scientists used plane-mounted thermal infrared detectors to gauge the extent of oil spills.

The potential for using space-based platforms for reconnaissance of earthly phenomena was first established with the success of the Nimbus weather satellites. In July 1972, the National Aeronautics and Space Administration (NASA) launched its first Earth Resources Technology Satellite (ERTS). In 1978 the National Oceanic and Atmospheric Association, in conjunction with NASA, launched an ambitious satellite that used microwave surveillance to gather global synoptic data on ocean topography, waves, temperature, and wind. Although the satellite failed after three months, enough data was collected to keep about fifty oceanographers busy for five days at a conference at the Jet Propulsion Laboratory. These were exciting days for oceanography, but satellite data had one crucial limitation: it could not penetrate the surface of the sea.

To truly understand the world's oceans as a global system, remote sensing had to be deployed in the ocean itself, an initiative that stemmed from two general events. First, modern computer advances were making oceanic modeling more sophisticated. Second, in the early 1980s, scientists the world over were becoming more interested in climate research. In 1984 nine teams of international researchers came together at Liège, Belgium, to compare modeling

Satellite imaging of natural phenomena, like this depiction of water temperatures in the Pacific, are produced by remote sensing devices that have dramatically increased the amount of data accumulated by explorers. Remote sensing and satellite imaging represent the latest phase of technological non-human exploration. (Time Life Pictures/Getty Images)

strategies relating to the tropical atmosphere's influence on the Pacific Ocean and the ocean's influence on the atmosphere. The teams were composed of meteorologists and oceanographers who puzzled over the oceanic and atmospheric conditions necessary for El Niño and other southern oscillation events, a hotbed of research in the 1980s, especially after the remarkable 1982 season. The efficacy of the modeling techniques made clear that a more precise understanding of global weather depended on a rigorous system of real-time data collection of both the atmosphere and the subsurface ocean temperatures and currents. That same year, international scientists responded with the Tropical Ocean-Global Atmosphere (TOGA) program that, over the next ten years, moored seventy buoys in arrays that stretched along the equatorial Pacific. Individually, the buoys gathered information on ocean temperatures and currents as well as atmospheric conditions and sent the information in real time to central data centers via polar-orbiting satellites. When collectively assembled, the data became a metaphorical oceanic blood-pressure cuff that provided, and continues to provide, massive amounts of dynamic data that helps both researchers and forecasters to understand oceanic and atmospheric phenomena. The data from TOGA dovetails with other international exploratory efforts such as the World Ocean Circulation Experiment (WOCE), both of which are initiatives given force by the World Climate Research Programme, an international program established in 1980 to explore climate prediction and the possibilities of anthropogenic climate changes.

In the 1970s, oceanography took on new uses and new forms of exploratory activity to track the environmental degradation of the oceans. Some precedent existed, especially in the United States. Fueled by the public outcry concerning atmospheric nuclear testing, in 1955 the NAS convened the Committee on the Effects of Atomic Radiation on Oceanography and Fisheries. The committee's initial report considered the dumping of atomic wastes in the ocean and recommended further study of ocean mixing and ecology before the United States committed itself to a policy of wide-scale ocean dumping. They also pointed to an extraordinary opportunity for ocean scientists. The tracing of radioactive isotopes could be used as an exploratory method for studying ocean currents, circulation, and ecology; such tools were becoming a standard ecological method in the 1950s. Between 1946 and 1962, dumping had become fairly routine as the United States disposed of some 86,000 containers of nuclear waste. In 1962, the committee recommended strict limits on the practice. Scientific data proved not at all powerful, and international policy was regulated to politics. But scientists were able, now and again, to lend some credibility to public fears that dispersion of nuclear waste in the ocean could prove harmful. If oceanographers played a marginal role in the 1960s debates on the safety of disposing radioactive material on the ocean floor, they were well poised to gain from environmental politics in the 1970s.

As usual in the world of environmental politics, it was the highly visible nuisance of an oil spill off the Santa Barbara coast in 1969 that made ocean scientists acutely aware of some of the dangers that postwar economic growth posed for the ocean. Western countries had been extracting petroleum from the seafloor in ever-increasing amounts since the end of World War II. Spills and leaks such as the one near Santa Barbara and the one from the tanker *Torrey Canyon* (1967) pushed oceanographers to think seriously about the impact of oil on marine ecosystems. In 1973 the Southern California Academy of Science hosted a number of government scientists, industry experts, and concerned marine and physical oceanographers at a conference to establish a baseline study of the outer continental shelf marine environment off the coast of California. In a manner similar to the oceanographic work done at Operation Crossroads, which measured the effect of atomic explosions on the Bikini environs, such baseline studies were meant to establish sound data on relatively pristine ecosystems so that the effect of anthropogenic pollutants could be readily determined.

Over the past thirty years, oceanographers have been at the leading edge of scientific and public campaigns that seem to demonstrate rapid anthropogenic climate change and a rather dire prognosis on the state of the globe's oceans. Institutionalized, internationalized, and equipped with highly sophisticated remote-sensing technologies, these scientists have charted unprecedented and— fifty years ago—barely imaginable methods and techniques in oceanic exploration. Reports from commissions in the late 1990s and early 2000s contain little of the enthusiasm compared to their predecessors some forty years ago. Owing to the immense oceanic and climatological problems that the world faces, the development of oceanic exploration is poised for dramatic growth in the twenty-first century.

Bibliographic Essay

Susan Schlee's *The Edge of an Unfamiliar World: A History of Oceanography* (1973), though more than thirty years old, remains the definitive text in synthesizing the development of oceanography and ocean exploration. More focused on the history of the marine sciences is Margaret Deacon's *Scientists and the Sea 1650–1900: A Study of Marine Science* (1971). Regrettably, historians of science are today a bit reluctant to provide such sweeping and comprehensive histories. We can only take solace in the fact that the era of historical specialization was preceded by the synthetic vision of historians such as Schlee and Deacon. Two historiographical treatments of the ocean sciences that are helpful, even if a bit dated, are Eric Mills, "The Historian of Science and Oceanography after Twenty Years" (1993), and Mott Greene, "Oceanography's Double Life" (1993). Two collections of essays dealing with the history of oceanography are very helpful in appreciating the scope and depth of the field: Keith Benson and Philip Rehbock, eds., *Oceanographic History: The Pacific and Beyond* (2002), and Mary Sears and Daniel Merriman, eds., *Oceanography: The Past* (1980). These volumes contain wonderful essays written by historians, oceanographers, and people otherwise involved in oceanography. Their chief virtue is that they represent oceanography's wider history and therefore stand as a corrective to the Anglo-American emphasis above.

Matthew Fontaine Maury has, for the most part, evaded the attention of most professional historians except for Chester Hearn's *Tracks in the Sea: Matthew Fontaine Maury and the Mapping of the Oceans* (2003). An account of the work of whalers in the creation of the science of cetology is Lyndall Baker Landaurer, *From Scoresby to Scammon: Nineteenth Century Whalers in the Foundations of Cetology* (1982). The text should be more widely published because it provides an excellent analysis of the links between science and commerce.

William Roger Stanton's *The Great United States Exploring Expedition of 1838–1842* (1975) still remains the definitive account of the expedition as a scientific endeavor, but Nathaniel Philbrick's popular *Sea of Glory: America's Voyage of Discovery—the U.S. Exploring Expedition, 1838–1842* (2003) provides a gripping and balanced account as well. American contributions to the marine sciences in the nineteenth century by way of Louis Agassiz and his son Alexander are covered in Mary Winsor, *Reading the Shape of Nature: Comparative Zoology and the Agassiz Museum* (1991). A wonderful chapter in Philip Pauley, *Biologists and the Promise of American Life: From Meriwether Lewis to Alfred Kinsey* (2000), provides a glimpse into the work of the U.S. Bureau of Fisheries. On the latter, see also Jeffrey Brosco, "Henry Bryant Bigelow, the US Bureau of Fisheries and Intensive Area Study" (1989).

One of the more exemplary studies of the *Challenger* Expedition that puts the venture into the wider context of British and U.S. oceanic exploration at the end of the nineteenth century is Helen Rozwadowksi's *Fathoming the Ocean: The Discovery and Exploration of the Deep Sea* (2005). For an analysis of the mingling of military and scientific cultures, see Rozwadowski's "Small World: Forging a Scientific Maritime Culture for Oceanography" (1996). A fun and accessible book is Richard Corfield's *The Silent Landscape: The Scientific Voyage of HMS Challenger* (2003).

For all its importance, oceanography from Scandinavia has not garnered much historical research. One helpful book is Helen Rozwadowki's definitive history of ICES, *The Sea Knows No Boundaries: A Century of Marine Science under ICES* (2002). Although the focus is on ICES, the start of the analysis does consider some of the interesting developments in the dynamic method established by Scandinavian scientists. Another helpful book is Eric Mills, *Biological Oceanography: An Early History, 1870–1960* (1989). Mills's account is focused more on biological oceanography and gives an exemplary history of the Kiel school of early plankton specialists, but he also necessarily covers parts of the Scandinavian story because of proximity in time and space. Other links between seaside biological laboratories and their links to oceanography are Susan

Schlee's *On Almost Any Wind: The Saga of the Oceanographic Vessel "Atlantis"* (1978), which provides a wonderful ship analysis that revolves around the WHOI, and Elizabeth Noble Shor, *Scripps Institution of Oceanography: Probing the Oceans, 1936–1976* (1978), which is the West Coast analog.

By far the most detailed and perhaps insightful treatment on the relationship between oceanography and the U.S. Navy from World War I to about 1960 is Gary E. Weir's *An Ocean in Common: American Naval Officers, Scientists, and the Ocean Environment* (2001). While a general knowledge of most of this topic has been established by other historians, Weir's legion research has turned up troves of insightful detail, bolstered by analysis, of what the key participants were thinking and doing. Not to be missed is Chandra Mukerji's *A Fragile Power: Science and the State* (1989), which analyzes postwar American oceanography to make a case for the new patronage relationship between the federal government and science. Her analysis of Operation Crossroads can be supplemented with Johnathan M. Weisgall, *Operation Crossroads: The Atomic Tests at Bikini Atoll* (1994).

Ronald Rainger has also detailed some of this story in several articles, including "Constructing a Landscape for Postwar Science: Roger Revelle, the Scripps Institution and the University of California, San Diego" (2001), which gives the story of how Revelle's wartime experience with the federal government and the military were enlisted to establish basic oceanographic research at La Jolla and UCSD. Rainger's "Adaptation and the Importance of Local Culture: Creating a Research School at the Scripps Institution of Oceanography" (2003) demonstrates an alternative to the top-down model of research tradition development through the analysis of Norwegian scientist Harald Sverdrup at the helm of the nascent SIO. Naomi Oreskes, widely recognized for her history of continental drift research, has also taken on the subject and has provided some insightful readings on gender in oceanic exploration. See her "Laissez-tomber: Military Patronage and Women's Work in Mid-Twentieth-Century Oceanography" (2000) and "Objectivity or Heroism? On the Invisibility of Women in Science" (1996). Oreskes has also teamed up with Rainger to write "Science and Security before the Atomic Bomb: The Loyalty Case of Harald U. Sverdrup" (2000), which reveals much of the cultural insecurity that existed in wartime science.

Jacob Darwin Hamblin has written fairly extensively on the international culture of the ocean sciences. His "Visions of International Scientific Cooperation: The Case of Oceanic Science, 1920–1955" (2000) and "Science in Isolation: American Marine Geophysics Research, 1950–1968" (2000) are helpful in contextualizing ocean science within the wider initiatives of U.S. international diplomacy in the post–World War II period. His "Environmental Diplomacy in the

Cold War: The Disposal of Radioactive Waste at Sea during the 1960s" (2002) is less focused on scientific exploration than international diplomacy. Nevertheless, it remains the best analysis of the highly controversial issue of the dumping of radioactive waste in the ocean. Christopher Vanderpool, "Marine Science and the Law of the Sea" (1983), also provides an insightful analysis of the influence of international politics on marine science.

7

Human Exploration under the Sea

No earthly environment has presented so daunting a challenge as has undersea exploration. In many ways, the reluctance of the oceans to grant easy access has dealt many a blow to the daring and adventurous plans of explorers, military strategists, mineral prospectors, and scientists alike. Despite setbacks, cost overruns, and the loss of human life, industrialized nations became quite adept at wresting the secrets of the sea, a feat made possible through the fast pace of technological innovation in the twentieth century. Only space travel compares when considering the all-important need for industrialized technology in the field of underwater exploration. But developing ingenious technologies was only one of the impediments to scientific exploration of the undersea realm. Even those scientists who successfully marshaled funds, technology, and research teams faced an uphill battle when it came to justifying large costs with sometimes rather modest scientific dividends. Indeed, to many within the scientific community, underwater research seemed just a bit too much like a stunt to merit serious consideration. Diving scientists were sometimes an uneasy mixture of showmen and serious explorers.

There are basically three separate groups of technologies that shaped undersea exploration. The first dealt with those technologies that enabled individuals to spend long periods of time under the sea such as diving helmets and scuba. The second allowed teams of divers to spend days or even weeks under the sea through the use of underwater living habitats such as SEALAB and the less famous, but perhaps more important, Tektite Project. The third was the development of underwater submersibles especially the recently retired flagship research submersible *Alvin* that was largely responsible for human exploration of the deep sea.

Historians have been slow to take up the story of underwater exploration for the same reason that scientists sometimes criticize the effort. Even Cindy

Lee Van Dover, an *Alvin* pilot, professional oceanographer, and articulate advocate of human underwater exploration, noted that "there is no reason why a robot equipped with multiple cameras and manipulators could not accomplish the tasks *Alvin* can do. . . . There is an indefinable advantage to seeing the ocean bed with one's own eyes. I think a creative force and a passion develop that are unattainable from watching a mere image, no matter how good that image might be."[1] It is beyond question that the technological tools outlined in the previous chapter outstripped by far the scientific results produced from human underwater exploration. This fact is underscored by the recent decommissioning of *Alvin*, a research submersible that served as one of the only means for deep-sea research for just under forty years. Rather than being measured in the quantity of scientific publications, the value of undersea exploration has been about opening a new and alien environment for the world to see. The heroic tradition of exploration, as valid today as it was during the days of James Cook and Meriwether Lewis, holds that a place simply is not a place until it has been visited by a human being.

Cindy Lee Van Dover (circa 1955–present)

Cindy Lee Van Dover's exceptional career has been devoted to exploring the ecology, biodiversity, and biogeography of deep-sea vents as a scientist and an *Alvin* pilot but always as a naturalist. In so doing she has become one of the preeminent authorities on the peculiar fauna of thermal vent ecosystems. She represents one of the hundreds of scientists who made a living within the protection of a titanium sphere, but she stands out as a naturalist who went to extraordinary lengths to spend large amounts of time in these eerie seascapes. She probably represents, more than anyone else, the passion of ocean exploration, and she was able to parlay that passion into a prestigious scientific career, no mean feat by any standard.

　　Van Dover took her undergraduate degree in zoology at Rutgers, and instead of moving on to graduate school, she spent several years as a technician in Ft. Pierce, Florida, at one of the many American biological laboratories that mushroomed along the edges of the continent after World War II. In 1981 she edged her way onto an *Alvin* expedition at the East Pacific Rise. Having succumbed even further to the deep-sea bug, she earned a master's degree at UCLA before earning her Ph.D. at the MIT/WHIO joint program in oceanography. She then made an unconventional decision. "The deep sea is a compelling place, and being just a passenger on an occasional dive to the seafloor did not satisfy. Becoming an *Alvin* pilot seemed to me a logical next step in my career as a deep-sea ecologist. Not only do pilots dive far more frequently and in more places than any scientist, they

Personal Exploration: Helmets and Scuba

Throughout the nineteenth century, diving suits and helmets were routinely used by sponge divers in the Mediterranean. The technology was magnificently simple. Usually constructed of heavy copper and plate glass, the device was hoisted onto the diver's shoulders, thus providing a small sanctuary of air. A rubber hose tethered the helmet to a ship or boat equipped with a pump that continuously fed air into the helmet. A diver with a helmet never really swam but rather walked on the ocean floor. There is precious little evidence that naturalists endeavored to don the weighty equipment to explore shallow waters for substantial research purposes. If a nineteenth-century oceanographer wanted to see what was under the water, he would use seine nets, trawls, and dredges to bring the seabed to the surface for examination. As oceanic depths became a matter of increased scientific scrutiny in the early twentieth century, a number of naturalists began using the equipment to view the ocean realm *in situ*. Even exploring with a diving helmet required considerable resources, and so it is not

have the ultimate control over the dive. What better way to see the seafloor?"[2] It may have been a "logical" decision, but it was nonetheless an usual decision. *Alvin* pilots were U.S. Navy personnel and a male-dominated group of folks at that. Van Dover went through the pressures of pilot training that were made even more arduous because of the sometimes oppressive machismo social environment. She became a qualified *Alvin* pilot in 1990 and over the next year-and-a-half piloted forty-eight dives before heading back to academia, first at Duke, then the University of Alaska, and then to the College of William and Mary where she held the prestigious Marjorie S. Curtis Associate Professorship in the Biology Department, and then back to Duke.

Van Dover's deep-sea exploration has manifested itself in more than sixty peer-reviewed articles and the textbook *The Ecology of Deep-Sea Hydrothermal Vents* (2000). She has also donned the cap of the literature-loving general naturalist and has written one of the most popular and artful descriptions of deep-sea life, *Deep-Ocean Journeys* (1996), a text that bears a resemblance to William Beebe at his best. She has also been involved in many public awareness campaigns aimed at revealing the secrets of the undersea world to America's younger generation. And in the tradition of modern explorers, Van Dover has also become something of an environmental advocate, describing how even deep-ocean caverns have been marked by humanity's sometimes carelessly wasteful ways. In 1996 she wrote, "We are only beginning our attack on the blue waters, with the coastal waters and the continental shelf doubling as our beachheads and communal sewers. I think we could screw it up badly—maybe not in this generation, but give us time."[3]

altogether surprising that a number of New York naturalists were able to make good use of the technology.

Roy Waldo Miner, for instance, was an expert in invertebrates and eventually held the position of curator of marine life and living invertebrates at the American Museum of Natural History in New York City. Miner earned his degrees from Williams College and Columbia University and joined the museum in 1905 as assistant curator, quickly focusing his attention on the exploration of the Lesser Antilles. In the 1920s he turned his attention to several group exhibits for the new Hall of Ocean Life that opened in 1930.

Beginning in 1924, Miner went on his first of a series of expeditions to Andros Island in the Bahamas. Andros Island was home to beautiful tropical and coral environments and also the site of John Ernest Williamson's photosphere. In the first decade of the twentieth century, Williamson designed and operated a device that consisted of a long tube with a viewing chamber mounted on the end. The tube stuck out of the bottom of a barge and was lowered to the ocean floor, where Williamson took moving pictures such as *Thirty Leagues under the Sea* (1914) and one of the first film versions of Jules Verne's *Twenty Thousand Leagues under the Sea* (1916). It is not entirely clear how many naturalists and oceanographers descended Williamson's photosphere. More certain is that Miner's work with the technology, as well as a similar expedition in the late 1920s for Chicago's Field Museum of Natural History, made possible some of the first substantial museum exhibits of the underwater realm.

Miner began by sitting at the bottom of the cramped viewing chamber to make routine observations of the coral barrier reef. He also used Williamson's diving helmets to get an even closer view of ocean life. Miner took photographs with a specially designed underwater camera; he sketched the form of living corals and the movements of reef fish with a specially designed underwater pad and oil paints; and he selected, secured, and raised, with the help of sledges, drills, and dynamite, more than forty tons of coral that were then sent back to the museum for display. Over the course of three expeditions and seven years, and with the help of several museum staff members, numerous wealthy patrons, Williamson's generosity, and the labor of numerous inhabitants of the Bahamas, Miner was able to reconstruct in downtown Manhattan the Andros Island barrier reef. The exhibit allowed visitors to stand "on the sea bottom and [gaze] into a tangled forest of coral," just as Miner had done with the diving helmet.[4]

Miner followed up the success of the Andros Island exhibit with a new group display on the pearl divers of the South Pacific. In 1936, the San Francisco industrialist Templeton Crocker offered his yacht, *Zaca*, and use of his diving equipment so that a small crew could explore the Tongareva Island lagoon. Miner and his assistant, staff artist Chris Olsen, explored the bottom of

Opening of the Hall of Ocean Life at the American Museum of Nature History in New York, 1933. These preserved and recreated specimens gave urban audiences a glimpse into the underwater realm and the fruits of underwater exploration. (Bettmann/Corbis)

the lagoon, observing the rich diversity of coral and reef life, especially the pearl-producing oysters scattered over the lagoon floor. They also observed the practice of indigenous pearl divers who could hold their breath for about three minutes. Olsen devised a new underwater canvas and easel and used a pallet knife to capture the color of the living environment. Miner spent some of his time experimenting with a bang stick, essentially a dynamite cap that exploded at the end of a stick. The explosion would stun nearby fish, and Tau, one of Miner's befriended indigenous divers, swam freely with a net to collect the fish, which were then plaster-cast for re-creation in New York. The expedition returned with paintings, specimens, knowledge, and several tons of coral that went into the Pearl Divers Group at the museum. The exhibit opened in 1941, shortly before the displayed region was the site of cataclysmic warfare.

Another New York figure who was even more famous for his diving exploits was William Beebe, the director of the Department of Tropical Research (DTR) at the New York Zoological Society (NYZS), home of the Bronx Zoo.

Beebe is most often remembered for his bathysphere dives, but before these famed dives, he made much use of a diving helmet. He had arrived at the Bronx Zoo in the first decade of the twentieth century and, through the NYZS and its network of wealthy patrons, went on several exciting expeditions to Asia to examine the ecology of pheasants. He then turned to the tropical jungles of South America and developed a series of famous field stations. Numerous scientists used these stations as home bases from which to conduct research in tropical ecology. Starting in 1923, Beebe fell under the good graces of a number of patrons who made it possible for him to study the tropical ocean. The most notable expedition was his 1925 trip to the Sargasso Sea and the Humboldt Current aboard the *Arcturus*, a steam yacht that was retrofitted with $250,000 worth of oceanographic equipment. With a scientific staff of fourteen and a complete manifest of fifty-one, the *Arcturus* spent six months exploring what Beebe often called a "wilderness of water." The Sargasso phase of the expedition proved to be something of a disappointment; the Sargasso weeds were not present in the dense mats that Beebe had expected. But the exploration of the Humboldt Current and especially the waters of the Galápagos Islands proved especially exciting.

The *Arcturus* was equipped with a diving helmet, the use of which Beebe believed was the most important discovery of the expedition. He recalled "trembling with terror, for I had sensed the ghastly isolation" while struggling against a bad swell on the steep slope of Tagus Cove. Beebe's explanation for writing these "personal digressions" was "to make real and vivid in the mind of the reader, the unearthliness of the depths of the sea."[5] He compared the depths to interstellar space, then to the moon and Mars. When Beebe pulled out of these soliloquies, he described the clarity of the water, beautiful lava-sculpted undersea mounts, encounters with tiger sharks, and a host of life forms. He brought down bags of bait and let tropical fish feed from his hand. Occasionally he made use of dynamite caps to stun specimens long enough for easy retrieval. Beebe enjoyed this more than any other activity; it was his true medium, experiencing nature's wonders *in situ*.

Beebe's *Arcturus Adventure* (1926) was both a narrative of the expedition and a collection of essays that highlighted the main events of the trip. Reviews of the volume were mixed, and they highlight some of the problems that scientists encountered when using new technology such as the diving helmet. Many gave the typical thanks to Beebe for making science palatable and enjoyable for a lay audience. Of special interest were his accounts of helmet diving, but it was not his description of the coral-edged Galápagos environment that drew attention. Instead, commentators concentrated on his encounters with sharks. While such courage received approbation from one point of view, it also had the potential to bite back on account of the sensational nature of such displays. One literary critic who had nothing but praise for Beebe's other books leaned toward ambivalence when reading of a new species of deep-sea fish, *Diabolidium arcturi Beebe*. The reader "begins to dread the worst: that his trusted guide has succumbed to the enemy's snare at last, and now bids for popularity by purveying sensation." The trouble with Beebe, according to a tropical naturalist, was that he looked at "everything in nature as an 'adventure,'" a trait that the reviewer thought was distasteful. He also challenged Beebe's tendency to exaggerate the dangers of exploration. In short, "Too much poetry and too little science, too much adventure and too little calm thinking . . . are the most evident faults of Beebe's writings."[6] Despite such criticisms, Beebe made the diving helmet an indispensable part of the DTR's arsenal of scientific equipment. It was especially valuable in Bermuda, where Beebe spent much of the 1930s. The politicians, industrialists, bankers, and other scientists who visited the Bermuda station could count on a dive under Beebe's direction.

Beebe's Bermuda field station was just one among many marine laboratories that kept a helmet at the ready. Throughout the 1940s and 1950s Carl Hubbs, an ichthyologist at the Scripps Institution of Oceanography (SIO), routinely brought his students to the kelp beds of southern California. And in the mid-1940s, the University of Miami opened its marine station near Miami Beach; its director, Dr. F. C. Walton Smith found the time to take Rachel Carson on an all-too brief dive while she was writing *Sea Around Us* (1951).

In the end, the diving helmet probably did not have a direct influence on scientific research. Miner, Beebe, Smith, and Hubbs all produced scientific papers that made little use of diving equipment. But the virtue of the exploration was more of the ineffable kind. The experience touched these scientists with a sense of awe and beauty that went to the heart of their popular books, articles, movies, slide shows, and museum exhibits. This form of underwater exploration played a vital role in informing the world about the complex beauty that characterizes subsurface ocean environments. The history of the diving helmet also offers an instructive lesson that stands roughly true for most forms of human exploration of the subsurface ocean. Helmets were used for a wide range of activities: commercial exploitation, motion pictures, leisure and recreation, and even military operations. Scientists were just one of many groups who had a common interest in walking on the bottom of the ocean. As such, scientific exploration of the ocean was less a solitary endeavor hatched from a brilliant and inventive explorer than a complex social activity that brought scientific explorers into various nonscientific circles through the common use of a shared technology. This was true for virtually all technologies of underwater exploration.

Despite the rare glimpse into the ocean world that diving helmets offered, naturalists were nevertheless relatively anchored on the ocean floor and enjoyed only limited mobility. The development of the aqualung changed naturalists' engagement with the ocean in fundamental ways. Jacques Cousteau is widely recognized as the co-inventor of the regulating device that made it possible for humans to spend long durations in a variably pressured liquid environment. Cousteau's aqualung emerged out of his recreational skin diving activities in the Mediterranean as well as the war demands of 1940s' France, and its first uses were largely militaristic. After the war, Cousteau set to the work of commercializing the new technology, and he also began an extraordinarily energetic campaign to put scuba to use in as many fields as possible. This campaign took place largely on the deck of the *Calypso*.

With British and French patronage, Cousteau purchased a furrowed American minesweeper, stripped it down to the hull, and refitted it with a sonar apparatus, an aluminum flying bridge, a diving station off the stern, a false prow with windows, and an interior diving well for easy access to the ocean. The *Calypso*'s maiden expedition began on 24 November 1951. The ship's manifest listed a few members of the Undersea Research Group (a government-funded organization), a small crew for the *Calypso*, and at least five French naturalists; their destination was the Red Sea coral island of Abu Latt off Saudi Arabia. The naturalists went about the business of observing the diverse flora and fauna of the tropical sea. The expedition's success was not so much with the equipment's scientific value but rather with its ability to capture on film the colors of the undersea world. Cousteau and Frederick Dumas conducted experiments with color emulsion film and with flash photography to illuminate the brilliant colors of their eerie studio. The expedition was written up in *National Geographic* along with sixteen photographs. Certainly, the photographs removed a bit of the dense sea curtain, but the primary goal of the article was to highlight Cousteau's technological marvels.

While Cousteau is remembered as an ocean explorer and an important environmentalist, the fact that he often called himself an "oceanographic technician" usually escapes the attention of most writers. It is true that his articles, books, films, and television series did much to educate the public about the ocean realm. But just as important, Cousteau was educating the public about the aqualung and the endless stream of marvelous accoutrements that further advanced human exploration of the oceans. A number of naturalists and archeologists benefited from accompanying Cousteau on various expeditions. But the interesting stories to tell about scuba and ocean exploration usually happened far from the filming crews.

Scuba, according to two SIO oceanographers, offered an unparalleled opportunity for observation, sampling, and measurement of inshore phenomena. As opposed to the strict use of mechanical devices, a scuba-equipped scientist "becomes part of the medium. Subject to the same forces, he partakes of the feelings of the other underwater animals as he is swaying by passing waves and drifts along, weightless, with the ocean currents." By 1953, SIO scientists were using the technology to study the ecology of kelp environments, calibrate and operate undersea instruments, make surveys of otherwise inaccessible ocean canyons, map seabed terrain, and observe physical oceanic processes. The key difference between previous underwater technologies such as diving helmets was the freedom inherent in the system that made observations and manipulations relatively easy. In an odd hybrid of human and machine, SIO scientists noted that the "investigator equipped with the SCUBA is thus the paragon of instruments."[7]

One example of the advantages of *in situ* human exploration made possible with scuba was its adoption by Eugenie Clark at the Cape Haze Laboratory (now the Mote Laboratory) in the 1950s. Clark was a New York-trained ichthyologist whose early underwater exploration consisted of a single helmet dive with Hubbs and some extensive fieldwork in the Caribbean, Hawaii, the Red Sea, and Micronesia, where she primarily used skin diving equipment. She established the Cape Haze Laboratory in 1955 and immediately began exploring the coastal waters of western Florida with scuba. Among her purely natural historical research was an investigation of the functional hermaphroditism of a small grouper prevalent in coastal communities of western Florida. The sexual anatomy and behavior of *Serranus subigarius* was still something of a mystery in the mid-1950s, and Clark was somewhat confused by the presence of large colonies of these organisms, all of which had large bellies and thus appeared to be comprised solely of females. Clark brought the organism into the laboratory to conduct anatomical studies and learned that *Serranus* was a functioning hermaphrodite, possessing the sexual organs of both males and females.

It was a modest beginning. The core of this research, after all, took place inside the laboratory. But the research staff at Cape Haze continued to grow, and scuba took on the status of an indispensable technological apparatus for exploring the undersea world. Clark became a celebrity for her multiple expeditions investigating shark behavior that were published in *National Geographic*. Shark research eventually became the primary focus of Cape Haze research. But scuba exploration was not limited to the observational activities of Cape Haze naturalists.

Eugenie Clark (1922–)

An ichthyologist by training who began with guppies and eventually moved on to shark research, Eugenie Clark was a pioneer of *in situ* exploration through the use of scuba. Like Jacques Cousteau and Hans Hess, Clark made the familiar journey from skin diving to scuba in the postwar period. Her work represents the thousands of marine biologists who worked in biological laboratories that enjoyed rapid growth after World War II. She was also a popular naturalist whose work was instrumental in unpacking the many myths surrounding sharks. More than anything else, she played an important role in democratizing ocean exploration by showing how it could be an activity enjoyed by persons of humble means and training.

A child of New York City, Clark was schooled at Hunter College and then received her doctorate from New York University while working with Charles Breeder, a notable ichthyologist at the American Museum of Natural History. After defending her thesis on the mating behavior and the significance of sexual isolation in platys and swordtails, shortly after the end of World War II she accepted a fellowship with the Pacific Science Board to conduct fieldwork in Micronesia, where, among other things, she became an aficionado of skin diving. She then conducted similar research in the Persian Gulf. These two experiences were the subjects of a very popular book, *Lady with a Spear* (1953), that launched Clark into the limelight. The book was responsible for her connection to William and Albert Vanderbilt, who provided the land and resources for the 1955 establishment of the Cape Haze Marine Laboratory in Placida, Florida (today known as Mote Lab). There she turned her attention to shark research that entailed field research in the Gulf of Mexico with scuba as well as behavioral projects in seaside enclosures. Other than her behavioral studies, which showed how sharks responded to stimuli in interesting ways, she also aided in the search for an ever-elusive shark repellant. Much of this work was chronicled in her second popular book, *Shark Lady* (1969). She left Cape Haze in 1968 and eventually secured a position at the University of Maryland's department of biology.

From the earliest days of her career, Clark was always a popular and stunning figure. After the success of *Lady with a Spear*, she would go on to write twelve beautifully illustrated articles for *National Geographic* and appeared in documentaries by Cousteau and in National Geographic's popular television exposé, *The Sharks* (1982). The fact that Clark was a woman explorer also led to compelling coverage by *Sports Illustrated* and *Ms. Magazine*. She usually speaks humbly of her scientific output, but her work on the sexual behavior of many marine fauna should not be understated. Her wider impact, however, was on the sport and hobby of recreational diving. The activity would grow steadily through the 1960s but then exploded in the 1970s. Clark's claim that her most significant contribution has been to dispel the notorious reputation of sharks should be interpreted more broadly. She was one among many who introduced the ocean as a lovely and safe place to visit and explore.

John Randall, an ichthyologist at the University of Puerto Rico in the early 1960s, was effective in using scuba for observation, but he was equally effective in using the technology to turn the seabed into a kind of laboratory through the manipulation of field variables. Throughout the 1950s, Randall used scuba in tropical environments in Hawaii and the Caribbean to observe the grazing behaviors or reef fish. Starting in 1958, Randall began an extensive survey of tropical fish off Saint John in the Virgin Islands. On the ocean-side of Saint John's reef is a band of bare sand that fringes the entire reef. Just beyond the sand, again continuing toward the direction of the open ocean, are extensive fields of sea grasses. This is a natural geographical phenomenon common to many of the world's coral reef environments.

Conventional wisdom at the time had it that the bare strips of sand were sections of coarse and shifting material ill-suited for sea grass growth. Through a series of ingeniously simple field manipulations, Randall showed that the barren strip was actually the result of grazing activities of tropical fish that used the reef for security. The fish would leave the security of the reef to graze on the sea grass but only to a certain point, thus resulting in the strikingly linear nature of the sand–sea grass boundary. The hypothesis called for experimentation, and scuba was indispensable. Randall dug up a strip of sea grass and placed it next to the reef. It was almost completely consumed within a day. Then he placed a strip of sea grass that stretched across the band from the reef to the fields. The strip of grass succumbed to grazing beginning at the site of the reef and, over time, finishing at the boundary of the sea grass field. He then assembled an artificial reef out of cylinder blocks that extended from the reef, across the band, and into the field of sea grass. The reef fish used the holes in the cylinder blocks for protection and moved along the artificial reef and into the field of sea grass. A band of bare sand soon developed in the sea grass field around the artificial reef. At the time that his studies were published, Randall believed that his ecological understanding of fish–sea grass interactions had some implications for the resurrection of the green sea turtle, chiefly led by Archie Carr in the 1950s, for the economic utility of humans. Less anticipated was that Randall's studies became foundational for establishing artificial reefs.

Outside of the field of biology, scuba also presented unique opportunities for archeologists. While Cousteau's early investigations of various Mediterranean wrecks sometimes teetered on a line between salvage and science, his widely publicized activities highlighted the potential advances for archeology. In the 1950s, most undersea archeologists simply extended terrestrial methods into the ocean. Innovators such as George Bass, however, began to develop new methods and new technologies designed specifically for the underwater environment. In 1960, with the support of several universities, museums, and equipment manufacturers and the newly formed Council of Underwater Archeology, Bass led a team of investigators to an ancient wreck off the coast of Turkey. The running narrative of this and other expeditions was published in his *Archaeology under Water* (1966), a book that also contained a useful treatment of underwater techniques and technologies. There was some criticism within the scientific community regarding the legitimacy of the new field, which, in the estimation of one archeologist, "probably engendered as much hostility and suspicion as new subjects generally do."[8] Bass's text, according to this reviewer, added rigor and legitimacy. Bass's activities were bolstered by H. E. Edgerton, whose numerous technological innovations, such as underwater strobe photography and sonar detection equipment, played a pivotal role in marine archeology. By the mid-1960s, archeologists were investigating sunken wrecks, ports, and cities all over the world, and more recently, the subject has enjoyed widespread popularity and patronage thanks to the discovery of the *Titanic*.

Underwater Habitats

One of the more exciting, and sometimes the most fruitful, uses of scuba was at underwater facilities. It only made sense to build an underwater habitat, after all, if its inhabitants could leave the structure with scuba to undertake an assortment of tasks. The chief virtue was that a diver could go on repeated deep dives without needing to spend the required time decompressing on the way up, a protocol necessary to prevent a case of the bends. This kind of saturation diving allowed humans to dive deeper and longer than they had in the past. Underwater habitats were creatures of the 1960s, a period in which many politicians, industrialists, and scientists were increasingly enthusiastic about the exploitation of the ocean through modern technology. It was an experimental decade in which the primary research goal was to see what could be done, and no scientific field benefited more than human physiology. As the cost and danger of deep facilities became increasingly apparent, the rush to the bottom was moderated by a more humble approach. Beginning in the 1970s, habitats were generally shallow structures that enabled naturalists and scientists to conduct research in the surrounding area, mostly in coral environments. It was at this point that scientific exploration of the sea through the use of underwater habitats began to pay off hoped-for dividends.

Cousteau was at the leading edge of the habitat craze. In 1963 he announced the potential arrival of a new species, *Homo aquaticus.* This was a man who could live and breathe underwater, a man, according to James Dugan, who "would dwell among his kind in submerged towns and swim about on his daily labors in the open depths."⁹ Such an evolutionary jump, or reversal, needed to be helped along by technological know-how, a kind of "surgical transformation" spurred, in Cousteau's words, "by human intelligence rather than the slow blind natural adaptation of species." He noted that NASA was constructing an "artificial gill" that would allow the oxygenation of an astronaut's blood without the pesky nuisance of breathing. The gill, "fitted under an arm and linked with the aorta by surgical manipulation," could easily be adapted for underwater exploration, though given the pressure of ocean depths, it would be necessary to alter the pack significantly.¹⁰

Shortly before the announcement of the new species, Cousteau had sent two men 33 feet down to reside in a 17-foot cylinder filled with compressed air for a week. His overall goal, he wrote in *Popular Mechanics*, was to "free man from the slavery of the surface, to invent ways and means of permitting him to escape from natural limitations . . . to move about, react and live within the sea and take possession of it."¹¹ The project was officially called *Conshelf I*, which represented Cousteau's desire to turn Earth's continental shelves into an extension of terra firma, a place for human labor, prospecting, and fish farming. *Conshelf II* took place in the Red Sea at a depth of 33 feet. The new Starfish House was equipped with all of the comforts of home, including a garage for Cousteau's deep-sea submersible. Five divers spent their time collecting marine specimens and sending them topside for study. *Conshelf III*, some two years later, took place at a remarkable 355 feet beneath the surface of the Mediterranean Sea. On this occasion, Cousteau's divers served as oil prospectors. A mock oil rig was sent down to the ocean floor, and the divers tested their dexterity with routine roughneck tasks. Despite their successful manipulation of the rig, the divers were constantly hampered by mechanical failures. Cousteau came face-to-face with the stunning limitations of a technological universe, and he then decided to forgo plans for *Conshelf IV* and *V*.

Cousteau's efforts, which were supported by both France and oil interests, were evenly matched by the United States and its three underwater habitats. For the United States, the military took on the role as the primary patrons. The U.S. Navy had long employed divers to perform the tasks of salvage and hull repair, and the threat posed by German submarines increased the need for diving research during and after World War II. One of the foremost sites for this kind of research was the Medical Research Laboratory, a U.S. Navy submarine base in New London, Connecticut. In 1957, George Bond, a young doctor from the backcountry of North Carolina, joined the New London staff as a member of the Undersea Medical Service. For five years, he pioneered the field of saturation diving through a series of animal and human tests called the Genesis Project. By 1962, Bond had proven, in experimental conditions, that divers could stay submerged for long periods of time as long as they followed a tightly orchestrated decompression protocol. Bond proposed field tests in submarines, but it was tragedy, more than anything else, that breathed life into the SEALAB program. The failing in 1963 of the USS *Thresher*, a nuclear attack submarine, made it imperative that a method be developed for rescuing trapped sailors, or at least recovering nuclear technology. A report by the Deep Submergence Review Group called for a dedicated commitment to deep-sea technologies that included submersibles and SEALAB. Accordingly, the Office of Naval Research (ONR) approved Bond's SEALAB proposal, and plans for the habitat were drawn up at the U.S. Navy Mine Defense Laboratory in Panama City. By the following spring, Bond had assembled his team for the first trial, close to Argus Island off the Bermuda coast.

Late in May, SEALAB came to a gentle rest some 192 feet beneath the surface of the ocean. Bond assembled a team of four aquanauts who were, in his estimation, an elite group of bold explorers whose only interest was to advance humanity. Despite Bond's modesty, he was well aware that SEALAB needed to be a popular event similar to the nascent space program that was culling comparatively massive federal funding. Throughout the 1960s ocean explorers found themselves in a kind of battle with space explorers, and the latter had a decisive advantage. Bond played the game brilliantly and saw to the participation of Scott Carpenter, the astronaut who piloted *Aurora 7* of the *Mercury* program for three orbits around Earth in 1962. Bond was clearly interested in cashing in on the social cachet of Project Mercury. These were, after all, the men with the "right stuff." At one point, Bond called Carpenter his "astronaut shield of protection," by which he meant a heroic and familiar face for public relations.[12] Carpenter had a motorcycle accident and was scrubbed from SEALAB I, though he played an instrumental role in the success of SEALAB II.

The first trial was a relative success, although it only lasted eleven of the sixty planned days. Navy officials were impressed enough to ask Bond to direct the medical side of the Man-in-the-Sea program. SEALAB II was thus brought fully into the fold of naval support by falling under the Deep Submergence Systems Project. The second facility was lavish in comparison, much like Cousteau's Starfish House with living quarters, a laboratory, an equipment room, a kitchen, and open access to the sea. At $1.8 million, SEALAB II hosted three teams composed of nine aquanauts for a duration of fifteen days per team. Located off the La Jolla coast in southern California, the habitat was lowered to 205 feet just southeast of Scripps Canyon. The mother ship, the *Berkone*, was a catamaran that was instrumental in launching the new *Polaris* submarine. The purpose of SEALAB II, according to Bond, was to open the continental shelves for naval exploration. And it took every bit of energy from 400 staff members, which included countless technicians and engineers, navy personnel, private consultants, and academic experts. New telemetry devices were being tested to monitor the physiological stresses of the aquanauts. Communication networks were created, tested, and reengineered on the spot. The deck of the *Berkone* very much resembled NASA's mission control.

Like America's space program, SEALAB II was observed by the American press with similar enthusiasm. Reporters, photographers, and camera operators jostled among the technicians and engineers to record and publicize the heroic mission. Press conferences and television interviews consistently dogged Bond and other leaders of the project. There were some special flourishes to the daily routine of the trial. Tuffy, a porpoise trained by navy personnel, acted as a courier between the surface and the facility, one of the more benign uses of marine life by the U.S. Navy. Carpenter also had dramatic conversations with other frontier explorers. He spoke with several of the oceanauts in Cousteau's *Conshelf III*, which coincided with SEALAB II. And perhaps most stunningly, he exchanged greetings with Colonel L. Gordon Cooper Jr., who was in *Gemini 5* passing 100 miles above Earth.

For the most part, the true objective of SEALAB II was to observe the physiological and psychological stresses of humans working underwater. The work of the aquanauts themselves played second fiddle. One team performed experimental salvage operations and installed an underwater weather station, while another team included a number of oceanographers from the SIO. After the trial, the biologist Arthur Flechsig noted that the type of science conducted was fairly routine, "old-fashioned biology—like going on a nature walk."[13]

The success of SEALAB II marked the apex of the era of oceanic enthusiasm. Half of this enthusiasm was militaristic, as war planners envisioned a network of undersea laboratories along the Atlantic Ridge. Such forts could serve as undersea listening stations to monitor submarine traffic, and they might also serve as provisioning stations for nuclear submarines, thereby extending their time underwater. John P. Craven, head of the Navy Deep Submergence Systems Project, reinforced that spirit in a 1966 article titled "Sea Power and the Sea Bed." Since 1955, Craven had worked as a scientist for the navy's Special Projects Office, an outfit of military and civilian war planners who were primarily responsible for envisioning the deterrent potential of undersea warfare, beginning with the *Polaris* initiative and thence to deep-sea rescue and salvage. At one point they were considering the creation of hardened silos built into the seabed.

SEALAB III was America's most extravagant push to inhabit the ocean. The $10 million project led to a 57-foot, 340-ton bright yellow habitat. Five teams were to spend twelve days apiece at a remarkable 610 feet off San Clemente Island. The Miami Seaquarium obtained a contract to train eight porpoises for the effort. The affair was scheduled for the spring of 1968 but was delayed a year while the leadership team dealt with a barrage of technical complications. Air leaks, flooding chambers, malfunctioning transports, and other problems continued as the team lowered the facility to the ocean floor. On 17 February a team of four divers went down to fix some of the problems, but early in their work, Berry L. Cannon began to seize. The other three divers brought him quickly to the surface decompression tank, but Cannon died on the ascent. An investigation found that Cannon's breathing apparatus was missing the vital chemicals that scrubbed out carbon dioxide. There were even charges that a saboteur was wreaking havoc on the team.

SEALAB III came to a quick end, as did the SEALAB program. Captain William Nicholson thought that Cannon's death raised serious doubts about the ability of humans to function in deep waters for long periods of time. A *New York Times* editorial admitted that the economic and military potentials of deep-sea exploration were enormous, but "Cannon's death sounds a warning that these gains may be much harder to wrest from the sea than has hitherto been assumed."[14] Bond began to publicly state that the SEALAB habitats did not have military value; instead, they would be valuable as teaching institutions and for private industry.

Thus, in the late 1960s and early 1970s, the engineering bravado of underwater habitats gave way to some exciting developments in underwater research, most notably the Tektite Project. To call Cousteau or Bond scientists or naturalists is slightly misleading; their explorations were technological. For example, only five of the twenty-eight aquanauts in SEALAB II were scientists who were more interested in the ocean than the habitat itself, and four of them were still graduate students. But the entire structure of engineering impulse increasingly joined hands with the scientific community in the late 1960s and early 1970s. Scientists in the Sea, as distinguished from the Man-in-the-Sea initiative, was a program that met such an objective. The initiative was sponsored by the U.S. Navy, the Department of the Interior, and NASA. The Tektite habitat was constructed by General Electric and consisted of two metal cylinders, four meters in diameter and six meters high, connected by a flexible tube. It was a fairly typical habitat, with living and laboratory spaces and a wet-room with open access to the ocean. But the unit was lowered to a depth of only fifteen meters in the beautiful tropical waters of St. John's Lameshur Bay, still within the boundaries of the U.S. Virgin Islands National Park.

NASA and the U.S. Navy were interested in Tektite in order to test the group psychology of people under the stress of living in tight places and hostile environments. The entire habitat was covered by cameras wired topside to a closed circuit television. Graduate students from the University of Texas monitored the movements and attitudes of the aquanauts. NASA and the navy were interested in seeing if everyone could get along for long periods of time, but they were also interested in assessing the quality of their scientific work.

Tektite I had four scientists living in the habitat for two months. The scientists spent some time conducting a census of the geology and biology of the reef, but much of their time was occupied with physiological testing. *Tektite II*, in contrast, was more ambitious. Scientists proposed their research projects, and fifty of them were selected and grouped into ten teams of five (four scientists and one engineer). Each team spent between ten and twenty days in the habitat. This situation was perfect for scientists who needed to spend long periods of time observing underwater life and structure. Sylvia Earle, one of the participants, believed that the extended time spent in the study area was crucial for understanding oceanic ecology. The focus of the research, more than anything else, was coral reef ecology. Some of the scientists tinted their work with an environmental spirit that was becoming increasingly popular in the early 1970s. They were taking what ecologists call baseline data of a pristine coral reef environment that had not suffered abuse from human interaction. This data would prove valuable when compared to that of stressed environments. One reporter was able to succinctly distinguish Tektite from the earlier Man-in-the-Sea program. "Tektite's program is a big step in another direction, the exploration of the undersea environment with a view to peaceful and ecologically sound use of its resources."[15] With the technological problems sorted out and with a new environmental agenda at hand, underwater habitats had found a niche in scientific fieldwork. By 1975 an additional fifty underwater research habitats had mushroomed in the world's oceans.

Deep-Sea Submersibles

The history of deep-sea submersibles parallels the histories of personal exploration and underwater habitats. Each mode of technological exploration initially went through a heroic phase before gaining legitimacy as a scientific activity that produced rigorous and compelling scientific knowledge. Underwater submersibles were initially a form of dilettantish exploration; the scientists who rode along often had to settle for the mere aesthetic beauty of going and seeing the underwater realm for the first time. It was not until the 1970s, with the active use of *Alvin* and a few other vehicles, that the technology began to be truly seen as an important scientific activity.

Sylvia Earle (1935–)

Sylvia Earle came of age right at the climax of the engineering enthusiasm of the 1960s, at about the same time as a new environmental consciousness was sweeping the nation. Engineering and environmentalism became the twin vectors of her career as an ocean explorer. She also logged time in an underwater habitat and was part of one of the first teams that did truly compelling scientific research while living underwater. Earle, like Clark, was also a popular writer and bureaucrat whose work has been instrumental in the articulation of an ocean ethic, an extension of Aldo Leopold's land ethic, into wetter regions of the globe.

Earle spent her teenage years in Florida and developed a love for the ocean through her personal exploration of the Gulf of Mexico. She earned her Ph.D. at Duke University in 1966 specializing in marine algae. She then enjoyed directing Cape Haze lab for a brief spell. Through a connection with Edward Link, she then became a part of the U.S. Navy's Scientists in the Sea program, which eventually led to a coveted position on the *Tektite II* project. Earle spent two weeks in a shallow underwater habitat in the Virgin Islands with a team of four other women. There she drew on her extensive knowledge of ocean algae and investigated the ecology of reef fish grazing behavior. In the 1970s she held various research posts at Berkeley, the California Academy of Sciences, and the Natural History Museum of Los Angeles County. All the while, she continued to participate in various expeditions and even had the opportunity in 1979 to dive in a "Jim" suit, a rather robotic-looking container in which she was able to dive to 1,250 feet and walk around the ocean floor. The experience brought her into contact with Graham Hawkes, an ocean engineer who designed equipment for marine industries. In the 1980s Hawkes and Earle joined forces to establish Deep Ocean Engineering and Deep Ocean Technologies—companies that designed and built ocean submersibles. They had some success in introducing a number of remotely-operated vehicles and then the Deep Rover, a small one-person submersible that combined the virtues of the deep-diving submersible with the freedom of the solitary diver. In 1990 Earle was appointed by President George H. W. Bush to be chief scientist of the NOAA.

Earle's environmental ethic goes clear back to her early work with *Tektite II* when she noted that knowledge of reef ecology "will provide a model against which to compare—and, one may hope, to correct—disturbed and unbalanced parts of the undersea environment."[16] Her advocacy was further elaborated in popular books such as *Deep Frontier* (1980), where she called for an ocean ethic: "A century ago the ocean was wilderness. Before we are gone, we may choose to keep what remains of the sea of Eden, or cause—and witness—paradise lost."[17] In the 1990s she kept up a busy schedule of consulting, research, and public speaking; all the while her environmental message became even more pointed. Her popular *Sea Change: A Message of the Oceans* (1996) was thus a wonderful blend of history and autobiography as well as a call for an ocean ethic. More recently she has become a National Geographic explorer in residence, and between 1999 and 2003

Marine Botanist Sylvia Earle with "Jim," the submersible suit she wore to the depth 1,250 feet in 1979. "Jim" was one of many designs engineered for deep-sea exploration since the 1960s. (Bettmann/Corbis)

she headed up the joint NOAA/NGS Sustainable Seas Expeditions. Over a five-year period, Earle coordinated teams of diving scientists in the survey and investigations of many of America's National Marine Sanctuaries. More than just another scientific survey, Earle's work once again blended technological savvy and an environmental message into a highly popular educational opportunity for our world's newest generation.

The story begins with Beebe. He and his research team were way ahead of the curve and engaged in an exciting, and terribly dangerous, form of scientific activity. Since his *Arcturus* Expedition (1926), Beebe had received many proposals for designing technologies for deep-sea exploration. He had relocated his DTR on the island of Bermuda in the late 1920s and was soon visited by Otis Barton, an engineer and amateur naturalist and the inheritor of a sizable fortune, who presented a feasible design with an offer to pay for the vehicle's construction. The bathysphere was cast and fitted with three eight-inch-thick windows of fused quartz, the clearest glass ever produced, as Beebe incessantly told the press. A new tug was chartered and rigged to safely lower and retrieve the six-ton sphere. Beebe designed a meticulous protocol for the DTR staff and tug crew that ensured everyone's safety. By May 1930, the DTR was ready to integrate bathyspheric research into its wider program.

Naturalist Dr. William Beebe and Engineer Otis Barton pose with their invention, the bathysphere, in Bermuda, ca. 1934. Beebe and Barton are still famous for a series of dramatic deep-sea dives made in the 1930s off the coast of Bermuda. (Ralph White/Corbis)

A considerable amount of time went into preparations, but the dives themselves took up a scant part of three weeks in the entire 1930 season. The sphere was lowered fifteen times over the course of seven days. It was equipped with a high-energy light, provided by General Electric, and a Bell telephone through which Beebe would relay his observations to Gloria Hollister, who took notes from the tug's deck. Indeed, Beebe often received this equipment free of charge so long as the companies had free rein to advertise the usefulness of their products half a mile under the sea. The modest nature of the bathyspheric research was reported by Beebe who, understandably, emphasized findings of his more traditional deep-sea and Bermuda shore research conducted with trawls and nets. Newspapers did not immediately seize upon the fantastic nature of the dives until it was announced that Beebe and Barton had reached a depth of 1,426 feet.

The bathysphere was put into service during the 1932 and 1934 field seasons. Coverage of the events included the book *Half Mile Down*, numerous newspaper and magazine articles, and a live radio broadcast. Some critics believed that Beebe was engaged in a stunt that had little scientific value. Other critics immediately thought they were witnessing a contest for a world record and minimized the descent by citing that miners had been working in the bowels of the world as far down as 5,200 feet. Beebe was quick to disabuse the public and indicated that he was not out for a record. He began what would be one of many argumentative maneuvers that portrayed the dives as a serious scientific endeavor. Shortly after the first dive to 1,426 feet, Beebe reported that he had seen many fish, "attesting to the scientific value of the apparatus," and upon his return to New York he remarked that "the importance of this deep dive is not the fact that it is the furthest man has ever been under the sea but the great value to science in its study of deep-sea inhabitants."[18] At first, there was little response from fellow naturalists. Henry Fairfield Osborn, a friend of Beebe's, was easy to convince and wrote an approving letter to *Science* attesting to the scientific value of the bathysphere. Beebe was no doubt concerned with winning over others. After reading the *National Geographic Magazine* account of Beebe's 1930 dives, E. J. Allen wrote a glowing letter of praise.

Beebe was constantly trying to find ways to improve the scientific value of bathyspheric dives. For instance, he made inquiries to physical scientists for advice on possible experiments so as to put the bathysphere to good use. E. O. Hulbert of the Naval Research Laboratory responded by giving a long list of experiments that would investigate brightness, spectrum analysis, polarization, and cosmic rays. He also suggested that Beebe contact Robert Millikan, of cosmic ray fame, for further possibilities. Beebe wanted to mirror the stratosphere flight's use of a spectrograph, and he would have done so had it not been for the bathysphere's physical limitations.

One way of proving the scientific value of the bathysphere, or of disproving the stunt, was to stay at the bottom of each dive for an extended period of time. Beebe did much to publicize the extended periods of observation that these dives required. More than a quick jaunt for a record, Beebe was hoping "to stay in the depths of the ocean for five or six hours at a time, photographing and studying the inhabitants."[19] Then he came up with a new way to use the technology. Towed by a ship, the bathysphere could swing low over Bermuda's underwater plateau, and thus Beebe could construct an accurate topographical map of the seafloor. Unfortunately, the winch was not able to raise and lower the sphere fast enough for these contour dives to have any value. But the innovative use of the bathysphere illustrates Beebe's desire to undertake research that would become commonplace some forty years later. Clearly, Beebe was deeply concerned that the dives appear unsensational and thoroughly scientific. But he still could not rise above one insurmountable problem that is absolutely key to the practice of the natural historian: the collection of actual specimens. For taxonomists such as Beebe, the pickled specimen in hand was necessary to do the work of science. In the bathysphere, he was little more than a voyeur. Beebe even attached baited hooks to the outside of the sphere but with no success. When he tried to name a new species of fish that he had identified through the bathysphere's quartz window, a host of naturalists cried foul.

The bathysphere dropped out of scientific business in 1934, and it was not until the 1950s that others took up the call. In the 1950s, most underwater vehicles were built not so much with the aim of scientific research but rather with the aim of going deeper. They were in the business of record breaking. The contest began in earnest in 1953 with the launching of two new vehicles, France's *FRNS-3* and Auguste Piccard's *Trieste*. The vehicles were called bathyscaphs, and their advantage over the bathysphere was the lack of a shipside tether. The crew continued to inhabit a pressure-resistant steel sphere, but atop of it sat a large structure that was filled with aviation gasoline. The gasoline provided buoyancy, and lead shot served as ballast. The principal objective of the bathyscaph was to descend to the bottom of the sea, cruise along the bottom at the very slow speed of one knot, release the ballast, and float home. The vehicles were primarily used in the Mediterranean, and when not racing for a record, a number of European scientists took the time to do a bit of exploring.

Piccard traveled to the United States in 1956 to sell the bathyscaph idea to American oceanography. The following year, a number of American scientists traveled to Europe to observe the *Trieste* in action. Shortly after, Columbus Iselin, the director of the Woods Hole Oceanographic Institute (WHOI), recommended that the United States acquire the *Trieste*. The ONR bought *Trieste*, and a few extras, for $1 million and then initiated Project Nekton in 1959. The

goal, again, was to get to the bottom, and this they accomplished in 1960—
35,805 feet in the Challenger Deep in the Marianas Trench. Navy Lieutenant
Don Walsh and Piccard's son, Jacques, were the pilots. On the bottom they saw
a flatfish, perhaps a flounder, then they returned.

The *Trieste* made its home in San Diego and throughout the early 1960s
became a part of southern California's oceanographic fleet, providing especially
fruitful opportunities to the oft-marginalized biological sciences. But the bathyscaph idea, with its inherent danger and limited mobility, gave way to radical
new designs. Again, Cousteau, the ocean technician, played a leading role. The
late 1950s and early 1960s were triumphant days of fame and technological prowess for the Cousteau team. It constructed a deep-sea submersible that brought a
crew of two to the depths of 1,000 feet in a kind of underwater flying saucer. To
play host to the saucer, in 1961 Cousteau christened *Amphitrite*, a Zodiac-built
vessel composed of a nylon and neoprene sandwich, with new forms of fiber
glass, foam plastic that was lighter than wood, and a new stainless tubing of
aluminum and magnesium alloy. Other than Tupperware, it is hard to imagine
a more significant emblem for the postwar industry of synthetic materials. The
lightweight ship was part of Cousteau's new Airborne Underseas Expeditions, a
unit that could be flown around the world in order to dispatch the diving saucer.
These extraordinary technological devices set the stage for Cousteau's experiments in saturation diving, his grandest techno-utopian project.

While Cousteau's diving saucer certainly provided impetus for Americans
to respond with submersibles of their own, it was more likely the mandate from
the Deep Submergence Systems Review Group that provided the largest push.
An interesting conjuncture occurred around 1964 that brought together the military's desire for deep-sea recovery vehicles, the oil and mining industries' wish
to access the continental shelf, oceanographers' goal of catching a glimpse of
the undersea world, and military industrialists' desire to cash in on it all. This
resulted in a flurry of engineering activity that produced a spate of submersibles. All were beautifully named. The navy's *NR-1* and *Trieste II*, Lockheed's
Deep Quest, North American's *DOWB*, Electric Boat Company's *Star* series,
Reynold's *Aluminaut*, and Westinghouse's *Deepstar*. Between 1961 and 1967,
according to a report in the October 1967 edition of *Science*, the United States
had built twenty submersibles. The same source counted approximately 600
research dives between 1964 and 1967 for about 150 scientists.[20]

From resource extraction to undersea warfare to the Man-in-the-Sea initiative, oceanology, or ocean engineering, provided the language, concepts, and
tools that shaped much of America's postwar relationship with the ocean. But
this general air of ocean enthusiasm should not be confused with consensus.
By the late 1960s, a number of critical voices began to dampen the spirits of

ocean pioneers. The litany of technologies gone wrong, including the Texas Towers, the *Thresher*, SEALAB, and *Johnson-Sea Link*, did more than underscore the fact that undersea exploration was a dangerous endeavor. The events spurred a discussion on the limitations of the human invasion of the ocean. President Lyndon Johnson weighed in on the issue in 1968 when he addressed a Geneva audience and pleaded that the ocean floor be barred as a location for hiding nuclear missiles, just as it had already been agreed to ban weapons from space and Antarctica. Johnson's voice joined with civilian criticisms of oceanological optimism. *Wall Street Journal* writer Kenneth Slocum noted in 1968 that "despite considerable hoopla in recent years about the great underwater wealth presumably about to be tapped as technology enables men to conquer what some enthusiasts call 'the earth's last frontier,' there's little indication that the dream of harvesting vast quantities of food, drugs, minerals and petroleum from the oceans is anywhere near fruition."[21]

One piece of technology that came out of this era had enormous implications for scientific exploration. *Alvin* was very much a creature of the technological optimism so rife in the 1960s, and as with most technologically sophisticated gadgets of the Cold War era, it required the collaboration of science, government, and industry. With the ONR's wetted appetite, given the success of *Trieste* and the exposure to Cousteau's saucer, the ONR sought to contract with Reynolds Aluminum and Metal Company a research submersible named *Aluminaut*. The deal fell through, and the ONR chose to let WHIO engineers, who were contracted to General Mills, design and build their own submersible, with the stipulation that the U.S. Navy could continue to enjoy access. *Alvin* was launched in 1964. The vehicle was fully equipped with propulsion, a mechanical arm, various surveying devices, and a sophisticated array of communications devices. But, ironically, the structure that would create a safe haven for the crew of three continued to resemble Barton's and Beebe's bathysphere. The heart of *Alvin* was a sphere six feet, ten inches in diameter and two inches thick.

Alvin was tested in 1965, and one of its first missions in 1966 was to help in the recovery of a submerged hydrogen bomb from a tragic B-52 crash over the coast of Spain. *Alvin* received credit for spotting the bomb, at 2,450 feet, but it was a navy-owned remote operated vehicle that retrieved it. *Alvin* continued to serve in a military capacity, monitoring and servicing deep-sea sonar arrays that listened in on submarine traffic. Indeed, in the early years of its use, the details of prioritizing navy use and scientific research had not been ironed out, a matter of some frustration to scientists. This and the fact that the use of *Alvin* entailed enormous costs meant that deep-sea submersible science was not in high demand.

Scientific exploration during *Alvin's* early years was quite humble. A number of marine geologists and biologists had undertaken a modest amount of research in 1966 and 1967. Then, for about a year, in 1968 through 1969, *Alvin* rested on the bottom of North America's continental shelf, 5,200 feet under water, a casualty of a rusted cable. By the time the vehicle returned to service in 1971, the spirit of oceanological enthusiasm had been doused. *Alvin*, however, survived and increasingly became the flagship submersible for science. The timing could not have been better. A host of oceanographic research in the postwar period was responsible for building a credible argument for the theory of plate tectonics. Much of this evidence involved knowledge of the world's mid-oceanic ridges, the point of genesis for new ocean floor. Marine geologists had explored the region through traditional surveying mechanisms, but in 1973 and 1974, American scientists joined hands with their French colleagues for a joint expedition of a section of the mid-Atlantic Ridge. Both *Alvin* and the French submersibles *Archimede* (a bathyscaph) and *Cyana* were enlisted in the venture. The expedition, named Project FAMOUS, brought back water samples, 3,000 pounds of rock, and more than 100,000 photographs. The submersibles had proven their value, and it became increasingly difficult to argue against the theory of plate tectonics.

The following year, *Alvin* investigated the Galápagos Rift, another ocean spreading zone, that resulted in the discovery of hydrothermal vents. Just as important, *Alvin* aided in the discovery of a new ecosystem. Communities of organisms had evolved to harness the energy from the chemicals that emerged from the vents. This exciting discovery resulted in a flood of biologists who lined up along with marine geologists to request precious time in *Alvin*. In 1978 and 1979, *Alvin* participated in expeditions that discovered the presence of black smokers, hydrothermal vents that emitted superheated water and a large host of precipitating chemicals. Physical oceanographers who specialized in ocean chemistry were added to the manifest of submersible science.

There is something about the cachet of *Alvin*, and human deep-sea exploration in general, that tended to minimize the contributions made by other nonhuman ocean probes. All of the above expeditions relied on a large number of mechanical devices. Military topographical maps that had previously used the Sing Around Sonar System were essential for laying out the rough contours of the ocean floor. Side-scan sonar systems, such as the SIO's Deep-Tow, used equipment attached to tethered sleds to gain a better picture of seabed topography. It was only after SIO oceanographers had investigated the Galápagos Spreading Zone that *Alvin* was used. Towed camera systems were always employed to extensively photograph potential dive locations. A lot of traditional oceanographic research was already done by the time biologists or geologists were allowed to go down to take fine-tuned measurements and samples.

Conclusion

By 1976, according to one report, approximately 350 submersible projects had been published. These publications, however, were written by a small core of scientists; half of the papers were written by 12 percent of the total authors. *Alvin*'s yearly operational budget in 1976 was about $1.2 million, paid in the form of grants by the ONR, the National Science Foundation, and the National Oceanic and Atmospheric Association (NOAA). This was a modest investment for such compelling dividends. Many scientists hoped that more submersibles would be built that would increase scientific exploration of the oceans. But starting in the 1980s, the rapid development of computer, communication, and fiber optic technologies made the remote operating vehicle (ROV) a true competitor, and by the 1990s most scientists were investigating the midocean ridges with ROV technology.

One of the leading developers of ROVs was Robert Ballard who, despite his long history with *Alvin* and other submersibles, primarily used towing sleds and a prototype of the famous Jason ROV to explore the wreck of the *Titanic*, although *Alvin* made twelve dives in order to test the ROVs' performance. Jason and a host of other ROVs spurred a familiar debate over the costs and benefits of human exploration of the deep sea. In 1993, Ballard noted that "the paradigm shift has begun. I just don't believe manned systems are competitive with ROVs."[22] He predicted that human-operated vehicles such as *Alvin* would become obsolete within the decade. *Alvin* offered constrained bottom time, and utilizing it was a costly operation due to the safety protocols necessary to sustain human life in the deep sea. ROVs solved some of these problems, but many scientists continue to attest that the human view simply cannot be replaced by a television screen. *Alvin* advocates continued to maintain that there was no substitute for the fine-tuned operations of human exploration. The increasing sensitivity and accuracy of ROVs, however, made this an increasingly difficult position to hold. After forty years of service and more than 4,000 dives, *Alvin* was decommissioned in 2004. Its successor is scheduled to launch in 2008.

While the history of human exploration of the deep sea is certainly a story of great achievement, daring bravado, engineering genius, and scientific success, it also reveals many of the social, economic, and cultural problems beset by those people who wish to visit environments that are hostile to human life. From a purely economic perspective, such efforts carried a heavy cost that rarely met expectations. From the perspective of scientific exploration, a slow curve existed from skepticism to acceptance. In some ways, it almost seems that technological innovation came first and that scientific uses of this technology followed. More importantly, the social status of the scientific explorer was consistently questioned as he or she moved into the public light. The ocean was not the only place where scientists had to deal with such problems.

Bibliographic Essay

Piecing together the story of underwater exploration presents numerous challenges for the historian. The first issue is its very recent history; underwater exploration did not become widespread until the 1960s. One of the consequences is that the history of undersea exploration is widely chronicled in popular books and articles that are intended for the general audience. Many of these treatments are written by the explorers themselves as memoirs, some of which are characteristically self-lionizing. Despite the drawbacks, these writings provide evidence of the interesting dance between science and showmanship, and they can be read as fascinating primary sources. Several studies of underwater exploration have been published by professional historians. In general, however, the field remains ripe for historical inquiry.

There are two among many general histories of human diving that are worth mentioning. For a delightful treatment of diving history up until 1934, turn to the first four chapters of William Beebe, *Half Mile Down* (1951). Beebe comes from the tradition of the true naturalist who is versed in history and literature. These chapters, like much of Beebe's literature, are full of history, insight, and speculation on the earliest human endeavors to study the subsurface ocean. The second is Sylvia Earle and Al Giddings, *Exploring the Deep Frontier: The Adventure of Man in the Sea* (1980). The book is really a coffee table text intended to highlight Earle's 1979 dives in a "Jim" suit, but it should not be dismissed as a popular treatment. Through words and wonderful illustrations, Earle and Giddings bring us through the history of human exploration of the ocean and provide especially fruitful discussions of underwater habitats and deep-sea submersibles. Two very prominent figures in this chapter are Beebe and Jacques Cousteau. While substantial research has been done independently for the purposes of writing this chapter, it nonetheless is built on two historical works: Robert Henry Welker, *Natural Man: The Life of William Beebe* (1975), and Axel Madsen, *Cousteau: An Unauthorized Biography* (1986).

There is a great need for a history of the scientific use of helmet diving. In the meantime, one can turn to Roy Waldo Miner's articles: "Forty Tons of

Coral" (1931), "Diving in Coral Gardens" (1933), and "Coral Castle Builders in Tropical Seas" (1934). On Ernest Williamson and his photosphere, see C. Moffett, "Taking Moving Pictures at the Bottom of the Sea" (1915), and Williamson's autobiography *Twenty Years under the Sea* (1936). Brian Taves, a film scholar, has done some of the best work on Williamson; see Taves's "With Williamson Beneath the Sea" (1996). William Beebe's *The Arcturus Adventure: An Account of the New York Zoological Society's First Oceanographic Expedition* (1926) should be consulted for Beebe's use of the diving helmet in the Galápagos, but the helmet sections are fairly modest. For a Beebe text that highlights the helmet, see his *Beneath Tropical Seas* (1926).

The best route through the history of science and scuba is through primary sources. The best brief and general introduction is Willard Bascom and Roger Revelle, "Free-Diving: A New Exploratory Tool" (1953). For Eugenie Clark, see her *Lady with a Spear* (1953) and *The Lady and the Sharks* (1969). For Cousteau, see his *Silent World* (1953), and be sure to watch the same-named documentary. On John Randall's studies, see his "Overgrazing of Algae by Herbivorous Marine Fishes" (1961) and "Grazing Effect on Sea Grasses by Herbivorous Reef Fishes in the West Indies" (1965). For the influence of scuba on archeology, see Froelich Rainey and Elizabeth K. Ralph, "Archeology and Its New Technology" (1966). Finally, there are hundreds of memoirs written by the first generation of scuba enthusiasts. One that is truly worthy of mention is David Powell's *A Fascination for Fish: Adventures of an Underwater Pioneer* (2001).

Literature on underwater habits is sparse. On Cousteau's Conshelf experiments, see his *The Living Sea* (1963). There is no better source on the history of SEALAB than George F. Bond, *Papa Topside: The Sealab Chronicles of Capt. George F. Bond, USN* (1993). Project Tektite is discussed by Sylvia Earle in *Sea Change: A Message of the Oceans* (1995) and in her *Exploring the Deep Frontier*.

On deep-sea submersibles, turn first to Beebe's *Half-Mile Down* for his accounts of the bathysphere in the 1930s. Supplement this with Otis Barton's *The World Beneath the Sea* (1953), which also includes some of Barton's other submersible exploits. On the *Trieste*, see Jacques Piccard's *Seven Miles Down: The Story of the Bathyscaph Trieste* (1961). For a very good autobiographical example of the spirit of ocean engineering enthusiasm that dovetailed with America's contribution to the Cold War, see John Piña Craven, *The Silent War: The Cold War Battle Beneath the Sea* (2001). Edwin Link's contributions to submersible science are available in Susan Van Hoek and Marion Clayton Link, *From Sky to Sea: A Story of Edwin A. Link* (2003).

Gary Wier's *An Ocean in Common: American Naval Officers, Scientists, and the Ocean Environment* (2001) includes a chapter that discusses the U.S. Navy's involvement with the *Trieste* in the late 1950s and early 1960s. Bruce Robison's "Submersibles in Oceanographic Research" (2002) provides a brief but informative overview of the topic. His conclusions are worth noting, even in brief: "Submersibles have played a relatively small but significant role in twentieth-century oceanographic research. In most cases, technology evolved first and methodology followed."[23] Robison is more sanguine about future prospects. For a mid-1970s evaluation of the technology, see J. R. Heirtzler and J. F. Grassle, "Deep-Sea Research by Manned Submersibles" (1976).

A wonderful and quick entry into *Alvin* literature can be found in Naomi Oreskes, "A Context of Motivation: US Navy Oceanographic Research and the Discovery of Sea-Floor Hydrothermal Vents" (2003). Here Oreskes highlights how oceanographic science followed military objectives instead of the other way around. Historians need to tackle other areas of undersea exploration with the same diligent seriousness. Popular treatments of *Alvin*'s history, usually written by people close to its operation, are also helpful. Robert Ballard with Will Hively, *The Eternal Darkness: A Personal History of Deep-Sea Exploration* (2000), provides a nice chronology of *Alvin*'s heyday years, but the book also makes an argument for ROV development. A more literary treatment is Cindy Lee Van Dover's lovely and informative *Deep-Ocean Journeys: Discovering New Life at the Bottom of the Sea* (1996). If *Alvin* can be said to have a biography, then its Victoria Kaharl's *Water Baby: The Story of Alvin* (1990). Its only vice is its frustrating lack of source citations. It is helpful to consider *Alvin* from the perspective of people who directly and indirectly supported submersible science through their own nonhuman exploration techniques. See, for instance, Kathleen Crane, *Sea Legs: Tales of a Woman Oceanographer* (2003). *Sea Legs* also provides the clearest articulation of the problems and issues that women face in the generally male-dominated field of ocean research.

Human Exploration
of the High Frontier

During the nineteenth century, at the same time that European and American explorers were exploring and mapping the globe, they also began to investigate upward from Earth's surface. Two new technologies enabled exploration of the high frontier: balloons in the nineteenth century and rockets in the twentieth. Motives for this exploration were varied. While exploration of the American West, Africa, and Central and South America had often been motivated by the hope of territorial or resource gains, motives behind the exploration of the atmosphere and space were varied. Individuals were interested in scientific discovery, setting new records, and personal glory. Nations, particularly during the space race between the United States and the Soviet Union, were far less interested in scientific knowledge than in gaining technological superiority and national prestige, although they were careful to justify their efforts in the name of science. By the end of the twentieth century, humans had reached outward to land on the moon before retreating to more permanent habitation in low Earth orbit.

Exploring the Atmosphere

Two Frenchmen, Jacques and Joseph Montgolfier, built the first successful hot air balloon in 1783. To make certain that a human could survive the flight, the two sent aloft a duck, a sheep, and a rooster. They then sent up a scientist, François Pilâtre de Rozier, who was interested in investigating the air's temperature and pressure. In the eighteenth century, the atmosphere was a largely unexplored frontier, and no one knew much about it. Its constituent elements, its circulation, and its temperature and pressure variation with altitude were not well understood. Scientists also did not know how well the human body would function as the air grew thinner and colder. Exploration of the atmosphere, and eventually exploration of space, required scientists to first explore the limitations of the human body.

Within a month of Rozier's first flight in the hot air balloon, chemist Jacques Alexandre Charles flew the world's first piloted hydrogen balloon. Hydrogen was a more powerful lifting gas than hot air, and Charles's balloon reached much higher altitudes. On his first flight, which took place on 1 December 1783, he rose to 9,000 feet. Balloons evolved very rapidly for a time, then stalled technologically as balloonists reached the limitations imposed by their own bodies. The cold, thin air beyond 15,000 feet or so was beyond most aeronauts' capabilities, and for the duration of the nineteenth century only a few dared to seek higher altitudes, and they did so with mixed results. In 1804, the French scientist Joseph Louis Gay-Lussac reached 23,000 feet, a record that held for more than fifty years. Americans Henry Coxwell and James Glaiser barely survived a flight to 30,000 feet in 1862; Coxwell lost consciousness from lack of oxygen, and Glaiser saved them both. Only one of a trio of Frenchmen survived an ascent to 28,000 feet in 1875. It was not until 1901 that a human survived an ascent beyond 30,000 feet, when the German meteorologist Arthur Berson survived a flight to 35,424 feet.

Jacques Alexandre César Charles departs Nesle, France, after the landing of the first hydrogen balloon flight from Paris on December 1, 1783. Marie-Nöel Robert (on the ground at left), who accompanied Charles on the flight, takes the statements of the witnesses. (Library of Congress)

The danger and expense of reaching high altitudes with piloted balloons limited their usefulness to scientists for most of the nineteenth century. Unpiloted balloons were of little use, too, until meteorologists developed automatic recording instruments to keep records of temperature, pressure, and humidity as balloons ascended and descended. Scientists first flew balloons equipped with these automatic instruments in France in 1892, initiating a revolution in meteorology. Because they were far less expensive than piloted balloons, could reach higher altitudes, and did not risk human lives, these sondes, as automated balloons were named, quickly became the primary means of carrying out high-altitude research. Using sondes, in 1899 the French meteorologist Teisserenc de Bort discovered that the atmosphere had different layers. His instrument package found that above about 36,000 feet, the atmosphere stopped getting colder and its temperature stabilized. He named this region of constant low temperature the stratosphere. Scientists had expected the atmosphere to simply keep getting colder until it reached absolute zero. The fact that it did not drove other scientists to question why the stratosphere did not behave as they had expected.

The availability of balloon sondes and the development of other types of recording instruments opened other kinds of balloon-based scientific research. Charles Greeley Abbot, the director of the U.S. Smithsonian Astrophysical Observatory, equipped sondes with pyrheliometers to study solar radiation at different altitudes in 1914. Prior to this, in 1912, physicists in the United States and Europe also began using sondes to carry out research into a very high altitude radiation that kept the recently named stratosphere slightly ionized. An intense scientific controversy drove this line of research. A Viennese lecturer had argued that this odd radiation originated from an extraterrestrial source; indeed, he thought that they originated from beyond the solar system. Eventually called cosmic rays, their existence called into question the dominant Western theory of the universe's origin. Because that theory had its roots in Christianity, the challenge that the putative cosmic rays posed was not to a mere scientific theory but to deeply held religious beliefs. If cosmic rays originated from deep space, then they were evidence that the universe was far, far older than the age specified by Christian theology.

The controversy over cosmic rays was one issue that led to a resurgence in interest in piloted scientific ballooning. Piloted ballooning promoters used the issue to argue that humans could take better instrument readings than machines. But the most important element driving a race to the stratosphere during the late 1920s was international competition for prestige. Aeronautics had become a venue in which nations demonstrated their technological prowess. Nations competed in annual events to set new records for speed, altitude, and distance. The Frenchman Jean Callizo set a new altitude record in a balloon, 40,820 feet, in 1926. The same year, balloonists from the U.S. Army and Goodyear won first and second place at the International Gordon Bennett Race, starting a six-year period of American dominance. The next year, U.S. Army balloonist Hawthorne Gray, who had taken second place in the Gordon Bennett classic, made three attempts to beat the existing world altitude record. Gray died on the third attempt, after having reached 42,470 feet. While army spokesmen claimed that Gray's flights to the stratosphere were for scientific purposes, he carried no scientific instruments. Instead, prestige motivated the race to the stratosphere.

One piece of knowledge did come from Gray's death. Aeronauts no longer attempted the stratosphere without pressurization. Gray had relied on an oxygen mask and warm clothing during his flights, and his death on the third try was a result of his clock freezing. Without the clock, he had no way of gauging his oxygen supply, and he eventually ran out of air. Future human explorers of the stratosphere ascended in sealed, pressurized gondolas that insulated their occupants from the harsh outside conditions. These sealed capsules were the technological forerunners of the space capsules that eventually carried men to the moon.

Auguste Piccard, a University of Brussels physicist who was interested in cosmic ray research, was the first to build a successful pressurized gondola. Financed by the Belgian government, Piccard designed an eighty-two-inch aluminum sphere capable of holding two people and cosmic ray instruments. Named *FNRS*, for the National Fund for Scientific Research that had financed it, Piccard's gondola made its first ascent on 27 May 1931. It reached 51,775 feet that day, a new record, and set another record the following year. In 1933, Piccard's brother Jean came to the United States to build a still larger balloon, named the *Century of Progress*, for another record-setting flight. This time, the competition was a balloon from the Soviet Union. Soviet military balloonists had constructed the largest balloon built to that point, the *U.S.S.R.*, and flew it to 60,590 feet on 30 September 1933. Jean Piccard's *Century of Progress* balloon did not successfully ascend until 20 November but nonetheless took the record back from the Soviets. The *U.S.S.R.* tried again the following January with a flight to 72,178 feet, but the crew of three died during the descent. They descended too quickly, and the excessive speed destroyed their balloon.

This process of trying to score records with ever-higher balloon flights came to an end in 1935 with the flight of the American *Explorer 2*. Funded jointly by the Army Air Corps and the National Geographic Society, it reached 72,395 feet, a record that held for twenty years. These piloted expeditions had been enormously expensive, and sondes had already far surpassed this altitude. In addition to the older automated recording instruments, scientists interested in using sondes for meteorology, cosmic ray research, and other endeavors developed instruments that transmitted their data back to the ground by radio, making recovery of the instrument packages unnecessary, although still desirable to save money, and permitting the sondes to be tracked precisely by their signals. These achievements meant that men were superfluous. The only reason to send men to high altitudes was to test life-support systems. When piloted scientific ballooning began again in the 1950s, testing life-support equipment for piloted spaceflight was the primary motivation. The one thing that sondes could not do was to explore the limits of the human body. For the rest of the twentieth century, humans were the centerpiece of piloted exploration.

Getting Man into Space

The brief balloon competition between the United States and the Soviet Union over altitude records foreshadowed the competition that emerged after World War II. The two nations had been allies during the war, but their alliance was built on the need to defeat a common enemy, Germany. Once that enemy had been destroyed, the alliance ended. By 1948, the United States and the Soviet Union were bitter enemies who dared not attack each other directly. Instead, they financed proxy wars in the Third World, carried out extensive espionage activities against one another, and divided the globe into two armed camps. This conflict-short-of-war was called the Cold War, and it inspired the next phase of human exploration of the high frontier: the space race. From 1957 through the early 1970s, the two nations carried out a competition to see which could first put men into space, then into Earth's orbit, and finally on the moon.

The space race drew its technological origins from the liquid-fueled rocket research of the 1920s and 1930s that had been conducted by a small group of space enthusiasts in Germany. Calling themselves the German Rocket Society, they had completed 87 flights and more than 200 ground tests by 1932, when many of the group's members joined the German army. The German rocket team, led by Werner von Braun, was eventually located at Peenemünde, an island in the Baltic Sea, where the rockets could be safely fired over the ocean. Their rocket weapon, the V-2, required an enormous amount of effort to bring to fruition, and von Braun's group managed its first successful launch late in 1942. By the end of the war in 1945, 6,000 V-2s had been built, primarily by slave labor in an underground factory. As Germany collapsed, most of the members of von Braun's group voted to abandon their facility and surrender themselves to the U.S. Army instead of the closer Soviets, believing that they would be far better off in American hands. Quick work by American forces resulted in the extraction of 100 V-2s from the production facility before it was captured by the Soviets, and thus the United States acquired the best part of the German Rocket Society's legacy. Most of von Braun's group wound up in Huntsville, Alabama, at the U.S. Army's Redstone Arsenal, where they became the core of the Army Ballistic Missile Agency and, eventually, of the National Aeronautics and Space Administration (NASA) Marshall Space Flight Center.

The Soviet Union came to rocket technology through a somewhat different route. Konstantin Tsiolkovskiy had theorized extensively about rockets prior to the founding of the USSR in 1917, and although he died in 1935, his work inspired other Russian engineers to pursue rocketry. The Soviet Union's political leaders, however, did not immediately perceive the value of this technology, and in keeping with their tendency to execute or exile perceived dissidents, they decimated the ranks of domestic rocketeers. The few who managed to prove useful to the regime survived to the end of the war, when they were put to work analyzing captured German rocket technology and collaborating with the German rocket engineers who had not left with von Braun. The Soviets quickly embarked on a project to copy the V-2. In 1947, the state established a design bureau under space enthusiast Sergei Korolev to complete a V-2–derived rocket, the R-1, and develop more advanced longer-range rockets.

Both the United States and the Soviet Union sought rockets primarily as weapons. Married to the atomic bomb, the rocket would become a devastating weapon. Because the atomic bombs of 1949 were far larger than a V-2 could carry, both the United States and the Soviet Union used their V-2s to develop procedures for launching rockets, to carry out research to improve future rockets, and for scientific research in the upper atmosphere. These scientific flights were the public face of rocketry in both nations through the 1950s while both pursued larger nuclear-armed rockets in the deepest secrecy they could manage.

Men began to return to the stratosphere during the 1950s, too, as a way to experiment with pressure suits that might eventually be used in piloted spacecraft. Projects Stratolab and Man High, funded by the U.S. Air Force and the U.S. Navy, set new altitude records for piloted balloons. Man High, in particular, provided important physiological data for eventual human space flight, a goal that the air force was working toward clandestinely in a project called Dyna-Soar. The Dyna-Soar vehicle was winged and piloted, and it was to be launched into space by a rocket and then be flown back to a runway by its pilot. It would have been the first piloted spaceplane had it been completed. Instead, it was a victim of a stunning accomplishment by the USSR's Korolev, who unexpectedly orbited an automated satellite in October 1957.

The year 1957 was supposed to have been when the United States was to showcase its scientific and technological prowess. It had been declared the International Geophysical Year in 1950, and during it the United States, many

European nations, and the Soviet Union were to collaborate on a number of global scientific research efforts. In 1954, American representatives announced their intention to launch the world's first artificial satellite. The Soviet rocket program had concentrated solely on ballistic missiles to this point, but using the American declaration that it would launch a satellite as justification, Korolev got permission in early 1956 to develop a simple satellite. On 3 October 1957, he secretly launched a 183-pound satellite into Earth's orbit; once the Soviet government was satisfied that it was working properly, it revealed the coup to the world as *Sputnik*. The Soviet Union had beaten its rival to space.

In the United States, there were two very different reactions to *Sputnik*. President Dwight Eisenhower knew what few others did, that the U.S. ballistic missile program was proceeding very well and that the nation soon would be orbiting a satellite. The Army Ballistic Missile Agency's Jupiter C rocket had just completed a series of reentry tests, and it could easily have orbited a small satellite had Eisenhower directed it to do so. In fact, *Sputnik* gave Eisenhower something he badly wanted. International law prohibited nations from sending aircraft into each other's airspace without permission, which made overflights by intelligence-gathering aircraft illegal. It was not clear whether space would bear the same kind of restriction. By launching *Sputnik*, which overflew the United States repeatedly, the USSR unintentionally gave Eisenhower the legal precedent he wanted. Satellites would not be held to the same restriction as aircraft, so they were free to overfly anyone, and a highly classified Central Intelligence Agency program to develop spy satellites could go ahead.

The public reaction to Sputnik, however, was disastrous for Eisenhower. Because reporters knew very little about the ICBM program and because Eisenhower's political opponents, particularly supporters of the U.S. Air Force, fed the press scare stories of Soviet bombardment and American destruction, reporters and political commentators created a media firestorm around America's failure to beat the Soviets into space. Eisenhower's program was simply insufficient to those who saw space as the ultimate high ground for military operations. Within the three branches of the U.S. military, there was intense competition to gain control of the nation's eventual space program, whatever it turned out to be. In both the U.S. Army and U.S. Air Force, there were strong factions promoting the development of very large rockets whose sole purpose was to launch piloted vehicles into space. Humans, advocates insisted, would be needed to operate military space stations, first designed to be observation platforms and later to be used as orbiting weapons. Piloted spaceplanes would be necessary for space combat, for destroying enemy space stations and satellites, and for rapid, global deployment of combat troops. Sensing the huge costs of such weapons, Eisenhower had resisted the demands of the military space enthusiasts and kept his program directed at robotic flight as well as he could. He was, after all, the general who had orchestrated the American victory over Nazi Germany, and his critics had been afraid to challenge him on military issues. But *Sputnik* gave his critics an opening that they were happy to exploit.

On 2 November 1957, just in time for the fiftieth anniversary of the Soviet state and only a month after *Sputnik*, Korolev launched a dog named Layka into orbit. To this point, the Eisenhower administration had been able to defend its position adequately, but *Sputnik 2* produced a revolt within the mass media. Henry Luce, the publisher of *Life*, the largest circulation newsmagazine in the nation, openly attacked Eisenhower's policy in a series of editorials in which Luce called for a vast expansion of the American space effort. Other editors, essayists, and pundits followed, producing enormous pressure on Eisenhower to produce a new space plan. The media storm was reinforced by the failure of the first U.S. attempt to launch a satellite on 6 December 1957 that exploded on the launch pad in full view of the press, and was only slightly mollified by von Braun's Army Ballistic Missile Agency, which succeeded in launching a satellite the following month. That same January saw one of Eisenhower's most prominent political foes, Senate majority leader Lyndon Johnson, take up the space enthusiasts' cause. The control of space was necessary for American victory in the Cold War, Johnson argued. Failure to win the space race, he claimed, meant the end of the United States as a world power.

Eisenhower's reply to his critics and to the Soviet challenge came in July 1958 with his approval of legislation designed to create a unified civilian space agency. NASA was given control over the nation's rocket programs and its civilian satellite efforts, although responsibility for intelligence and military satellites remained in the hands of the Central Intelligence Agency and the Defense Department. Eisenhower also authorized a manned space program designed to quickly orbit a human. While he personally disapproved of what he believed was essentially a feat of engineering with little real value, he understood the political necessity of restoring Americans' faith in themselves through technological spectacle. His man-in-space program became Project Mercury, and although he attempted to limit its scope by restricting its budget, in February 1960, Senator (and presidential aspirant) Johnson arranged for a major increase in space-related spending.

During the election campaign of 1960, Johnson and fellow Democratic candidate John F. Kennedy ran against Eisenhower's candidate, Richard M. Nixon, using charges that Eisenhower's refusal to take the rocket race seriously had resulted in a missile gap. They claimed that during Eisenhower's presidency, the Soviet Union had built a huge lead in ICBMs that required a crash investment in ballistic missiles and space. Republicans, they asserted,

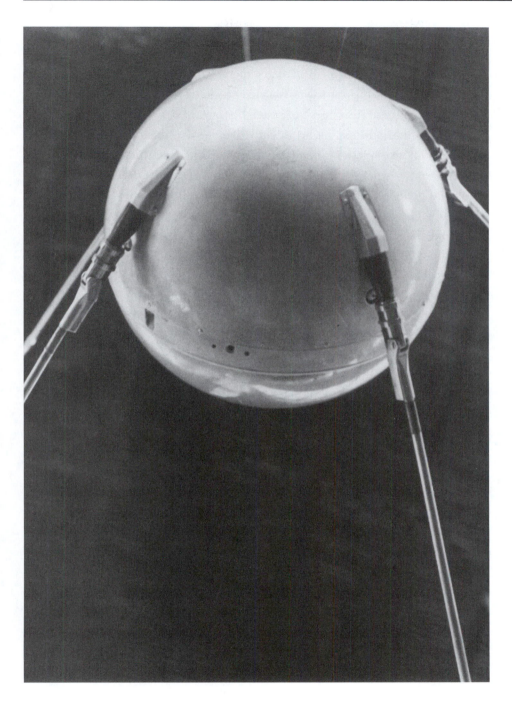

The Soviet satellite Sputnik 1, *the first human-made object in space. Its launch in 1957 shocked the world and gave the Soviet Union a solid lead in the space race. (NASA)*

were thus endangering the nation. While the charge was false, aerospace companies who hoped to profit from large increases in weapons spending amplified it throughout the nation in advertisements and via carefully placed opinion pieces. Kennedy, with Johnson as his vice presidential candidate, won the 1960 election and altered the course of space history.

In his final address as president, Eisenhower reflected on the growing power of defense-related business within the American political system. He warned that the American government had to guard against the rise of a military-industrial complex. Eisenhower had seen firsthand how supposed threats to national security could be used to justify otherwise unsupportable programs and projects, and the space race was one of these. He believed that there was nothing useful for men to do in space, and putting them there simply served the interests of corporations seeking to profit from the public treasury. But the Kennedy administration, indebted to the aerospace industry for aiding its victory, took little time in launching a vastly expanded human spaceflight program, one targeted at the moon.

Prior to the creation of NASA in 1958, the Soviet Union had not possessed an organized space program. *Sputnik* had been a response to the American International Geophysical Year-related satellite effort, not the beginning of an organized space program. The Soviet Union was a relatively poor country with many demands on its limited resources, and space advocates had not had much success selling an expensive human spaceflight program to the leadership. The R-7 had been available to launch *Sputnik* because it had been developed as a weapon, and therefore the *Sputnik* effort had been relatively inexpensive. A space program oriented around manned spacecraft required a much larger purpose-built rocket that would necessitate a huge investment of national resources. The Soviet leadership was well aware of restructuring and expansion of the American space effort in *Sputnik*'s wake, however, and Soviet Premier Nikita Khrushchev believed that an expanded effort would be necessary to prevent his nation from rapidly falling behind the United States. In June 1960, he approved a space program that included a goal of orbiting men.

The key Soviet advantage in 1960 was Korolev's R-7, which could lift a small one-man space capsule into low Earth orbit. No proven American booster could match this performance. The R-7 therefore allowed the Soviets to score a number of quick victories in the early space race. The first unpiloted flight of its crew capsule, named Vostok, occurred in May 1960. After a second flight containing two dogs exploded in July, a third flight, containing two more dogs and a variety of other animals, succeeded in August. Those two dogs, Belka and Strelka, were the first living beings to be safely recovered from space. The spacecraft had performed flawlessly on this flight and led to approval for the first launch of a human cosmonaut, which was to take place at the end of the year.

The Soviet drive to launch a cosmonaut in December was delayed by a succession of failures, including an enormous tragedy that the Soviet Union kept secret for decades. A new ICBM designed by one of Korolev's competitors exploded on 23 October 1960, killing 126 people, including the head of the nation's Strategic Missile Forces, Marshal Mitrofan Nedelin. Work on Korolev's programs stopped while the tragedy was investigated. Then an unpiloted Vostok carrying two dogs reentered incorrectly on 1 December and was destroyed by its onboard self-destruct system. An unpiloted launch on 22 December suffered a launch vehicle failure, although this time the escape system successfully rescued the dogs on board. Permission for the piloted flight was temporarily revoked due to these failures, and two more automated flights were launched in early 1961 to restore confidence in the vehicle.

Fortunately for the Soviet effort, the U.S. program had encountered delays of its own. The principal setback came from the innate conservatism of von Braun's rocket engineers, who managed the launch vehicle development for NASA. The booster chosen for the first set of piloted suborbital flights was judged unready for Project Mercury's scheduled launch on 26 April 1961, and von Braun had recommended that another automated flight be made to fully validate the vehicle. NASA administrator James Webb accepted the recommendation, pushing the first American attempt off until May. This provided Korolev with the reprieve he needed to gain another victory for the Soviet Union.

The two remaining automated Vostok launches occurred flawlessly in March 1961, removing the cloud of doubt that had engulfed the program at the end of the previous year. On 8 April 1961, the cosmonaut Yuriy Gagarin was chosen for the first piloted flight and was launched into orbit on 12 April. After nearly two hours, he was back on Earth and in excellent health. His single orbit flight was not matched by the United States until after two Mercury astronauts, Alan Shepherd and Virgil Grissom, flew suborbital flights in May and July. Finally, the astronaut John Glenn flew an orbital mission on 20 February 1962. The first lap of the space race was over, and the Soviets had won.

Shooting the Moon

The November 1960 presidential election in the United States brought about major changes in the nation's approach to the space race. Eight days after Gagarin's triumphant orbit, President Kennedy assigned his vice president,

Yuri Gagarin (1934–1968)

Yuri Gagarin was born on 9 March 1934 on a collective farm west of Moscow and became the first human to orbit Earth in April 1962. Gagarin had been a top graduate of the Soviet Union's Air Force Academy in 1957, received fighter pilot training, and joined the new Soviet cosmonaut corps two years later. Gagarin received his cosmonaut training at Svësdni Gorodok, or Star City as it became known in the United States, a highly secret closed community east of Moscow. His mission lasted 108 minutes, and its purpose was to demonstrate that humans could survive in space and, of course, to demonstrate Soviet technological capacity. Unlike the later American astronauts, Gagarin had no control over his spacecraft. Soviet engineers designed the control systems to be fully automatic. The first American astronauts had been able to get a similar decision overridden, but the cosmonauts had not. Gagarin's capsule also did not have a soft-landing system, and when it reached about 20,000 feet, Gagarin ejected and, using a conventional parachute, landed near the village of Smelovka. His success made him a hero within the Soviet Union and helped motivate President John F. Kennedy's decision to send Americans to the moon. Gagarin never flew in space again. He became assistant director of the cosmonaut training facility and died in the crash of a MiG-15 trainer on 27 March 1968.

Lyndon Johnson, the former Senate majority leader who had been such a forceful promoter of an expansive space program, the task of formulating a new set of space goals. In May, Johnson proposed a grand program of exploration. Believing that the nation's self-image was collapsing under the weight of successive Soviet victories, Johnson argued for an accelerated space program aimed at restoring American confidence and prestige. On 25 May 1961, before a joint session of Congress, Kennedy announced the new goal: before the end of the decade, the United States would land an American on the moon and return him safely to Earth.

The Soviet leadership did not take Kennedy's pronouncement seriously because such speeches in the Soviet Union were merely propaganda, as they often were in the United States. Kennedy's proposal, however, was quickly backed by congressional action. Congress provided comparatively vast funds for the new effort almost immediately, and NASA constructed two new projects to prepare the necessary technologies, Gemini and Apollo. The United States made available financial, managerial, and technological resources far greater than those available to Korolev in the Soviet Union. To make things worse for the USSR, factions within its own space effort began to challenge Korolev for leadership in the wake of the Vostok successes, throwing the overall program into chaos at the moment that the United States began organizing its mission to the moon. As a result, the Soviet Union was barely in the moon race at all.

To put men on the moon, a much larger launch vehicle than either nation possessed in 1961 was necessary. The rocket NASA chose was named Saturn. In its ultimate configuration, the Saturn could put 130 tons (260,000 pounds) into Earth's orbit, dwarfing the capabilities of anything else in the world. NASA named the piloted lunar program Apollo, and several other programs supported it. NASA also established Project Gemini to develop and validate the technologies and procedures needed for the complex series of maneuvers and dockings that the moon flight required. A preexisting robotic lunar exploration project, Ranger, was redirected to support Apollo by photographing potential landing sites. The agency also constructed two other robotic projects, Surveyor and Lunar Orbiter, to provide scientific data for Apollo. The Surveyor spacecraft were designed to land on the lunar surface and report on the surface's strength, radiation levels, micrometeoroid activity, and other data that the engineers needed for the design of the piloted lander. The Lunar Orbiters, derived from a classified spy satellite program, were designed to photograph almost the entire surface of the Moon, including the far side, again to help select potential landing areas.

In the Soviet Union, the moon program Korolev wanted to undertake in response to Apollo did not get approved until late in 1964, leaving him nearly three years behind the United States. The Soviet leadership was comfortable with the USSR's apparent lead over the United States because, through the end of 1964, the American program seemed to have stalled. Project Mercury ended in early 1963, and not a single American was scheduled to fly in 1964. Soviet cosmonauts had accumulated much more time in space than had American astronauts, and the Soviet Union had twice orbited two Vostok spacecraft simultaneously. It also orbited the first woman, Valentina Tereshkova. To overshadow the upcoming two-man American Gemini flights scheduled for 1965, Korolev was tasked with turning the one-man Vostok spacecraft into a three-man Voskhod, which flew in October 1964 to more accolades. Soviet leaders had good reason to be proud of their successes.

But Soviet complacency did not survive 1964, because the more deliberately paced American program started to produce results commensurate with the vast resources it had been granted. The accomplishment that drew Soviet attention and finally gained Korolev money for his own moon rocket, the N-1, was the arrival in orbit of the first Apollo capsule, a dummy sent aloft on a Saturn I booster on 28 May 1964. While the Saturn I, with only a seventeen-ton orbital capacity, was a pale shadow of the Saturn V that would eventually put men on the moon, it was still more powerful than any existing Soviet rocket. Further, the Apollo capsule it carried was new, designed to sustain three men for a ten-day lunar mission, and was state-of-the-art, while Voskhod was just a quick modification of the Vostok capsule that could barely sustain three men without space suits or experimental apparatus for one day. The contrast was obvious. The Americans had, in fact, been serious in their moon race pronouncements, and they would win the race to the moon without a similar national commitment. Korolev finally received party support for his N-1 in August 1964. In the end, this would not matter. The year 1965 marked NASA's ascendance and, after this, leadership in the moon race.

During 1965 and 1966, NASA's Gemini program successfully orbited ten piloted missions and several different target vehicles. Because Gemini's purpose was to advance the necessary technologies, its pacing was cautious and its accomplishments lacked the spectacle nature of the Soviet program. But these were critical to Apollo's success. Gemini missions 4, 5, and 7 were aimed at extending knowledge of how the human body would perform and how well life-support systems would sustain humans during moon missions of up to fourteen days. Geminis 8, 10, 11, and 12 made repeated rendezvous and docking maneuvers to prove that a human pilot could accomplish a docking and to show that the docked spacecraft system could then be safely maneuvered. All of the Gemini missions had as goals the refinement of recovery techniques so that American astronauts did not suffer prolonged waits for recovery. Soviet cosmonauts routinely waited for a couple of days to be picked up, often in freezing weather, because Soviet controllers could not accurately estimate where the space capsules would land. By the end of the Gemini program, NASA could project touchdown to within 3 miles and position U.S. Navy ships to make an almost immediate pickup.

Gemini's success at validating the maneuvering and docking techniques necessary for the lunar mission was matched during 1965 and 1966 by success with the Saturn rocket. Gemini had used a modified U.S. Air Force ICBM, the Titan II, to reach orbit. Saturn was unlike the Mercury and Gemini vehicles in that it had more than one stage and used clusters of several engines in the first two stages. The use of multiple engines produced dynamic stresses on the engines and supporting structure that had to be investigated, and the use of multiple stages required the development of control systems capable of precise timing and of systems capable of jettisoning spent stages safely. To accomplish all of the engineering developments that Saturn required, NASA actually developed three Saturn variants. The first version, Saturn I, was a two-stage rocket that used clusters of engines intended to be used in the second- and third-stage moon rocket, the Saturn V. All ten launches of the Saturn I were successful, demonstrating that NASA had finally surmounted the unreliability problem that had plagued the early space program.

The second Saturn variant, Saturn IB, was the first version that was intended to carry human beings. First launched in February 1966, the Saturn IB used the Saturn V's third stage as its own second stage. This allowed NASA to verify that the stage could carry out its fairly complex mission. Unlike the other Saturn stages, this one had to start, shut down, and then relight when in space. After three successful unpiloted flights, NASA had prepared to launch three astronauts—Virgil Grissom, Edward White II, and Roger Chafee—on 21 February 1967 using the Saturn IB. Unfortunately, this effort ended in a disaster. During a prelaunch checkout on 27 January 1967, an electrical short within the Apollo capsule's systems sparked a fire, and the pure oxygen atmosphere of the capsule caused the fire to spread with extraordinary speed. The three astronauts could not open the hatch quickly enough to escape and burned to death, with the helpless flight controllers listening to the tragedy unfold. The deaths of these three men were the first in the American space program, and the tragedy set off an orgy of recriminations and investigations that set back the lunar exploration program by a year.

After the Apollo tragedy, NASA redesigned the capsule, adding an instant-opening explosive escape hatch, and launched its first three Apollo astronauts—Walter Schirra, Donn Eisele, and R. Walter Cunningham—into orbit on 11 October 1968. Named Apollo 7, the mission was the final test of the Apollo capsule. The redesigned capsule performed flawlessly, as did its launch vehicle. Apollo 7 was the interim Saturn IB's last flight, however. It had only been intended as an interim vehicle to get NASA to the Saturn V, and von Braun's monster was ready.

The Saturn V's first flight had occurred eleven months prior to Apollo 7, on 9 November 1967. Named Apollo 4, it represented an enormous gamble by NASA. Originally, the agency had intended to flight test each of the rocket's stages separately, as it had with the third stage using the Saturn IB. But this approach would not have gotten men to the moon by 1970, President Kennedy's arbitrary deadline for putting an American on the moon. The Apollo program director told von Braun to drop the conservative step-by-step approach and test the rocket in its entirety. Von Braun, unable to prove that this approach would not work, accepted the decision with substantial doubts. In return, he got the most extensive ground-testing program to which any rocket had been subjected. The Saturn V experienced several problems in its ground tests but virtually none when it finally flew. Apollo 4's flight was flawless.

Americans would not reach the moon for another eighteen months, but Apollo 4's flight marked the effective end of the space race. The Soviet program achieved no comparable successes; instead, it suffered repeated tragedies while Gemini and then Apollo soared. The first disaster was the unexpected death of Korolev, who had been the driving force behind the Soviet space program.

Korolev, suffering from intestinal bleeding, had been admitted to the hospital in January 1966 for surgery. His surgeon found that Korolev had a large malignant tumor that had gone unnoticed. The surgeon removed the tumor, but Korolev died shortly afterward. The Soviet Union's space visionary was given a state funeral on 18 January 1966.

Korolev's successor, Vasiliy Mishin, had a very difficult task before him. The first flight of the new Soyuz space capsule, which was planned to replace the old Vostok/Voskhod vehicle, occurred in November 1966 and suffered multiple failures in orbit. The second attempt a month later to launch Soyuz automatically aborted on the launch pad when the capsule exploded. The launch complex was destroyed, and one person died. A third attempt placed a Soyuz capsule in orbit in January 1967, but again the vehicle suffered multiple failures. Despite these failures, a piloted mission was scheduled for April in response to the enormous pressures produced by NASA's Gemini and Saturn successes. On 23 April, the cosmonaut Vladimir Komarov went into orbit aboard *Soyuz 1*. He planned to rendezvous and dock with a second vehicle carrying two more cosmonauts, which would score another Soviet space success, but his Soyuz capsule suffered multiple failures. The second Soyuz launch was canceled as the controllers tried to bring Komarov back alive. They directed Komarov to manually initiate the de-orbit engine burn, and he did so successfully. But when the search team located his capsule, they found smoking wreckage. The parachutes had stuck in their containers, and the capsule had plunged through the atmosphere out of control, exploding when it hit the ground. The second Soyuz spacecraft had the same flaw as Komarov's. If Mishin had not canceled the second launch, the two cosmonauts aboard it would likely have died, too. Soyuz did not have a fully successful automated flight until August 1968, and it did not carry a crew into orbit until October 1968.

The Soyuz capsule was not the Soviet lunar vehicle, however. It was only intended to ferry the cosmonauts to the true lunar vehicle, which was to be launched separately. The lunar vehicle was carefully hidden from the Western press, and its tests had the same code name—Zond—given to robotic flights. The lunar vehicle, first launched in September 1967, also suffered repeated failures. A booster explosion destroyed the first attempt. The second, which reached space as Zond 4, failed during reentry. Two more attempts failed before Zond 5, carrying two tortoises, completed a successful orbit of the moon after a 15 September 1968 launch. Mishin had hoped that one more automated mission would permit a piloted attempt to orbit the moon, but Zond 6, launched in November 1968, failed during reentry. In any case, Soyuz had already suffered its own return to failure. *Soyuz 2* and *Soyuz 3* had been launched into orbit in late October to validate the rendezvous and docking technique, which Gemini had accomplished more than two years before, but the docking attempt had failed. What little chance the USSR had of beating Apollo to the moon vanished with these failures.

On 21 December 1968, NASA launched Apollo 8 toward the moon with three astronauts aboard: Frank Borman, James Lovell, and William Anders. Its mission was to complete the rendezvous procedure while orbiting the moon. The astronauts of Apollo 8 passed behind the moon on Christmas Eve and became the first humans to see the moon's far side. They also witnessed and photographed the first Earthrise, bringing back with them a spectacular picture of Earth hanging above the moon's surface. For the next three years, NASA would deliver the same sort of humiliation to the USSR that *Sputnik* had caused in the United States. Each Apollo mission built on the last, and with a single exception, the missions were enormously successful.

Apollo missions 9 and 10 continued validating the performance of the vehicle and its systems. Apollo 9 was the first to have a fully functional Lunar Excursion Module (LEM), and it carried out docking and crew transfer experiments while still in Earth's orbit. Apollo 10, launched in May 1969, carried out LEM experiments while in orbit around the moon. During this mission, the LEM with its crew of two separated from the Apollo command module and descended toward the moon's surface. Instead of landing, however, it jettisoned its descent stage and ignited its ascent engine to ensure that it would get the astronauts back into lunar orbit and safely docked again with the command module. Apollo 10 completed NASA's validation of the Saturn-Apollo complex. The next mission would land the first humans on the moon.

The landing site for the first landing mission, Apollo 11, was chosen after extensive surveying of the moon by robots. By the end of 1967, five Lunar Orbiter spacecraft had produced photographic maps of almost the entire surface. Based on these maps, the Apollo mission planners had selected a number of potential landing sites that were then rephotographed at higher resolutions by the Apollo 8 and 10 astronauts. The performance of the Lunar Module, in particular its descent speed and its control precision, was the final factor in selection of a landing site. Mission planners chose a spot in the Sea of Tranquility as Apollo 11's landing site and set a launch date of 16 July 1969.

Exploration and Lunar Science

Apollo 11 left on schedule, carrying Neil Armstrong, Michael Collins, and Edwin Aldrin into space and into history. On 30 July 1969, Armstrong and Aldrin piloted their LEM, named *Eagle*, to the lunar surface, touching down at 4:18 P.M. EDT. Armstrong was the first down the module's ladder and was followed a few

Apollo 11 astronaut Edwin Aldrin stands facing the U.S. flag on the moon, 21 July 1969. (NASA)

minutes later by Aldrin, becoming the first humans to reach the moon. Their presence on the Sea of Tranquility and their safe return to Earth on 24 July represented an enormous engineering achievement by NASA and its legions of contractors. It also marked the beginning of the piloted space program's contributions to lunar science. Prior to Apollo 11, scientists had served the lunar mission, helping engineers design and validate the lunar vehicle and helping mission planners select suitable landing sites. With the *Eagle*'s arrival on the moon, Apollo could finally begin producing a scientific return.

Apollo 11 had been equipped with a limited package of experiments, including a core sampler to take subsurface samples, a solar-powered seismometer to detect moonquakes, and a Laser Ranging Retro-Reflector that consisted of a series of mirrors that would reflect a light beam back to Earth. Geologists

Neil Armstrong (1930–)

Neil Armstrong was born on 5 August 1930 in Wapakoneta, Ohio. The first of twelve American astronauts to explore the moon's surface, Armstrong had been a naval aviator prior to joining NASA's Lewis Laboratory as a test pilot in 1952. He later transferred to the agency's High Speed Flight Station in California and flew many experimental aircraft, including the hypersonic X-15. In 1962, he was accepted into the astronaut corps and flew his first mission in space aboard Gemini VIII, with David Scott, as the mission's command pilot. During this mission, Armstrong demonstrated that it was possible to dock with other space vehicles. Despite this success, the mission was nearly a disaster because a malfunction in the capsule control system caused it to spin wildly once docked. Armstrong succeeded in separating the capsule from the target, regaining control, and making an emergency splashdown. Because of his ability to respond well in crises, he was appointed mission commander for Apollo 11. Apollo 11's crew consisted of lunar module pilot Edwin E. "Buzz" Aldrin and command module pilot Michael Collins. On 20 July 1969, Aldrin and Armstrong landed on the moon, provoking Armstrong's famous statement, "That's one small step for a man, one giant leap for mankind." After their return to Earth, Armstrong became deputy associate administrator for aeronautics and resigned from NASA in 1971 to become a professor of aerospace engineering at the University of Cincinnati. In 1979, he entered private industry, serving as chairman of Computing Technologies for Aviation, Inc., then of AIL Systems, Inc.

hoped that the core sampler would permit them to understand the structure of lunar soil. Data from the seismometer would permit geologists to evaluate the moon's internal structure, especially after later Apollo missions deployed additional seismometers in other locations. Furthermore, NASA planned to create moonquakes for the seismometers to record by crashing the spent Saturn V third stages into the moon. The seismometer data would show how the impact's energy moved through the moon, further revealing lunar structure. Finally, the reflector experiment would allow extremely precise measurement of the distance between Earth and the moon. The seismometer and reflector stayed on the moon's surface near the landing module's descent stage when Armstrong and Aldrin lifted off on 21 July to rendezvous with Collins. The three men splashed down in the Pacific Ocean on 24 July and were picked up by USS *Hornet* almost immediately. They were then quarantined for a month to make sure they had not brought back any germs from the moon. They had not; the moon is sterile.

The next mission, Apollo 12, carried a much larger science package called the Apollo Lunar Scientific Experiments Package (ALSEP). In addition to a longer-lived seismometer and laser reflector, the ALSEP contained experiments to measure heat flow out of the moon, the moon's gravity, and its magnetic field. Launched 12 November 1969, astronauts Pete Conrad, Richard Francis Gordon Jr., and Alan LaVern Bean reached lunar orbit four days later, and Conrad and Bean were the next two astronauts to reach the moon. The three astronauts returned to Earth on 24 November.

The next Apollo mission was a failure. During the outbound voyage of Apollo 13 in January 1970, an oxygen tank in the service module exploded, crippling the spacecraft's electrical power system and leaving the three astronauts, James A. Lovell Jr., John L. Swigert Jr., and Fred Wallace Haise Jr., dependent on battery power. The command module's batteries were insufficient for the complete trip around the moon, back to Earth, and reentry, and to preserve the command module's power for reentry, the three men were forced to cram themselves into their two-man LEM, the *Aquarius*, and use it as a makeshift lifeboat. The *Aquarius* batteries could keep its temperature just above freezing on the long trip around the moon. But the *Aquarius* returned its astronauts safely to Earth's orbit, and the crippled command module *Odyssey* put them into the Pacific near the recovery ship on 17 January, attesting to the skill of the modules' designers. But if the success of the two previous missions had demonstrated that piloted exploration could produce scientific returns, Apollo 13 showed that it also carried significant risks. Television turned Apollo 13 into a dramatic object lesson on the continuing hazards of space exploration that was broadcast, in real time, into millions of American households.

Apollo 13's near disaster delayed Apollo 14 for a year while engineers from NASA and the service module's contractor investigated the failure. NASA had intended Apollo 15 to be the beginning of a much more substantial phase of lunar exploration. It introduced a modified landing module that could support its two astronauts on the surface for longer periods, a solar-and-battery–powered lunar rover that would carry its astronauts much farther from the landing module, an expanded group of scientific instruments, and new, more flexible spacesuits. The space agency had even more in mind for missions beyond Apollo 17. It had plans for dual-launch missions that would deposit an automated lander and housing unit that could support astronauts for two weeks. It wanted to deploy large science and engineering payloads on the moon to begin collecting the kinds of data needed to construct permanent bases. It had built and tested prototypes of crewed, mobile lunar geological laboratories and lunar flying vehicles to further extend the astronauts' exploration range. Finally, after the moon, the space agency wanted to send its astronauts to Mars.

But by 1970, Project Apollo had outlived its political usefulness. The United States had decisively defeated the Soviet Union in the space race, which had been the original motivation for the project. There was a new administration in the White House, the same Richard Nixon who had been defeated by Kennedy in 1960, and in his eyes Apollo's scientific returns were not worth the huge costs of its launches. The Mars project was even more absurd. Nixon had promised to cut government spending to balance the budget during his 1968 presidential run, and a recession that began in 1969 made the promise increasingly difficult to keep. NASA's space dreams were an easy and highly visible target with which to demonstrate his commitment to fiscal austerity.

In September 1969, Nixon's vice president, Spiro Agnew, proposed three options for a post-Apollo future. The first, a piloted Mars mission by the mid-1980s, included a permanent crewed Earth-orbiting space station, a reusable shuttle to reach it, and a Mars vehicle that would depart from the space station, at a cost of $10 billion per year. NASA's budget in 1969 had been $3.677 billion. The second option postponed the Mars mission for a few more years, reducing the cost to around $8 billion a year. Agnew's third option dropped a mission to Mars and kept the station and the shuttle at a cost of $4–5 billion a year. In March 1970, three months after the Apollo 13 near-disaster, a dissatisfied Nixon had created an option four for himself: a shuttle with no space station. Apollo missions beyond Apollo 17 were canceled, along with production of the Saturn V. The Saturn was too large for any post-Apollo future that did not include a piloted Mars mission, and Nixon did not want NASA keeping it around to prolong its dreams of a mission to Mars. Furthermore, the scientific community did not support a crewed Mars mission. By this time, scientific opinion was that the flexibility of human space exploration came at too high a cost. As had been the case in scientific ballooning, robots could do science well enough and for far less money. Human space exploration was suddenly dead, not even a year after Apollo 11's triumphant mission to the moon.

The last four Apollo missions flew in 1971 and 1972, generating little public interest. The twelve astronauts included the first and only scientist to reach the moon, Harrison Schmitt, aboard Apollo 17. The crews brought back hundreds of pounds of moon rocks for analysis, left behind experiments, and took photographs. Using the lunar rover, they traversed miles of the lunar surface during their sample-collection activities. The public, however, generally ignored these missions. If it had ever been interested in the Apollo missions as anything other than a space race with the Soviet Union, it had already turned to other concerns. When Apollo 17's astronauts returned to Earth on 18 December 1972, they came back as the last men to leave Earth's orbit in the twentieth century.

Contraction of the High Frontier: The Retreat to Low Earth Orbit

After Apollo, NASA's glory days were over, and the agency entered a long, troubled period in which it struggled to define itself and its purpose. In the Soviet Union, Mishin's program had suffered the same problem somewhat earlier. The Soviet moon rocket, N-1, never had a successful test. The first launch attempt, in February 1969, failed when the engine control system shut down all thirty-two engines only sixty-eight seconds into the flight. The second attempt in July of that year rose a few feet into the air, collapsed back onto the pad, and exploded. One more attempt to launch the N-1 failed the following year, terminating any hope the Soviets had of developing a heavy-lift vehicle to enable future crewed moon and Mars missions. Instead, the Soviet leadership turned to temporary Earth-orbiting space stations named Salyut. These stations were primarily laboratories for investigating human response to long-duration missions, but cosmonauts aboard them also performed solar and stellar observations, biological experiments, and metallurgical experiments.

NASA also built a temporary space station during the 1970s called *Skylab*. Built out of the shell of one of the remaining Saturn V third stages, *Skylab* was launched in 1973. It hosted three crews of three men each during its life, all launched during 1973. During stays of twenty-eight, fifty-nine, and eighty-four days, respectively, the three crews conducted repairs to *Skylab* and astronomical, biological, and materials research and underwent extensive physiological monitoring and experimentation. They demonstrated that people could live in microgravity conditions for extended periods of time and provided information on diet and exercise regimens necessary to preserve humans' health in space. After the final crew left in 1974, *Skylab* orbited empty until 1979. It was supposed to be rescued by an early mission of the agency's Space Shuttle, but delays prevented that from happening. *Skylab* burned up during reentry over the Pacific Ocean and Western Australia.

When NASA's Space Shuttle began operating in 1980, it became the only vehicle for NASA to conduct in-space experiments using humans. The shuttle could orbit for up to seventeen days with a crew of seven astronauts. In 1973, NASA had accepted an offer from the European Space Agency to build a space-based laboratory designed to fit into the shuttle's payload bay. The first major European contribution to human spaceflight, *Spacelab* flew twenty-five times between 1983 and 1998. Astronauts used *Spacelab* to conduct more than 400 experiments in life sciences, atmospheric sciences, and microgravity. It was retired in 1998 when NASA began to construct a new permanent space station.

By *Spacelab*'s retirement, the Soviet Union had been operating a permanent space station, named *Mir*, in low Earth orbit for many years. The successor to the Salyut series of temporary stations, *Mir* was composed of a series of individual modules, the first of which had been launched in 1986. The Soviet Union continued adding modules to *Mir* until the nation itself disintegrated in 1991; after this, the new Russian government and the United States agreed to carry out a Shuttle-*Mir* program to jointly operate and equip the station. An American docking adapter allowed the Space Shuttle to dock with *Mir* and supply crews. This program ran through 1998 as part of the preparation for building a new permanent space station, the *International Space Station*. Except for one brief period in 1989, *Mir* was crewed continuously from 1987 until shortly before its deliberate de-orbiting and destruction in 2001. Russian physician Valeri Polyka set a 438-day endurance record aboard *Mir*, demonstrating that humans could function in microgravity long enough to reach Mars.

Edwin Hubble (1889–1953)

Edwin Hubble was born in Marshfield, Missouri, on 20 November 1889 and was perhaps the twentieth century's greatest observational astronomer. Trained in mathematics at the University of Chicago and then in law at Oxford College in England, Hubble worked briefly at the Yerkes Observatory before serving in the U.S. Army in France during World War I. After the war, George Ellery Hale, founder and director of the Carnegie Institute of Washington's Mount Wilson Observatory in California, recruited Hubble. Mount Wilson's new one hundred-inch telescope was the world's most powerful at the time, and using it Hubble established that the scale of the universe was vastly greater than astronomers had believed. He convincingly argued that what many of the so-called nebulae that astronomers could see were actually other galaxies, beginning a long career of overturning accepted scientific "facts." Most importantly, Hubble and M. L. Humanson demonstrated that the light from many distant objects was redshifted. Just as a train whistle's tone lowers as it becomes more distant, light from receding objects shifts toward the lower, red end of the spectrum. Hubble's finding of redshift in the spectra of distant objects meant that those bodies were moving away from Earth. The universe was therefore expanding, not fixed in size. Hubble's theses became the basic foundation for cosmology, the study of the structure of the universe. Hubble died of a stroke in 1953 and thus did not live to witness the beginning of space travel. But due to his seminal importance to astronomy, the first of NASA's space-based observatories was named in his honor. The Hubble Space Telescope, launched in 1989, became astronomy's most powerful visible-light telescope after a repair mission in 1994 corrected a flaw in its 97.5-inch primary mirror.

Mir was replaced by the *International Space Station (ISS)*. Originally started as an American Cold War project named Space Station Freedom, after the collapse of the Soviet Union in 1991 it was merged with a Russian Mir 2 project, redesigned, and renamed. The *ISS* is the fulfillment of space enthusiasts' dreams of permanent space colonization. Built with American and Russian modules, the *ISS* received its first crew of three in November 2000, only a few months before engineers de-orbited *Mir*. But *ISS*'s future is highly uncertain. Originally projected to cost no more than $34 billion, the *ISS*'s costs have nearly doubled. Without the motivations provided by the Cold War in the form of technological competition to sustain political support for it, the project has been severely curtailed. It was to be built to sustain a crew of seven astronauts and scientists engaged primarily in research. After destruction of the Space Shuttle *Columbia* in 2003 and loss of its crew in a reentry accident, NASA and the Russian government reduced the crew to two, whose time is largely devoted to station maintenance, and if the *ISS* is not expanded to its intended complement, its scientific output will be negligible.

In January 2004, President George W. Bush announced that the United States would cease supporting the *ISS* after 2016 in order to focus its efforts on sending humans back to the moon and on to Mars, retiring the aged Space Shuttles in favor of a new vehicle after 2010. It remains to be seen what, if anything, will arise from this new initiative.

Conclusion

During the early twentieth century, space enthusiasts conceived grand dreams of human migration off Earth to other planets and eventually to other star systems. Constantin Tsiolkovsky, Herman Oberth, Wernher von Braun, Sergei Korolev, and many others believed in and worked toward human exploration and settlement of the high frontier. But their dreams came at enormous costs that governments supported only when human spaceflight served to advance their political and diplomatic goals. The Mercury, Gemini, and Apollo programs cost the United States about $25.4 billion in 1960s dollars and were in turn built on tens of billions of dollars spent on ballistic missile development. NASA's Space Shuttle cost $6.7 billion in 1971 dollars, the design studies for the never-built *Space Station Freedom* cost $8 billion, and the *ISS* will cost well beyond its budgeted $34 billion. There are no accurate accountings of the cost of the Soviet human spaceflight program. In the absence of the Cold War's technological competition, neither the United States nor Russia has had the political will to support an expansive human spaceflight agenda.

During the Cold War, the Russian leaders Khrushchev and Brezhnev and the U.S. leaders Kennedy, Johnson, and Nixon were solely interested in the political value of humans into space. As long as humans in space produced a positive political value, astronauts and cosmonauts had political support from the highest levels of government. Once piloted spaceflight became a political liability, it lost that support. Exploration of the rest of the solar system became the responsibility of increasingly sophisticated robots. Humans did not leave Earth's orbit again during the twentieth century.

Bibliographic Essay

For the ballooning age, see David DeVorkin's *Race to the Stratosphere: Manned Scientific Ballooning in America* (1989). For treatment of early American space-enthusiast culture, the most succinct analysis is Howard McCurdy's *Space and the American Imagination* (1997). Michael Neufeld's *The Rocket and the Reich* (1995) is the best English-language history of German rocket efforts before and during World War II. For the German Rocket Society, see Frank Winter's *Prelude to the Space Age: The Rocket Societies, 1924–1940* (1983).

The most thorough history of the Soviet-manned spaceflight effort in the 1960s is Asif Siddiqi's *Challenge to Apollo* (2000). This is the first major historical study in English that is based on original Russian sources. It should be paired with Walter McDougall's classic history of the space age, *From the Heavens to the Earth: A Political History of the Space Age* (1985). McDougall's work provides the overall frame for the space age, despite ending just as the space race began in earnest in 1964. More detail on the development of the American Saturn rocket can be found in Roger Bilstein, *Stages to Saturn: A Technological History of the Apollo/Saturn Launch Vehicles* (2003). William David Compton, *Where No Man Has Gone Before: A History of Apollo Lunar Exploration Missions* (1989), examines the entirety of the Apollo program. There are also detailed histories of the Mercury and Gemini programs: Loyd S. Swenson et al., *This New Ocean: A History of Project Mercury* (1966), and Barton C. Hacker and James M. Grimwood, *On the Shoulders of Titans: A History of Project Gemini* (1977). Finally, there are two dated, but still relevant, histories of space medicine: Mae Mills Link, *Space Medicine in Project Mercury* (1965), and John A. Pitts, *The Human Factor: Biomedicine in the Manned Space Program to 1980* (1985). And Margaret Weitekamp's *Right Stuff, Wrong Sex: America's First Women in Space Program* (2004) examines the inability of qualified women to become astronauts during the first decades of the space age.

The robotic explorations of the moon that preceded the Apollo landings are not as well documented as the manned missions. R. Cargill Hall, *Lunar Impact: A History of Project Ranger* (1977), examines the difficult early challenge to engineer spacecraft to hit the moon. Lunar Surveyor, the soft-lander series, is briefly examined in Clayton Koppes, *JPL and the American Space Program*, (1982). Lunar Orbiter, which provided the first complete photographic maps of the lunar surface, is documented in R. Cargill Hall, "SAMOS to the Moon: The Clandestine Transfer of Reconnaissance Technology between Federal Agencies" (2001), and Bruce Byers, *Destination Moon: A History of the Lunar Orbiter Program* (1977). There are no equivalents to the Soviet lunar robotics program in English, but the Soviet missions are summarized in Asif Siddiqi's *Deep Space Chronicle* (2002) and his *Challenge to Apollo*.

The primary history of the American *Skylab* project is W. David Compton and Charles D. Benson, *Living and Working in Space: A History of Skylab* (1983). Dennis Jenkins's *The History of Developing the National Space Transportation System* (1992) examines the development of the Space Shuttle, as does T. A. Heppenheimer's *The Space Shuttle Decision* (1999). Howard McCurdy's *The Space Station Decision: Incremental Politics and Technological Choice* (1990) documents the decision to build *Space Station Freedom*, the successor to *Skylab*. The redesign of Freedom into the International Space Station has not yet been made the subject of a history, but John M. Logsdon's *Together in Orbit: The Origins of International Participation in the Space Station* (1998) examines the evolution of the space station's international component. The utter failure of the Space Shuttle to deliver on its promise of inexpensive space access is examined by Roger Pielke Jr., "A Reappraisal of the Space Shuttle Programme" (1993). There is also valuable information on the Shuttle program in the *Columbia Accident Investigation Board Report*, released in August 2003 and available electronically or from the U.S. Government Printing Office.

9

Robotic Space Exploration

In 1894, Percival Lowell established an observatory in Flagstaff, Arizona, believing that the clear, stable air of the region would make it an ideal place to conduct planetary astronomy. He constructed the Lowell Observatory atop Mars Hill, at an altitude of 7,260 feet, and turned his twenty-four-inch telescope toward Mars. He was interested in that planet because the Italian astronomer Giovanni Schiaparelli had discovered what appeared to be channels in the surface of Mars. Lowell, during a year of intensive observation of Mars, came to believe that these channels were canals designed by intelligent beings to carry water from the poles to oases of vegetation that provided their food. In a series of books, *Mars* (1895), *Mars and Its Canals* (1906), and *Mars as the Abode of Life* (1908), Lowell made public his belief, arguing that the existence of a complex series of linear canals on Mars could only be the product of extraterrestrial intelligence.

Lowell's thesis was highly controversial for a number of reasons. Chief among them was that it challenged the dominant view that human intelligence was unique and that Earth was the only planet with life. In the waning years of the nineteenth century, most Americans held to the view that Earth had been specially created to hold life and that humans had been placed on it to serve as agents of the Creator's plans. Nothing in Christian theology suggested that other intelligent beings might exist elsewhere. Lowell's Martians did not seem to be part of the divine plan, and if they existed, they could undermine the very basis of the faith. It did not help Lowell's case that most other astronomers did not accept his claims either. While many could see Martian channels, few were willing to go as far as Lowell had in attributing them to nonhuman life forms. But Lowell's sensational arguments added to notion in the Western imagination that life might exist elsewhere. Like Jules Verne's earlier *Voyage to the Moon*, Lowell's fantasy proved enormously compelling.

While the space race that developed during the late 1950s between the United States and the Soviet Union was primarily aimed at getting men into

Konstantin Tsiolkovskiy (1857–1935)

Konstantin Tsiolkovskiy was born on 17 September 1857 in Ijevskoe, Russia. He lost his hearing at the age of ten to scarlet fever and received no formal education. Instead, he educated himself through reading. Inspired by Jules Verne's *Voyage to the Moon*, Tsiolkovskiy began to think about space travel and colonization. He eventually earned a teaching certificate and worked as a math teacher and then, as his theories on space travel became better known, a professor. His first space-related publication, "Investigating Space with Rocket Devices," was submitted in 1898 and published in 1903. In 1919, his prolific writings earned him membership in the Soviet Academy of Science, and he was granted a government pension in 1920. He wrote more than 500 scientific papers, all of which were based on mathematics. Unlike other rocket pioneers such as Robert Goddard and Werner von Braun, Tsiolkovskiy was not an experimentalist, so he never actually built a rocket. Nonetheless, he was able to analyze the action/reaction nature of rockets, including their ability to function in a vacuum; propose a variety of liquid propellants and analyze the efficiency that each could provide; and prove that multistage rockets would be necessary to leave Earth's orbit. By the time Tsiolkovskiy died in 1935, von Braun and Sergei Korolev were actively experimenting on his ideas.

space and then to the moon, there was also a significant element of competition to explore the sun and the solar system's other planets using robots. Like the space race, this competition flourished during the 1960s and died during the 1970s as the two competitors sought cooperation instead. But both nations continued sending out their exploring machines, albeit at a slower pace than during the golden age of the space race. Eventually, they were joined by the Japanese and European space agencies. By the end of the century, every planet but Pluto had received a robotic visitor, and Mars had been subject to a virtual bombardment of anthropogenic hardware designed to find out if there had ever been Martians.

The Golden Age of Robotic Exploration, 1960–1973

During space exploration's first decade, the United States and the Soviet Union each focused attention on the sun, Venus, the moon, and Mars. The moon received the most visitors, as both nations sent robots to prepare the way for manned landings, while perhaps the biggest scientific triumph and more than a few crushing disappointments came from Mars probes. The United States used three series of vehicles to carry out solar research during the decade.

Scientists pose with Explorer 1, *the first successful U.S. satellite, 31 January 1958. The satellite was built by the Jet Propulsion Laboratory (JPL), then an Army facility operated by California Institute of Technology. In December 1958, JPL became part of NASA. (NASA)*

The Orbiting Solar Observatories (OSOs) were placed into Earth's orbit above the atmosphere that limited ground-based observation, and they were used to study the sun directly through imagery taken at various wavelengths. OSOs contributed to understanding solar structure and activity and helped fill in gaps in the solar spectrum caused by filtering in the atmosphere. In addition to their solar studies, OSOs carried out early X-ray surveys of the galaxy. The National Aeronautics and Space Administration (NASA) put the first of these into orbit in March 1962 and the last, OSO-8, in 1975.

The second and third series of vehicles, the Explorer and Pioneer probes, carried out a wide variety of missions and also served as experiments in spacecraft design. The first U.S. satellite, *Explorer 1*, carried an instrument designed to determine the level of cosmic radiation above the atmosphere; its data, along with data from *Explorer 3*, enabled James Van Allen to discover the existence of belts of trapped radiation surrounding Earth. More than fifty Explorer-class satellites flew between 1958 and 1975, when NASA terminated the series of simple, inexpensive satellites.

The larger, more complex Pioneers were initially launched toward the moon and equipped, like many of the Explorers, to produce data on magnetic fields, charged particles, and gamma radiation. *Pioneer III*, launched in 1958, was the first to return scientific data and resulted in the discovery that Earth had two radiation belts. *Pioneer IV*, launched the following year, was the first U.S. probe to leave Earth's orbit, passing the moon in March 1959 and going into a solar orbit. *Pioneer V*, launched in 1960, was placed in an orbit between Earth and Venus to provide space weather reports. A series of launch vehicle failures delayed the next successful Pioneer launch until 1966, when *Pioneer VI* was launched into solar orbit to study the interplanetary environment. Three more Pioneers were also dedicated to this research during the remaining years of the decade.

The Soviet Union did not prepare missions designed specifically to carry out solar or interplanetary space research. Instead, as the United States would also do, it equipped planetary probes with instruments for studying the space environment while en route to their planetary objective. The first Soviet objective was to crash-land probes on the moon. Three attempts in 1958 never reached space due to launch failures, while a January 1959 launch publicly known as the Cosmic Rocket, *Luna 1* more formally, became the first Earth vehicle to reach escape velocity. But it missed the moon by a wide margin, going into solar orbit. *Luna 2*, the sixth Soviet attempt to hit the moon, successfully crashed on the lunar surface in September 1959, depositing a set of metallic spheres carrying Soviet emblems. *Luna 3*, originally launched as the Automatic Interplanetary Station in October, was the first major Soviet planetary triumph. Instead of crashing into the moon, *Luna 3* was a flyby mission, and it radioed back the first photographs of the far side of the moon. Because of the way the moon rotates in its orbit, only one side of the moon is ever visible from Earth, so *Luna 3* gave humanity its first view of the invisible side of the moon.

The United States also sought to crash robots into the moon, using a series of probes named Ranger and designed by the Jet Propulsion Laboratory

(JPL). The first two Rangers, launched in 1961, never left Earth's orbit. The third missed the moon due to guidance failures and incorrect ground commands, while the fourth suffered failure of its internal clock, never deployed its solar panels, and failed to return any scientific data. It did hit the moon, however. The first fully successful Ranger was *Ranger 7*, which launched in July 1964 and returned 4,316 images as it approached its impact on 31 July. *Ranger 7's* success was made more dramatic by the thirteen consecutive failures that had preceded it; in addition to the six failed Ranger missions, several Pioneer missions to orbit the moon had also failed.

After *Ranger 7*, the fortunes of the American lunar exploration program improved dramatically. Two more Rangers were completely successful, and the Ranger series was closed down in favor of two new lunar probes: Lunar Orbiter, a vehicle based upon a then-classified military reconnaissance satellite (SAMOS) intended to photographically map the moon, and Lunar Surveyor, a soft-landing craft designed to return information about the moon from its surface. Lunar Orbiter was stunningly successful, with all five vehicles orbiting the moon and returning hundreds of one-meter resolution photographs of potential Apollo landing sites. *Lunar Orbiter I* also returned one of the most famous pictures of the space race, a black and white photograph of Earth rising over the moon. Of the seven Surveyors, five successfully landed on the moon. *Surveyor 1* marked the first American soft landing attempt and success when it arrived on 2 June 1966 and followed the first Soviet soft landing by about five months; *Luna 9*, the twelfth Soviet soft-landing attempt, had landed on 3 February. Surveyor's primary mission was to generate engineering data for design of the manned lunar module, and several Surveyors had claws designed to determine the ability of the lunar soil to support weight.

Planetary Science

Oddly enough, the first unequivocal success in the American robotic planetary exploration program was a modification of the JPL's unhappy Ranger series that NASA sent to Venus. Known as *Mariner 2* (*Mariner 1* was destroyed in a launch vehicle accident), the vehicle was launched toward Venus in August 1962 and made its flyby of the planet on 14 December. *Mariner 2* returned forty-two minutes of data on the Venusian atmosphere and surface, permitting mission scientists to determine that the surface temperature was at least 425 degrees Celsius, hotter than molten lead, and there was very little difference in temperature between day and night sides of the planet. It also discovered a permanent cloud layer twenty kilometers thick surrounding the planet. Even this limited data made it clear that no life, at least not of a form like any on Earth, could be found on Venus, and the planet became much less interesting after this. A third Venus Mariner scheduled for launch during a 1964 launch window was canceled.

The first attempt to reach Venus had actually been the Soviet probe *Tyazhelyy Sputnik* in 1961, but it did not make it out of Earth's orbit. During the same July and August 1962 launch window that *Mariner 2* had used, the Soviets sent three missions to Venus (two landers and one flyby mission). Due to what would become a persistent problem for the Soviet program, none of the three made it out of Earth's orbit. In each case, the stage that was supposed to boost the vehicle out of orbit around Earth failed, and the three probes burned up in the atmosphere as their orbits decayed. During the next Venus launch window, in February 1964, the USSR sent three more missions to Venus, two of which did not make it out of Earth's orbit. The third, *Zond 1*, was put on the proper trajectory to Venus but failed along the way. Finally, during the next launch window in November 1965, the Soviet Union managed to get a robot explorer to Venus. *Venera 2*, one of three probes originally intended to fly to Mars the previous year, conducted its flyby on 27 February 1966. But the Soviet ground controllers were never able to retrieve the probe's data; apparently, after controllers ordered it to turn on its science instruments, it overheated. The second of the three converted Mars vehicles successfully left orbit, got most of the way to Venus, dropped its sterilized lander into the planet's atmosphere, and then again malfunctioned before controllers could retrieve any data. The third attempt in 1965 did not reach Earth's orbit.

The Soviet Union finally succeeded in reaching Venus in 1967 with a probe named *Venera 4*. Launched on 12 June 1967, this explorer's primary science mission was to place a lander on the planet's surface. It released the lander on 18 October 1967. The lander entered safely and, slowed by a parachute, transmitted for ninety-three minutes as it descended through the Venusian atmosphere. It ceased transmitting when crushed by the high pressures it encountered, seventy-five times those at Earth's surface. The lander reported that unlike Earth's atmosphere, Venus's was more than 90 percent carbon dioxide with no measurable nitrogen. It also found that Venus had a very weak magnetic field, which *Mariner 2* had missed because it had flown past Venus at too great a distance. The United States followed *Venera 4*'s successful mission with another Mariner, *Mariner 5*, that flew by Venus late in 1967 at a closer distance than *Mariner 2* had. *Mariner 5* verified *Venera*'s report of a weak magnetic field, and its data also suggested that Venus's surface pressure was actually a hundred times that of Earth's. This result indicated that *Venera 4*'s lander had been crushed well before actually reaching the surface. Several more Venera landers dispatched by the Soviet Union during the late 1960s were successful at returning more data about the Venusian atmosphere, and in 1970 the Soviets finally succeeded at depositing a lander on the surface. It functioned for twenty-two minutes, reporting that the surface temperature was 425 degrees Celsius and pressure was about ninety atmospheres, demonstrating that the American remote-sensing instruments had been essentially correct.

Venus had long been considered Earth's twin in the solar system because it was approximately the same size and mass, but the results produced by the Mariner and Venera explorers made it clear that Venus had developed in a radically different direction. The robots' data forced planetary geologists to revisit theories of the solar system's evolution to try to understand how two relatively similar bodies could diverge so enormously in their surface conditions. These missions did not provide definitive evidence to support any one new theory, however, and future missions were equipped to provide more information on the composition, temperature, and circulation of the Venusian atmosphere and to produce radar maps of the surface.

Mars, the planet that had been the target of Percival Lowell's investigations and that held the greatest hope for life beyond Earth, had also been the first target of planetary exploration and the source of the greatest disappointments. The earliest attempts to send space probes to Mars were failures, establishing a pattern that would last four decades. In 1960, the Soviet Union tried twice to send robot explorers on flyby missions to the Red Planet. Both of these missions suffered booster failure during launch and burned up in Earth's atmosphere. During the 1962 launch window, the USSR tried to send two flyby missions and a lander to Mars, but all three failed. A single mission during 1964 also failed, and the nation did not try again until 1969, when it dispatched two Mars orbiters that also failed.

The first robot to successfully reach Mars was the JPL's *Mariner 4; Mariner 3*, an identical spacecraft, had gone into a useless solar orbit after failing to separate from its launch vehicle properly. *Mariner 4* flew past Mars in July 1965, sending back twenty-one photographs of a barren, moonlike surface pockmarked with large craters. The probe also sent back data indicating that the Martian atmosphere was very thin, with a surface pressure of 4–7 millibars, or 0.1 psi (pounds per square inch); consisted of mostly CO_2; and had an estimated daytime surface temperature of –100 degrees Celsius. Two more Mariner flybys in 1969 reinforced these findings, demonstrating that *Mariner 4*'s readings had been accurate. These temperature and pressure measurements made it very clear that no life survived on Mars's surface, and there was no liquid water. Certainly no form of life known to Earthlings could live under those conditions.

There was clearly no life currently on Mars, but scientists could not rule out the possibility that there had never been life there. The photographs returned by the Mariners provided no indication that there had ever been intelligent life; instead, they demonstrated that the so-called canals were natural ravines, crevasses, and canyons, not engineered structures. But from orbit scientists could not disprove the idea that microbial life might have existed. To accomplish that, scientists needed to land on the surface and test samples of the Martian soil, and the effort to land robots dominated the Mars programs of both the United States and the Soviet Union during the 1970s.

A Decade of Retrenchment amid Exploration: The 1970s

The spectacular American victory in the moon race of 1969 signaled a dramatic reduction in the space efforts of both the United States and the Soviet Union. Both manned and robotic exploration were cut back substantially, and there were far fewer missions launched during the decade. Nonetheless, the decade witnessed spectacular achievements by the robot explorers, including soft landings on Mars and flybys of Jupiter and Saturn that produced magnificent photographs and substantial changes in human knowledge of these planets.

Just as the first missions to Mars had been launched by the Soviets, the first attempts to land on Mars were completed by the Soviets. *Mars 2* and *Mars 3*, both of which were launched in May 1971, each consisted of an orbiter to produce photographic maps of the Martian surface and a lander to descend to the planet surface. *Mars 2*'s lander crashed on Mars during descent, gaining for itself the distinction of being the first engineered artifact to reach the surface, but *Mars 3* landed successfully on 2 December 1971. Unfortunately, it stopped sending data twenty seconds after landing. The two orbiters recorded severe sandstorms in the lander's vicinity, and the Soviet engineers guessed that static electricity generated by the dust storms shorted out the lander's electrical system. During the next launch window in 1973, the USSR sent two more spacecraft, *Mars 6* and *Mars 7*, as well as an orbiter named *Mars 5* that was supposed to act as a communications relay for the landers. *Mars 5* entered planetary orbit successfully in February 1974, returning forty-three photographs. But it only functioned for three weeks, failing before the two landers arrived. The *Mars 6* lander, which descended to the Martian surface on 12 August 1974, disappeared right after it fired its landing rockets. The entry had appeared perfect, and the spacecraft's controllers never determined what happened to the vehicle. *Mars 7*'s lander, to finish the depressing tale, suffered a computer failure that prevented its retrorocket from firing, and instead of landing on Mars it went into orbit around the sun. After these failures, the Soviet Union did not send another explorer to Mars until 1989.

The United States had also launched a Mars probe, *Mariner 9*, during the 1971 launch window. This was the first spacecraft to orbit Mars instead of performing a flyby. Placed by the JPL's navigators into a polar orbit, it eventually provided photographic maps of the entire surface. But when it first arrived at Mars, the planet had been completely covered by a dust storm, causing the mission directors to reprogram it to wait for the dust to clear. In the meantime, atmospheric scientists used some of its instruments to examine the impact of the dust on the thermal structure of the atmosphere. Like Earth's, Mars's atmosphere is heated from below, by the surface, which is heated by sunlight. Cut off from the sun, the Martian surface cooled rapidly during the storm, while the suspended dust absorbed the incoming light and heated. This caused the atmosphere to warm to the same temperature as the surface, which helped sustain the dust layer for months. To the assembled scientists, this seemed to be very strange behavior and caused them to begin to examine the behavior of dust and other particles in Earth's atmosphere after *Mariner 9* ran out of fuel and was shut down.

For Mars enthusiasts, *Mariner 9*'s greatest result was that its photographs clearly revealed that water had once run freely on the Martian surface. *Mariner 9* imagery showed large-scale hydrologic structures in the surface, complete with runoff channels, outwash plains, and vast river deltas. The water itself, however, was entirely missing. Data from one of the spacecraft's instruments indicated that the polar ice caps were carbon dioxide ice—dry ice, not frozen water—leaving open the question of where the water went. This remained controversial for almost three decades.

The last American Mars program of the 1970s was named Viking after the earliest explorers known to have reached North America. It consisted of two orbiters and two landers. NASA had designed the program's schedule so that the two landers would reach Mars during 1976, the nation's bicentennial year. The orbiters carried cameras, sensors to detect water vapor in the atmosphere, and sensors to produce maps of the planet's heat distribution. The landers were much more complex. In addition to high-resolution cameras, the landers carried mass spectrometers to produce precise analyses of the atmosphere's chemical combination, a complete weather station, a seismometer, an extending arm to take soil samples, and the most important instrument of all, a miniature biological laboratory to look for signs of life in the soil samples. The landers also contained two radioisotope thermal generators to provide electric power instead of relying on solar panels. The Martian dust would quickly cover solar panels, and the design team wanted the landers to function for many months.

Viking 1 arrived at Mars on 19 June 1976 without incident. The next day, the orbiter started sending back images of the landing area that NASA's mission planners had chosen, Chyrse Planitia. The photographs indicated that the terrain was rougher than expected, and while the planners had hoped to land on Independence Day, the lander's descent was rescheduled while they chose a different site. On 20 July, the lander descended to its new target successfully and began sending back the first color photographs of Mars, revealing an unexpectedly yellow-red sky. Eight days later, the sample arm deposited soil samples in the biological laboratory. Unfortunately, the results of the analysis were mixed, producing neither proof of life nor definitive evidence that there was none. Like the current location of Mars's water, the results of this experiment remained controversial for decades. The *Viking 1* lander sent weather reports back until February 1983.

The second Viking mission went into orbit around Mars in August 1976. Its lander descended to the planet's surface the following month, landing 4,600 miles from *Viking 1*, in the Utopia Planitia region on the edge of the polar ice cap. The mission planners had chosen the location to improve the chances of finding life, since they believed that the ice caps, the surfaces of which consisted of frozen CO_2, might also contain whatever water Mars had under the carbon dioxide. At this site too, however, the biology instrument produced inconclusive results. The debate of whether life had existed on Mars thus continued unabated. *Viking 2*'s lander continued to operate until April 1980, while its orbiter died two years earlier. The orbiter passed within twenty-eight kilometers of the Martian moon Deimos in 1977, producing photographs greatly superior to those available from ground-based telescopes.

NASA's attention turned away from Mars for many years after the Vikings failed to find life, as its next scheduled missions were to the outer planets. Between the orbits of Mars and the next planet in our solar system, the gas giant Jupiter, lies a large asteroid belt that astronomers generally believe is a remnant of the solar system's formation billions of years ago. It has served as a convenient dividing line, permitting an easy definition of inner and outer planets. In the early space age, it also generated some fear that it would prevent exploration of the outer planets by posing a collision threat to the exploring machines, but this proved not to be. Instead, the small number of probes sent beyond Mars have been highly successful, sending back by radio images of enormous beauty and data that forced a reexamining of theories of the solar system's formation.

The first mission to the outer planets was *Pioneer 10*, a highly modified version of the same vehicle sent on solar explorations in the mid-1960s. Among the modifications to the vehicle was the replacement of solar panels with radioisotope thermal generators, making *Pioneer 10* the first nuclear-powered

Werner von Braun (1912–1977)

Werner von Braun was born on 23 March 1912 in Wirsitz, Germany. Like Tsiolkovskiy, von Braun was influenced by Jules Verne and by H. G. Wells and began to dream about space travel at a very young age. Von Braun became involved in the German Rocket Society in 1929, and in 1932 he joined the Wehrmacht to help develop ballistic missiles. He was also able to continue his education and completed the Ph.D. in 1934. During World War II, von Braun and his team of engineers successfully developed the A-4, also known as V-2, ballistic missile at Peenemünde. During 1944 and 1945, large numbers of these were manufactured by slave laborers at an underground facility called the Mittelwerk. Several hundred were fired at targets in Great Britain during the waning months of the war. The American Project Paperclip moved von Braun and several hundred of his engineers to the United States in 1945, where they became the core of the Army Ballistic Missile Agency and later NASA's Marshall Space Flight Center in Huntsville, Alabama. Von Braun was Marshall's first director, serving from 1960 to 1970. He was also the nation's most prominent advocate of an expansive space exploration program. He and his team successfully launched the first U.S. artificial satellite, *Explorer 1*, in January 1958; designed the intermediate-range ballistic missile Jupiter; and developed the Saturn I and V rockets to place humans on the moon. In 1970, with the cancellation of the Apollo program, von Braun moved to NASA headquarters in Washington, D.C., to aid the agency's long-range planning. He retired in 1972 and went to work for Fairchild Industries. Von Braun died in Alexandria, Virginia, on 16 June 1977.

American space probe. Its mission was to photograph the planet and its moons, map Jupiter's gravitational and magnetic fields, and investigate its atmosphere and interior. Launched toward Jupiter in March 1972, *Pioneer 10* passed safely through the asteroid belt and made its closest approach to Jupiter on 3 December 1973, sending back images better in quality than any Earth-based telescope could provide and demonstrating that Jupiter was primarily hydrogen. However, Jupiter was mostly liquid, which was a surprise. *Pioneer 10* also photographed the Jovian moons Callisto, Ganymede, and Europa, all visible from Earth with telescopes, but too small for astronomers to see in detail. After passing by Jupiter, it crossed the orbits of both Saturn and Neptune, demonstrating that Jupiter's magnetic wake extended all the way to Saturn's orbit. *Pioneer 10* officially left the solar system on 13 June 1983, and by 2005 it was more than 7 billion miles from Earth. NASA ceased routine contact with the vehicle in 1997 but maintained intermittent contact to collect engineering data from it. NASA engineers last heard from *Pioneer 10* on 23 January 2003 and do not expect to receive any more transmissions from the craft.

Pioneer 10's twin, *Pioneer 11*, was launched in April 1973 to conduct fly-bys of both Jupiter and Saturn. It reached Jupiter in December 1974, deliberately passing much closer to the surface and over the poles, which *Pioneer 10* had not. *Pioneer 11* also used Jupiter's gravity for a slingshot maneuver to change its direction and accelerate the craft toward Saturn, which it reached in 1979. It was the first space probe to reach Saturn, and it made several discoveries, including an unexpected additional ring and a new moon. The data it transmitted indicated that like Jupiter, Saturn was primarily liquid hydrogen. *Pioneer 11* passed out of the solar system in February 1990 still functioning. NASA's last communication with the vehicle was in 1995, at which point two instruments still worked. Earth's motion carried it out of sight of *Pioneer 11*'s antenna after that, permanently ending contact.

As the first travelers to leave Earth's solar system, the two Pioneers carried plaques designed to convey information about Earth in case they happened to be picked up by intelligent beings. The plaques were designed by astronomer Carl Sagan and carried a diagram of the solar system showing the location of those who had launched the vehicle and the Sun's relative position within its galaxy, a silhouette of the vehicle, and idealized drawings of its builders, a nude man and woman. It also contained a sort of "universal clock" that intelligent aliens could decode to figure out how long ago the Pioneers had been launched, as the two vehicles would not reach another star for several million years.

These two spectacular successes were followed by two more, a pair of craft originally envisioned as part of a five-vehicle grand tour of all of the outer planets. Named *Voyager 1* and *Voyager 2*, these spacecraft were significantly larger and more complex than the two Pioneer craft had been. Their design was derived from the Mariner series probes and, like the Pioneers, were nuclear powered. Although *Voyager 1* was launched after *Voyager 2*, it arrived at Jupiter first, entering Jupiter's moon system on 10 February 1979. It discovered that Jupiter also had rings like Saturn's although much thinner, and it provided detailed images of the five largest Jovian moons, Amalthea, Io, Europa, Ganymede, and Callisto. Io had not been imaged by the Pioneers, and it provided the most interesting discoveries of the mission. It proved to be highly volcanically active, unlike any other body explored by the robots. *Voyager 1* also discovered two new moons, which were named Thebe and Metis. The JPL's navigators redirected *Voyager 1* toward Saturn, where it arrived in November 1979. At Saturn it discovered five new moons orbiting the gas giant and another new ring, and it photographed six of the known moons, Titan, Mimas, Enceladus, Tethys, Dione, and Rhea. These, according to its data, appeared to be mostly water ice. *Voyager 1*'s controllers decided to redirect the probe to make a closer pass to Titan, where it discovered that this very large moon had an atmosphere that was mostly nitrogen, with a surface pressure about 1.6 times Earth's. Because of this redirection, *Voyager 1* could not be sent to Uranus and Neptune, however, and instead it began what NASA eventually decided to call the Voyager Interstellar Mission to gather data on the outermost limits of the sun's influence. In February 1998, *Voyager 1* became the most distant human-made object. In 2005, *Voyager 1*'s science team announced that it had finally passed through the termination shock, where the solar wind slows to subsonic speeds, at the edge of the solar system. It was ninety-four astronomical units—or 94 times the distance from Earth to the sun—from Earth.

Voyager 2's major contribution to planetary exploration occurred when it became the first probe to reach Uranus, although that planet had not been part of its original mission profile. Instead, the vehicle's very successful encounters with Jupiter and Saturn left it still fully functional and in a trajectory that could be modified to encounter Uranus in four and a half years. The controller team redirected it to Uranus without incident, and *Voyager 2* passed Uranus in January 1986 at a distance of 71,000 kilometers. It found ten new moons that were invisible from Earth, a pair of new rings, and a strangely oriented magnetic field. The vehicle's instruments also suggested that Uranus had a water ocean boiling 800 kilometers or so below the cloud surface, another apparently unique feature of the planet. With the probe still functioning, NASA redirected it again for a voyage to Neptune. When it reached Neptune in August 1989, all of its instruments still worked. *Voyager 2* reported that Neptune was mostly hydrogen, but it also contained a good deal of methane that gave it a bluish color. *Voyager 2* also found that Neptune's atmosphere was more active than expected given its distance from the sun. After its Neptune encounter, *Voyager 2*'s instruments were put in low-power mode to conserve energy, potentially permitting the vehicle to continue sending data until 2020, when its Radioisotope Thermoelectric Generators would no longer generate sufficient power to broadcast a signal across the 12 billion or so kilometers between the probe and Earth. Like its twin, *Voyager 2* was still sending data on interstellar conditions as of 2006.

The last four missions launched during the 1970s were to Venus, in the 1978 launch window. The United States and the Soviet Union each launched two probes. The American missions, *Pioneer Venus 1* and *Pioneer Venus 2*, consisted of an orbiter and a multiprobe lander. *Pioneer Venus 1* achieved orbit on 4 December 1978 and began using a radar system to map most of the surface. The south polar region could not be mapped from the craft's early orbit, and the radar was turned off in 1981. By 1991, the orbit had decayed in a way that permitted mapping the south pole as well, and the radar was switched back on. The orbiter burned up the following year. Its radar map of the planet showed that Venus was smoother and more spherical than Earth, further refining the image of Venus as very unlike Earth. The multiprobe, *Pioneer Venus 2*, consisted of a bus, a large probe, and three smaller probes. The bus released the large probe in November 1978 and the three small probes four days later. All five sections of the probe reached Venus on 9 December, with the large probe entering first. It descended on a parachute to impact, which it was not designed to survive. The three smaller probes, which did not have parachutes, fell more rapidly through the atmosphere, but two of them unexpectedly survived and continued transmitting data, the longest for sixty-seven minutes before its batteries died. The multiprobe returned a great deal of data on the planet's atmosphere, including the unexpected discovery that below thirty kilometers, the atmosphere was relatively clear.

The two Soviet missions in 1978 consisted of flyby bus/lander pairs and were both partially successful. Each of the landers carried color television cameras and soil sampling instruments in addition to instruments designed to take atmospheric readings. Both landers reached the surface and transmitted data, with *Venera 12* setting a record of 110 minutes of information. But the lens covers over the television cameras on both landers failed to open, so they did not produce much-desired pictures of the surface, and both soil samplers failed as well.

The Nadir: The 1980s and Robotic Exploration

Fewer missions were launched during the 1980s than in the decades before or after, but they carried a higher success rate. They also reflected an increase in the number of space-faring nations, with the European Space Agency (ESA) and Japan each launching their first "far travelers," to borrow a term from NASA's Oran Nicks, toward rendezvous with Halley's Comet. The missions of the 1980s reflected increasing cooperation, despite a return to harsh Cold War rhetoric, as space agencies struggled to make the most of shrinking budgets and reduced launch opportunities by seeking international partnerships.

During the first half of the decade, only the Soviet Union continued to launch probes, sending two more copies of the relatively successful *Venera 11* and *Venera 12* landers to Venus in 1981. *Venera 13* arrived on Venus in late February 1982, and its cameras and soil sampler both worked correctly, returning eight color panoramas of the area around the lander and demonstrating that the soil was similar in composition to basalt on Earth. *Venera 14*, which arrived in March 1982, also performed flawlessly, returning data that corroborated *Venera 13*'s while also showing that atmospheric conditions at its landing site were considerably different. During the next launch window in 1983, the USSR sent two radar mappers designed to extend the map of Venus produced by *Pioneer Venus 1*. These missions were partially successful, producing maps of the region north of thirty degrees north latitude to the pole, but their orbits did not permit mapping of the rest of the planet. They also carried a pair of East German infrared spectrometers that produced a temperature atlas of the atmosphere.

In 1984, the Soviet Union launched a pair of highly complex missions, *Vega 1* and *Vega 2*, that were designed to continue its Venus investigation while also addressing a new target, Halley's Comet. This was a cooperative project, involving six other Soviet bloc nations as well as France and West Germany. The two identical spacecraft were intended to deposit advanced landers on Venus, deploy two balloons into the atmosphere, and then use Venus's gravity to slingshot themselves into a rendezvous trajectory with the comet. The balloon gondolas were French and carried nephelometers to measure aerosols in the atmosphere. Both of these worked perfectly, drifting in the Venusian atmosphere and returning data for more than forty hours before their batteries failed. The two landers arrived in June 1985 and, like the balloons, returned a great deal of data. The two Vega spacecraft buses, after looping around Venus, proceeded to rendezvous with Halley's Comet, where they studied its gas and dust cloud during the comet's March 1986 encounter with Earth.

Japan and the ESA joined the Soviet Union in the Halley's Comet study with three more probes, whose missions had been coordinated in advance. Japan launched its first vehicle, *Sakigake*, in January 1985. This was a test launch and was not directed to a close pass of Halley's Comet; instead, *Sakigake*'s data was used as a reference for eliminating the effects of Earth's atmosphere on the transmissions of the two probes that followed it, the ESA's *Giotto*, and Japan's *Suisei*. *Giotto*, a spin-stabilized probe that was the ESA's first mission, was launched on 2 July 1985 toward a March 1986 flyby of Halley's Comet. *Giotto* had the closest flyby of the comet's core scheduled, and the probe had been protected against dust impact with a pair of thick shields. After course corrections based on tracking data from the two Soviet Vegas, *Giotto* passed within 605 kilometers of the comet. During *Giotto*'s approach, it was struck an average of one hundred times per second by dust particles but remained fully functional and returned 2,000 images of the comet and a great deal of other data. Later, the ESA steered it toward an encounter with Comet Grigg-Skjellerup, which it passed within 200 kilometers. The Japanese *Suisei*, launched in August 1985, was equipped to make ultraviolet photographs of the comet, the hydrogen gas cloud that surrounded the nucleus, of Halley's Comet. *Suisei*'s closest approach was on 8 March 1986.

In contrast to its extremely successful Venus missions earlier in the decade, the Soviet Union's last two Mars missions marked the continuation of its jinxed relationship with the Red Planet. This time, Project 5M, conceived in the early 1970s as a mission to land on Mars and return soil samples to Earth, resulted in two highly complex orbiter/lander pairs being sent to the Martian moon Phobos. Named *Fobos 1* and *Fobos 2*, the orbiters were to go into Mars's orbit to conduct long-term geophysical surveys, while the landers were supposed to descend to Phobos's surface. Each of the Fobos pairs carried twenty-two scientific instruments, making them the largest payloads ever sent to Mars. Unfortunately for the many scientists involved, the Fobos missions were no more successful than the USSR's earlier attempts. *Fobos 1* failed en route to Mars due to a programming error that disabled the orientation system, preventing the solar panels from recharging the spacecraft's batteries. *Fobos 2* went into Martian orbit late in January 1989 and operated successfully for two months. It was supposed to place the lander on Phobos in early April but did not respond to a routine communications session at the end of March. A weak signal received later that day suggested that *Fobos 2* had begun to tumble, preventing the craft from generating power or properly aiming its antennae at Earth. It was never heard from again.

Finally, in 1989, NASA returned to planetary exploration after eleven years of fiddling with its Space Shuttle. Its first mission, *Magellan*, was dispatched on 4 May 1989 by Space Shuttle *Atlantis* on a radar mapping mission to Venus. Due to the low energy available from a shuttle launch, *Magellan* did not reach Venus until August 1990. It carried out three cycles of mapping. Originally intended to image 70 percent of the surface at resolutions of less than 300 meters, *Magellan* actually achieved nearly 98 percent coverage of the Venusian surface. Its images showed evidence of vulcanism, turbulent surface winds, lava domes and tunnels, and tectonic movement, and they indicated that 85 percent of the surface consisted of volcanic flows. At the end of *Magellan*'s mission, it was commanded to enter the atmosphere to return information on aerodynamic drag, and it burned up on 13 October 1994.

Space Shuttle *Atlantis* launched a second mission in October 1989, this one the highly ambitious Galileo mission to Jupiter. *Galileo* was an orbiter/atmosphere-penetrator pair that was intended to study the planet's satellites, atmosphere, and magnetic field. It had been scheduled for launch in 1986, but due to the explosion of Space Shuttle *Challenger*, it had to be postponed and lost the most advantageous trajectory to Jupiter. After its 1989 launch, therefore, *Galileo* required an intricate series of gravity assists to reach Jupiter: a Venus assist and two Earth assists, which enabled a short Venus mission and an Earth mission. After looping around Venus in February 1990, it carried out a mission to Earth on 8 December 1990, where it clearly detected life, although it failed to comment on the life form's intelligence level (it detected human-generated pollution). It did a flyby of the asteroid Gaspra in October 1991 and then headed back to Earth for its second loop before finally heading off to its original mission in December 1992. *Galileo*'s high-gain antenna did not deploy properly after launch, probably as a result of damage from shipping the vehicle into storage during the delay, reducing the information it could return to about 70 percent of that intended. But it nonetheless achieved its first major scientific coup of the flight in July 1994, when it provided stunning pictures of the comet Shoemaker-Levy 9 hitting Jupiter.

Galileo released its atmosphere-penetration probe on 13 July 1995, still 80 million kilometers from the planet, and the penetrator entered the Jovian atmosphere on 7 December. It returned fifty-seven minutes of data, located an intense radiation field above Jupiter's perpetual cloud cover, found that there were almost no organic compounds in the atmosphere, and measured wind velocities up to 640 meters per second. The orbiter reached Jupiter the next day, beginning what was originally intended to be a twenty-two-month mission of surveying Jupiter and its moons Ganymede, Europa, and Callisto. But because the orbiter remained fully functional at the end of this period, its mission was extended two years to gain additional flybys of Europa, Callisto, and the volcanically active Io. Then, unable to give up on a functioning vehicle, NASA extended its mission to 2000, then extended it again to September 2003.

Galileo made striking discoveries during its eight years at Jupiter. It found that Jupiter has almost the same helium abundance as the sun, which makes helium through fusion. Four of Jupiter's moons, Callisto, Ganymede, Europa, and Io, have organic compounds—that is, compounds based on the carbon atom. *Galileo* presented evidence that there is a liquid water ocean under the ice surface of Europa, and it determined that Ganymede has a magnetic field, which no other moon in the solar system does. Because of the striking evidence produced by *Galileo*, many scientists began to think that life might exist under Europa's ice surface and that *Galileo* itself posed a risk of contamination to Europa when it ran out of fuel and became uncontrollable. NASA decided to eliminate this possibility by deliberately steering *Galileo* into Jupiter's atmosphere to burn up in September 2003, preserving whatever might exist on Europa until a properly sanitized lander could be built and sent there. Technologically, a robotic search for life under Europa's thick ice would be a daunting challenge and was well beyond NASA's budget expectations. But it gave space enthusiasts more hope that life existed beyond Earth.

The 1980s witnessed the fewest robotic missions of any decade since the space age began, but they were scientifically productive nonetheless. The several missions to Halley's Comet reflected exploration of a new class of interplanetary phenomena, and the highly successful *Galileo* and *Magellan* voyages allowed much more detailed examinations of Jupiter and Venus. The higher success rates

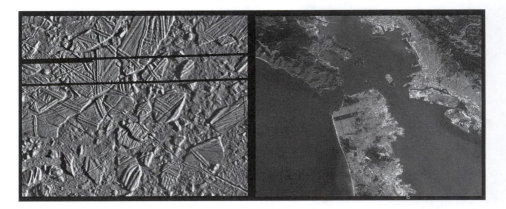

The left frame shows the surface of Europa taken by the Galileo probe, showing 'ice rafts' crisscrossed by stress fractures. The right frame is a Landsat photograph of the San Francisco Bay area. The two images are the same scale, revealing the huge size of Europa's ice blocks. (NASA/JPL-Caltech/Cornell)

of the vehicles represented a significant maturing of space technologies, and the growing number of participating nations indicated increasing technological capacity. Whereas the United States and the Soviet Union suffered repeated failures in their early space programs, the European and Japanese space efforts began with impressive successes.

The tiny number of missions during the 1980s, particularly American missions, also represented substantial problems for the space science community. The United States had gone ten years without new missions, making it virtually impossible to recruit new researchers to space science. It was unrealistic to expect graduate students to undertake space research if that meant spending ten or more years waiting for a launch opportunity, and the result was an aging and shrinking community of researchers. Because space research was funded through a political process, shrinkage of the research community meant potential loss of funding; without active researchers lobbying for new missions, there would be even fewer robot explorations in the future and thus further reduction in the size of the community. Reversing this vicious cycle became a major priority during the 1990s.

The Decade of "Faster, Better, Cheaper"

The year 1990 marked two relatively successful missions, one by Japan with its first lunar probe, *Hiten*, and the other the ESA's *Ulysses*. No launches graced 1991, and in 1992 NASA managed one launch that became the agency's nadir, the *Mars Observer*. Based on a military communications satellite, the *Mars Observer* had been intended to demonstrate cost reduction from synergy, the use of an existing robot bus applied to new missions. But it still cost nearly $1 billion, and it malfunctioned right before going into Martian orbit in August 1993, disappearing without a trace. *Mars Observer*'s failure permitted NASA's new administrator, Daniel Goldin, to force an agenda for change on the agency. Goldin believed that the agency had to get back to sponsoring smaller, simpler, less expensive, and most of all *more* robotic missions to restore vitality to the research community. During the 1960s, the agency had used the cheap Explorer series to carry out a substantial program of geophysical research, and this was the kind of thing that Goldin wanted to bring back. There was one more large, expensive mission left in the development pipeline, a joint U.S.-ESA mission known as Cassini/Huygens; other than this, Goldin intended to hold future planetary missions to $150 million. He named his new class of missions Discovery, and his management methodology became known as "faster, better, cheaper."

The first American mission after the *Mars Observer* debacle was a classified joint mission undertaken by the Defense Department's Ballistic Missile Defense Agency. Known as *Clementine*, this vehicle was the first American mission to the moon since *Explorer 49* in 1973. It was also the first demonstration of the "faster, better, cheaper" concept. *Clementine*'s purpose was to demonstrate a set of low-cost spacecraft technologies and instrumentation based on microminaturization. Despite weighing only 424 kilograms, it carried ten scientific instruments. It was launched in January 1994, and it became the first vehicle to image the entire lunar surface, including the poles. It became somewhat famous in the research community when the Defense Department revealed the mission in 1996. *Clementine*'s data suggested that the moon had water ice in a perpetually shaded crater in its south polar region, a discovery that was entirely unexpected.

A later mission in NASA's Discovery program, *Lunar Prospector*, was sent to the moon in 1998 to verify *Clementine*'s finding. *Prospector*'s instrumentation seemed to confirm that there were around 6 billion tons of water ice, and the probe also found that the moon appeared to have a very small, and entirely unexpected, iron core. *Prospector* was deliberately crashed into one of the craters suspected to contain water in hope of achieving further confirmation of the water discovery, as deep skepticism over the find still existed, but its crash failed to provide any further evidence. *Prospector* also delivered the remains of the planetary geologist Eugene Shoemaker, a central figure in the Apollo geology effort, to a very final resting place on the moon.

The 1990s also witnessed a renaissance of solar exploration in the form of the Inter-Agency Solar-Terrestrial Physics Program, encompassing the United

States, Japan, Russia, the ESA, and the Czech Republic. This project envisioned a total of twelve spacecraft, many of which were to study the interaction between Earth and the sun from the Earth-Sun Libration Point, or L-1. At L-1, the gravitational fields of Earth and the sun exactly balance, and a probe could therefore stay there potentially forever. The first of these robots, NASA's *Wind*, arrived at L-1 in November 1996 to study the interaction between the solar wind and Earth's magnetosphere in three dimensions. The second, ESA's 1995 *Solar and Heliospheric Observatory (SOHO)*, was aimed at investigating the sun's atmosphere and seismic activity and was still producing data in 2003. The third, NASA's *Advanced Composition Explorer (ACE)*, reached L-1 in January 1998 to study the composition of matter ejected by the sun and provide real-time warning of geomagnetic storms that might affect Earth-orbiting satellites. *ACE* was also providing data as of 2005. The robots sent to L-1 in this program are supplemented by a set of Earth-orbiting vehicles designed to study Earth's atmospheric reactions to solar phenomenon. These include the joint Japan-U.S. *Geotail* spacecraft; a set of four identical satellites placed into polar orbits by the ESA, called the *Cluster* mission; and NASA's *Polar*.

NASA also tried again at Mars, intending to make the Red Planet a focus of its efforts for the next two decades in preparation for a crewed mission that it hoped to launch in the early twenty-first century. In part, the agency's renewed attention to Mars was a product of new knowledge about Earth's oceans, which had been found to contain life that was not dependent upon the sun during the 1980s. These creatures lived off the energy released by volcanic vents in the seafloor, opening the possibility that life could still exist on Mars, deep underground where there might still be heat from the planet's core. The majority of Mars scientists believed that the planet's water was frozen underground, and volcanic heat might still be preserving subterranean environments of liquid water. These, in turn, might support struggling forms of Martian life like those surrounding Earth's deep-ocean vents.

Mars Global Surveyor, the first of the "faster, better, cheaper" Mars missions, was quite successful. Conceived in 1994 and launched in 1996, the *Mars Global Surveyor* was intended to inaugurate a new program of Mars research with launches every twenty-six months—in other words, at every conjunction of the Earth/Mars orbits. *Mars Global Surveyor* carried duplicates of five of the instruments on the lost *Mars Observer* and went into Mars's orbit in September 1997. One of its two solar panels failed to fully extend, complicating an atmospheric braking maneuver that was supposed to have placed it in a circular orbit synchronized with the movement of the sun across the Martian surface. Modifications to the maneuver produced the proper orbit but took a year longer, and its mapping mission did not start until March 1999. But the

Surveyor returned highly detailed images and produced the first reasonably direct evidence that water ice still existed near the surface. It was scheduled to complete its mission in January 2001 after one Martian year, but it was still fully functional and was kept in operation into 2006.

NASA launched a second Mars mission in 1996 called *Mars Pathfinder*, which consisted of a lander equipped with a tiny wheeled rover named *Sojourner Truth* and reached Mars on 4 July 1997. It returned the first images from the Martian surface since the Viking missions, and the rover used a miniaturized Russian instrument to sample rocks that it encountered. The rover was not designed to travel very far or very fast; in the eighty-three Martian days that it operated, it traveled a distance of only 101 meters at a top speed of 1 centimeter per second, never getting more than 12 meters from the lander. Nonetheless, it was enormously popular. The lander's cameras provided stereoscopic images of *Sojourner*'s efforts, sending back more than 16,500 very high-quality images. The Mars *Pathfinder* website received millions of hits during the probe's first week on the surface, and *Sojourner* achieved a cult following worldwide among schoolchildren, who followed the rover's slow journey around the landing site via the Internet. *Pathfinder* validated "faster, better, cheaper" and also the agency's focus on Mars. Even the spectacular *Galileo* had not drawn this level of public interest. In the annual Washington budget wars, missions such as *Pathfinder* were priceless, if short-lived, political capital.

The former Soviet Union's successor state, the Russian Federation, did not immediately copy NASA's new approach, and its last attempt at Mars was the most complex robot explorer ever built. Named *Mars 8*, it was launched in November 1996 on a Proton rocket. In addition to a large orbiter, the mission contained two landers and two penetration probes and a total of thirty-eight scientific experiments. During the launch sequence, however, the orbiter ordered one of the booster stages to shut down early, leaving it in an incorrect Earth orbit. Then the orbiter, its payload, and its propulsion system automatically separated from the booster and carried out the engine firing that was supposed to have sent it on the way to Mars. But because it was in the wrong Earth orbit, the firing had the unfortunate effect of injecting *Mars 8* into Earth's atmosphere, where it disintegrated.

Unfortunately for NASA, the next two "faster, better, cheaper" Mars missions were no more successful than *Mars 8*. In keeping with the new doctrine of a Mars launch every twenty-six months, the agency formulated two new Mars missions to be launched in the Mars window of December 1998–January 1999. The first of the two missions was named *Mars Climate Orbiter*, reflecting the nature of its scientific mission. It was scheduled to arrive in September 1999, shortly before the arrival of the second mission, the *Mars Polar Lander*. In addition to its own mission, the *Climate Orbiter* was designed to serve as a communications relay for the *Polar Lander* and future landing missions. The *Climate Orbiter* arrived on schedule, but during its orbital insertion engine burn, it disappeared. NASA's investigation of the loss revealed that a key piece of navigation software on the ground had been written using English engineering units, while the *Climate Orbiter* expected to be instructed in metric. The error placed it too deep within the Mars atmosphere, where it disintegrated.

The *Polar Lander* had been designed to land on Mars's southern polar ice cap and carry out extensive investigations of the local soil, digging for water ice with a mechanical arm. It arrived at Mars on 3 December 1999, but about six minutes before it was supposed to enter the atmosphere, the *Polar Lander* disappeared. NASA eventually attributed the loss to an incorrect computer program that caused its retrorocket to stop firing when the landing legs extended, instead of when they touched the Martian surface. The lander thus crashed instead of landing safely. The *Polar Lander* had also carried two penetration probes, collectively named *Deep Space 2*. These were lost as well. The software fault would have been caught if the lander had been properly tested before launch, a reality that caused a great deal of questioning of the entire "faster, better, cheaper" approach.

After the 1999 failures, NASA and the JPL, the agency's lead for Mars exploration, canceled the next set of missions, which were supposed to collect and return samples of Mars rocks to Earth, and replaced them with a somewhat less-ambitious set of orbiters and landers. These were to be built under the JPL's older and more expensive "test as you fly" philosophy to ensure that they actually worked. The public ridicule that NASA received over the twin failures convinced its leaders that it could not expect public support for sending humans to Mars if it could not get its robots there in one piece. These faster, better, but not cheaper missions were the 2001 *Mars Odyssey* orbiter and the 2003 *Mars Exploration Rovers*. They were to be joined by the ESA's first Mars mission, *Mars Express*, which also carried a British microlander named *Beagle 2*, a reference to HMS *Beagle*, which carried Charles Darwin on his renowned five-year voyage around the world more than 160 years earlier.

The 2001 *Odyssey* orbiter had one main mission: looking for Mars's missing water. One of its instruments had been designed to detect the energy emissions of buried ice down to a depth of about one meter, and in 2002 it confirmed a speculation by one of the Viking orbiter scientists that Mars's water was bound into the surface material as ice. This water ice appeared to be everywhere poleward of the fifty-degree latitude circles, with some also appearing at lower latitudes. Because the *Odyssey* instrument was very limited in the depth it could

Mars Exploration Rover Opportunity *took this image of its heat shield (left) on 28 December 2004. The circular feature in the foreground is the heat shield's impact crater. The heat shield is about 14 meters from the rover. (NASA/JPL-Caltech/Cornell)*

detect, the European *Mars Express* carried a radar-based instrument that its science teams hoped would be able to find water at much greater depths, where it might still be liquid. *Mars Express* arrived safely in December 2003, but the mission scientists did not order this instrument into operation until mid-2005 due to some concerns that its large antenna might damage other mission instruments. The Mars landers that followed *Odyssey* and *Mars Express* provided additional confirmation of Mars's wet past.

The *Beagle 2* lander did not reach the surface of Mars successfully, and because it had been built without the ability to send data back during its entry into the Martian atmosphere, its fate was never determined. The two NASA rovers, however, landed to great fanfare in January 2004. The first, named *Spirit*, had been aimed at Gusev Crater, which based upon surface appearances seemed to be a place that had once held a lake. The JPL's navigators targeted the second rover, *Opportunity*, at Meridiani Planum. Data from the still-operating *Mars Global Surveyor* suggested that hematite was present there, and hematite only formed in water. It also seemed an obvious place to look for signs of Mars's wet past. *Opportunity* turned out to be the first to find definitive indications of water, observing finely layered sedimentary deposits with its cameras and identifying water-related minerals with its other sensors. The layered deposits at Meridiani were similar to those found in shallow, temporary desert lakes on Earth, leading the rover's geologists back on Earth to hypothesize that Mars had repeated warmer, wetter periods in its remote past, but all within its first billion years or so. *Spirit* later found additional evidence of water near its landing site on the other side of the planet, making it clear that Mars had once been a much different place than it is now. Both rovers were still operating in late-2006, although they had greatly exceeded their design life spans of ninety

Layka (?–1957)

Layka the "space dog" was the first animal to orbit Earth. Layka had been a stray roaming the streets of Moscow before being captured and trained at the Soviet Air Force's Institute of Aviation Medicine for high-altitude flights aboard rockets. One of ten candidates for *Sputnik 2*, she was chosen for the flight due to her calm temperament. She was trained by reducing the size of her cage slowly until it matched the dimensions of the small, padded capsule. Surgeons attached sensors to her body to monitor her health during the flight. The capsule also held a television camera to return images and enough food and water for about ten days. But unlike the chimpanzees Ham and Enos that were used in the American space program, Layka was not expected to perform any tasks. The Soviet goal for her flight was simply to find out whether complex mammals could live in space for a short time, and the quickly designed and manufactured capsule did not include any provision to return her to Earth. On 31 October 1957, she was inserted into the capsule for her flight, and an R-7 rocket carried her safely into orbit from Tyura-Tam on 3 November. The capsule's telemetry systems functioned normally, but its environmental control systems were unable to prevent a gradual increase in temperature. Layka died of heat exhaustion a few hours into her flight, becoming the first Earth animal to both orbit Earth and die in space. *Sputnik 2*, dubbed "Muttnik" by the American press, burned up in the atmosphere on 14 April 1958.

Martian days. Their longevity on Mars, surviving even the bitter Martian winter and the incessant wear of ever-present dust, seemed to vindicate NASA's abandonment of "faster, better, cheaper" after the demise of the 1999 missions.

While the great Martian bombardment was taking place, what was supposed to have been the last of NASA's billion-dollar flagship missions was on its long voyage to the ringed planet, Saturn. A joint project carried out between NASA and the ESA, this was named Cassini/Huygens and, like the *Galileo* mission to Jupiter, consisted of an orbiter, *Cassini*, built by the JPL, and an entry probe, *Huygens*, provided by the ESA. *Huygens*, however, was not aimed at Saturn. Instead, *Cassini* would release it to land on Titan, Saturn's largest moon. Titan has a thick hydrocarbon atmosphere that prevented photography of the surface from space, and the mission scientists hoped that the probe could return images from inside the atmosphere. *Huygens* also carried instruments designed to study the composition of the atmosphere as it parachuted through it. *Cassini* reached Saturn on 1 July 2004 and released *Huygens* in December 2004. *Huygens* landed on Titan on 14 January 2005 and transmitted data for about an hour after reaching the surface. Its images revealed a strange terrain that resembled the desert mountains of coastal California but with temperatures far too cold for running water to carve the distinct canyons and river valleys. Instead, Titan seemed to have methane rain. *Cassini* itself was designed to operate until at least 2008, returning spectacular images of Saturn's rings, using a new synthetic aperture radar to map Titan's surface, and studying Saturn's magnetic and radiation fields.

Perhaps the most significant new finding of the late 1990s, however, was not returned by a robot explorer. Astronomers, first on the ground and then using the Spitzer infrared space telescope, began to find planets circling other stars. While this was not a surprise—it was impossible to believe, by this time, that Earth's solar system was the only one in the vastness of the universe—it was proof. The early planets detected were gas giants even larger than Jupiter. But the space agencies of the world began to design technologies capable of looking for ones of Earth's size in hopes of finding a second Earth sometime in the 2020s. Such planets would provide new destinations for humans to dream of sending their exploring machines to survey.

Conclusion

The competition between the United States and the Soviet Union in the Cold War drove the early years of space exploration for both human and robot explorers. Yet while the dramatic end of the Cold War seemed to doom humans to low Earth orbit due to the enormous cost of safely delivering humans to space and returning them, it witnessed a renaissance of planetary exploration as more robotic space probes, built by more nations and using increasingly advanced technologies, went out into the solar system. In part, the nationalist competition was replaced by an internal dynamic in which scientists claimed a share of national income for scientific exploration. But national governments also derived prestige from financing robotic exploration, gaining public accolades that played very well within the context of democracy. Hence, the desire of national leaders to be seen as visionary by their own populations was a major reason more nations joined the space-faring club during the 1980s. This strategy was not without risk, of course. Machines, like humans, sometimes failed.

The scientific knowledge produced by the exploring machines led scientists to think about both the heavens and Earth in new ways. The climates of Venus, Earth, and Mars had clearly evolved through time and to dramatically different results. This caused scientists to begin to think seriously about how climates evolved and how Earth's climate in particular was regulated. For regulated it clearly was, not having varied by more than ten degrees Celsius in at least a billion years despite dramatic changes in solar output. NASA scientists began to study Earth as an integrated system because that was how its robots had studied other planets. The agency went so far as to try to create a new scientific discipline, Earth System Science, for this purpose. Thus, robotic exploration of the other inner planets led back to a study of Earth. In the same way, discovery of deep-ocean life on Earth had helped fuel a new search for life on Mars. Increasingly, Earth and planetary science were converging in entirely unexpected ways.

Bibliographic Essay

The most complete survey of robotic space missions is Asif Siddiqi's *Deep Space Chronicle: A Chronology of Deep Space and Planetary Probes, 1958–2000* (2002). This work provides only a paragraph summary of a given mission, however. For more detail on Martian exploration, see William Sheehan, *The Planet Mars* (1996); the classic Edward and Linda Ezell, *On Mars: Exploration of the Red Planet, 1958–1978* (1984); and William Hartman and Odell Raper, *The New Mars* (1974). For the "life on Mars" questions as of the end of the Viking missions, the definitive work is Norman Horowitz's *To Utopia and Back: The Search for Life in the Solar System* (1986). However, recent work on life surrounding volcanic vents under Earth's oceans has reopened the question of life on Mars. The seminal work on this was Penelope Boston, Mikhail Ivanov, and Chris P. McKay, "On the Possibility of Chemosynthetic Ecosystems in Subsurface Habitats on Mars" (1992). Finally, the development of the science of life on other planets is discussed in Steven J. Dick and James E. Strick, *The Living Universe: NASA and the Development of Astrobiology* (2004).

Information on the 1996 *Mars Pathfinder* mission is provided in Donna Shirley's *Managing Martians* (1998) and Andrew Mishkin's *Sojourner* (2003). Oliver Morton, *Mapping Mars: Science, Imagination and the Birth of a World* (2002), provides an excellent nontechnical overview of scientific debates surrounding Mars's "missing" water. It also includes a brief discussion of the loss of the JPL's two 1998 Mars missions. For much more detail on the Mars water question, see Jeffrey Kargel, *Mars: A Warmer, Wetter Planet* (2004). For information on the *Mars 2001 Odyssey*, the *Mars Exploration Rover*, and *Mars Express* missions, there is very extensive material on their websites as well as in "On Mars, a Second Chance for Life" (2004) and S. W. Squyres et al., "The Spirit Rover's Athena Science Investigation at Gusev Crater, Mars" (2004).

Exploration of Venus has not received as much attention as has Mars. The study by Mikhail Ya Marov and David Grinspoon, *The Planet Venus* (1998), is comprehensive but very technical. Clayton Koppes, *JPL and the American Space Program* (1982), covers the early American Venus missions in a more accessible style. The final two American missions to Venus are treated in Richard O. Fimmel et al., *Pioneering Venus: A Planet Unveiled* (1995), and Henry Cooper, *The Evening Star: Venus Observed* (1993).

William E. Burrows captures some of the excitement of the *Voyager* missions in his popular *Exploring Space: Voyages in the Solar System and Beyond* (1990). A more recent history of the *Voyager* mission is Henry C. Dethloff and Ronald A. Schorn, *Voyager's Grand Tour* (2002). For a discussion of the possibility of life on Europa, see Committee on Planetary and Lunar Exploration, *A Science Strategy for the Exploration of Europa* (1999). There are, as yet, no histories of the *Cassini* mission to Saturn, but there are materials on the mission website. There is a single history of "faster, better, cheaper" to date: Howard McCurdy, *Faster, Better, Cheaper: Low-Cost Innovation in the U.S. Space Program* (2001).

Chronology of Significant Events

CHAPTER 1

1502	"Cantino's Planisphere," the first world chart to depict the tropics of Capricorn and Cancer, created by an unknown Portuguese cartographer.
1503	Phillip II of Spain establishes Casa de la Contratación (House of Trade) in Seville.
1543	Publication of Nicolaus Copernicus's *De revolutionibus*.
1600	William Gilbert publishes *De magnate* and experimented with magnets to aid navigation.
1627	Publication of Sir Francis Bacon's *New Atlantis*.
1662	Founding of the Royal Society of London.
1666	Founding of the Royal Academy of Sciences in Paris. Royal Society of London publishes "Directions for Sea-men, Bound for Far Voyages" in the *Philosophical Transactions*.
1675	Jean Richer's experiments with the pendulum in Cayenne.
1687	Issac Newton's *Principia mathematica* published.
1699–1701	Edmund Halley's voyages on the *Paramore*.
1714	Board of Longitude formed. Parliament passes the Longitude Act, offering £20,000 for a solution to the longitude problem.
1735	International scientific expeditions to determine the shape of Earth.
1753	Tobias Mayer and Leonard Euler's Tables of the Moon's Motion.
1759	John Harrison constructs H4, his most accurate marine timepiece.
1773	Constantine Phipps sets out for the North Pole.

CHAPTER 2

1577	Translation of Nicolas Monardes' *Joyful Newes Out of the Newe Found World*.

1626	Founding of the Jardin du Roi by Louis XIII.
1735	Publication of Linnaeus's *Systema naturae*.
1761	Measurements of the first transit of Venus.
1766	Voyage of Louis Antoine de Bogainville, first French circumnavigation of the globe.
1768	Captain James Cook's first voyage begins.
1769	Measurements of the second transit of Venus.
1771	Captain Cook's first voyage ends.
1772	Captain Cook's second voyage begins.
1775	Captain Cook's second voyage ends.
1776–1780	Captain Cook's third voyage.
1780–1820	Joseph Banks's presidency of the Royal Society of London.
1785	Ill-fated voyage of La Pérouse.
1789	Voyage of Alejandro Malaspina.
1791	Voyage of George Vancouver.

CHAPTER 3

1818	Parliament passes an act offering a substantial reward for the discovery of the Northwest Passage. Voyage of John Ross.
1819–1822	John Franklin's first overland expedition.
1819–1825	William Parry's three Arctic expeditions.
1830	U.S. Depot of Charts and Instruments established.
1845–?	John Franklin's last expedition.
1872	HMS *Challenger* Expedition begins.
1876	HMS *Challenger* Expedition ends.

CHAPTER 4

1799–1804	Travels of Alexander von Humboldt.
1831–1836	Charles Darwin's voyage on HMS *Beagle*.
1859	Charles Darwin's *On The Origin of Species* published.

CHAPTER 5

1803–1806	Meriwether Lewis and William Clark's Corps of Discovery.
1807	U.S. Coast Survey established by Thomas Jefferson.
1843	John C. Fremont expedition in the Rocky Mountains.
1848–1855	United States and Mexican Boundary Survey.

1853–1855   Pacific Railroad Survey.

1853–1856   North Pacific Exploring Expedition.

1857   Northwest Boundary Survey begins.

1861   Northwest Boundary Survey ends.

1868   Clarence King's 40th Parallel Survey.

1869   Powell's first Colorado expedition.

1870   Spencer Baird establishes the U.S. Fish and Fisheries Commission.

1925   Scripps Institution for Oceanography established.

CHAPTER 6

1838–1842   U.S. Exploring Expedition.

1849   Publication of the *Admiralty Manual of Scientific Enquiry: Prepared for the Use of Officers in Her Majesty's Navy; and Travellers in General.*

1875   British Arctic Expedition begins.

1876   British Arctic Expedition ends.

1882–1883   International Polar Year.

1902   International Council for the Exploration of the Sea (ICES) established.

1909   Climax of the race for the poles.

1926   Goddard launches first successful liquid-fueled rocket.

1930   Woods Hole Oceanographic Institution established.

1941   American entry into World War II precipitates federal funding for oceanography.

1944   Jet Propulsion Laboratory founded.

1948   Roger Revelle appointed as director of the SIO.

CHAPTER 7

1934   William Beebe bathysphere descent to 3,028 feet.

1960   *Trieste* descent to 35,800 feet.

1962   Jacques Cousteau and the Conshelf project.

1964   U.S. Navy begins SEALAB series. *Alvin* makes its first dives.

1968   Smithsonian-Link Man-in-the-Sea program.

1979   Chemosynthetic life discovered around deep-ocean vents.

1981   Sylvia Earle founds Deep Ocean Technologies, Inc.

1985   Robert Ballard finds the *Titanic*.

CHAPTER 8

1783	The Montgolfier brothers demonstrate the first hot air balloon.
1865	Jules Verne's *From the Earth to the Moon* published.
1895	Konstantin Tsiolkovsky proposes liquid-fueled rockets for space travel.
1903	First flight of brothers Orville and Wilbur Wright. Tsiolkovsky proposes liquid oxygen as a rocket fuel and multistaging to decrease rocket weight, and he calculates Earth's escape velocity.
1920	Robert Goddard proposes moon rockets and is ridiculed.
1927	Charles Lindbergh completes first solo transatlantic flight. The Verein fur Raumschiffart (Club for Spaceflight) is founded in Germany.
1928	Hermann Oberth first to propose multistaging outside the Soviet Union.
1932	Auguste Piccard, using a balloon, is the first human to reach the stratosphere. German General Staff hires Werner von Braun to develop bombardment rockets.
1933	Sergei Korolev builds first Soviet liquid-fueled rocket. Wiley Post makes first solo round-the-world flight, in seven days, eighteen hours.
1939	Pan Am initiates first scheduled transatlantic flights.
1942	Von Braun launches the first AS-4 (V-2 prototype).
1947	*Bell X-1* is the first piloted vehicle to safely exceed the speed of sound.
1957	The International Geophysical Year. The first artificial satellite (*Sputnik 1*) is launched. *Sputnik 2*, the first vehicle to carry a living passenger into space (Layka, a dog), is launched.
1958	Discovery of the Van Allen radiation belts surrounding Earth. First successful U.S. space launch. NASA founded.
1959	Soviet *Lunik 2* becomes the first Earth vehicle to reach the moon. It crash-lands deliberately.
1961	Cosmonaut Yuri Gagarin makes the first orbital flight.
1963	Apollo project to land men on the moon announced.
1967	Cosmonaut Vladimir Komarov and astronauts Gus Grissom, Edward White, and Roger Chafee die in space-related accidents.
1968	Apollo 8 successfully orbits the moon during Christmas.
1969	Apollo 11 launched to the moon. Astronauts Neil Armstrong, Edwin "Buzz" Aldrin, and Michael Collins reach the moon on 20 July.
1971	First docking with *Sayut 1*, first crewed space station.

1972	Final Apollo mission.
1973	*Skylab* launched.
1979	First European commercial rocket, *Ariane*, is launched from French Guyana.
1981	First Space Shuttle flight.
1986	Soviet *Mir* space station launched. Space Shuttle *Challenger* explodes after liftoff.
1990	Hubble Space Telescope launched.
1998	First module of the *International Space Station* launched in Russia.
2001	*Mir* de-orbited.
2003	Shuttle *Columbia* disintegrates during reentry.
2004	U.S. President George W. Bush announces "Vision for Space Exploration" to return humans to the moon.

CHAPTER 9

1962	*Mariner 2*, first probe to visit another planet, is launched to survey Venus.
1971	*Mariner 9* maps Mars. Soviet *Mars 3* lander reaches Mars but only operates for twenty seconds.
1973	*Mariner 10* is launched to visit both Venus and Mercury. *Pioneer 11* launched to visit Saturn.
1975	Soviet *Venera 9* sends back first photographs of Venus surface.
1976	*Viking 1* and *Viking 2* land on Mars to look for signs of life.
1977	*Voyager 1* and *Voyager 2* launched on grand tour of the outer planets.
1985	*Sakigake*, first Japanese deep-space probe, and *Giotto*, first European Space Agency deep space-probe, launched.
1986	*Voyager 2* first spacecraft to reach Uranus.
1989	*Galileo* orbiter/probe launched to Jupiter. *Voyager 2* reaches Neptune.
1990	*Magellan* orbiter maps Venus.
1995	*Galileo* reaches Jupiter.
1996	*Mars Pathfinder* and *Global Surveyor* launched.
1997	*Cassini/Huygens* launched. *Mars Pathfinder* lander and rover explore the surface of Mars.
1998	*Deep Space 1*, first ion-propulsion deep space mission, launched.

2000   Near-Earth Asteroid Rendezvous mission, first to study an asteroid, reaches Eros.

2001   Mars 2001 Odyssey launched.

2004   Mars Exploration Rovers *Spirit* and *Opportunity* reach Mars.

2005   ESA probe *Huygens* lands on Saturnian moon Titan. First measurements of comet interior materials. Penetrator from spacecraft *Deep Impact* hits comet Temple 1.

Aerobraking: A procedure in which a spacecraft uses atmospheric drag to modify its orbit instead of, or to supplement, the use of a retrorocket. This saves propellant weight and significantly reduces the cost of planetary spacecraft.

American exceptionalism: Historians use this term to refer to an idea popularly held since the Revolutionary period, and perhaps before, that America was distinct and superior when compared to European states. A common late nineteenth-century idea was that the West, and the Western experience, made America and Americans distinct from the rest of the world.

American Philosophical Society (APS): Scientific institution located in Philadelphia and created in 1743 by Benjamin Franklin. The APS is the American analog to European scientific societies that served as a library, a publisher of journals and correspondence, and a site for meetings and public experiments.

Anthropocentrism: The practice or belief that regards the creation and/or existence of human beings as the centerpiece of the universe. The idea is often criticized by people who espouse biocentric (life as the centerpiece) or ecocentric (the global ecosystem as the centerpiece) beliefs.

Astronomical Unit (AU): The distance between Earth and the sun, about 93 million miles or 150 million kilometers, used by astronomers as the standard measure of the universe.

Baseline data: A wide range of data accumulated in one location at constant intervals and over long periods of time. The purpose is to measure environmental changes over time, especially those changes caused by human-induced pollution.

Bends, the: Decompression sickness usually caused by a diver ascending too quickly from a dive. A quick ascent causes dissolved nitrogen to come out of physical solution and create bubbles, leading to joint pain, paralysis, and sometimes death.

Bestiaries: Medieval books that listed common and sometimes fantastic creatures. These books were often written by clerics, and the creatures often had moralistic qualities useful for writing sermons.

Cetology: The scientific study of ocean-dwelling mammals (whales and dolphins).

Chronometer: A sophisticated mariner's timepiece used to determine longitude at sea.

Continental shelves: The portion of a continental plate that is underwater and usually extends with a gradual slope into the ocean before reaching the steep continental slope. Many industrialized nations wished to explore, mine, develop, and even inhabit continental shelves in the 1960s. While the hope of the overly optimistic was quickly dashed, continental shelves did cede to oil extraction.

Core-periphery: A sometimes useful geopolitical dichotomy created by social scientists to characterize colonial and imperial processes. The core is defined as the place of greater social, economic, political, technological, and military power. The periphery is defined as those territories brought under partial control for the continued growth of the core. The distinction can also be used to refer to smaller geographies. For example, a city is the core, and its surrounding hinterland is the periphery.

Dead reckoning: The method of estimating a ship's position without using astronomical observations. To determine longitude, mariners combined distance traveled, obtained by throwing overboard a log attached to a rope, with the direction of the ship. The mariner then corrected for ocean currents, wind, and tides.

Discovery Program: A competitive awards program administered through NASA and intended to reduce mission costs and increase scientists' participation.

Dredges: A method of surveying the sediment and life of the seabed that consists of large metal scoops that are lowered to the seafloor and brought to the surface, usually through the use of mechanical engines. Widely used since the late nineteenth century, though less frequently now, dredges were an indispensable technology for oceanographers.

Economic botany: The practice of botany for economic gain by a government, usually through the acquisition and study of foreign species, which were then cultivated in local or other colonial environments, especially at Kew Gardens or other colonial botanical gardens.

El Niño: Oceanic phenomenon in the Pacific Ocean where a gradual buildup in ocean temperatures in the western Pacific suddenly shifts, thus bringing unusually warm water to the eastern Pacific. The oceanic process has a dramatic impact on weather patterns throughout the world. While Peruvian and Ecuadorian fishers have long been aware of the dramatic rains that come with an El Niño event, knowledge of the oscillation dramatically increased with the use of satellite and remote-sensing technologies.

Enlightenment: An intellectual movement that called for reform of thought, government, and society. It reached its peak in Europe in the eighteenth century, with roots dating back to the scientific advances of the seventeenth century. Its followers stressed the need for freedom of thought.

Ephemeris: A table showing the predicted or observed positions of astronomical bodies, used by mariners to compare with their own observations made during long-distance voyages in order to find longitude at sea.

Field stations: Usually located in the periphery, as distinct from a laboratory located at the core, a field station is usually a small laboratory and living quarters from where scientists explore the local environment. Ecologists, for instance, may spend a summer at a field station in Brazil to gather data on tropical rain forests.

Geodesy: The mathematical study of the shape of Earth's surface, including the shape of the entire planet as a whole.

Gravity assist: A means of increasing a spacecraft's speed through the use of a planet's or the sun's gravity field.

High-gain antenna: Planetary spacecraft generally have two communications antennae. The low-gain antenna has a very low data transmission rate but is not directional, allowing it to receive a signal from any direction. The high-gain antenna has a much higher data transmission capacity, but it must be very precisely aimed at Earth. The low-gain antenna's main purpose is to permit recovery of the spacecraft in the event it loses its lock on Earth, while the high-gain antenna is normally the primary antenna for the mission.

HMS *Beagle*: The British vessel, captained by Robert Fitzroy, in which Charles Darwin circumnavigated the globe. It was during the journey of HMS *Beagle* that Darwin collected much of his scientific data that he later used to formulate the theory of evolution through natural selection.

Humboldtian science: The type of science, popular in the late eighteenth century and the first half of the nineteenth century, that was characteristic of Alexander von Humboldt's approach to nature. It relied upon simultaneous observations over wide areas of Earth's surface to produce mathematical laws. The Magnetic Crusade was a highly visible Humboldian project.

Hydrothermal vent: A crack in the earth where magma is close to the surface. Water is then heated and rapidly shoots out of the vent. These cracks usually occur in seismically active places. When they are underwater they sometimes emit superheated water with dissolved minerals; these are called black smokers. Oceanic hydrothermal vents also can be host to unique ecosystems that derive their energy from chemosynthesis instead of the sun.

Ichthyology: The scientific study of fish.

In situ: Latin for "in place." A scientist is in situ when he or she is in the field studying natural phenomena, as distinct from using, for instance, satellites or remote-sensing technologies.

Intercontinental ballistic missile (ICBM): Intended to deliver atomic and later thermonuclear weapons, these rockets are suborbital. They do not place their payloads into orbit; instead, the suborbital weapon reenters Earth's atmosphere and explodes near its target. On early ICBMs, the warhead had no ability to maneuver once separated from its rocket. Lowering the payload weight enables conversion of an ICBM into an orbital launch vehicle. The American Thor, Redstone, Titan, and Atlas rockets all descend from ICBMs. The Saturn and Delta series do not.

Isogonic lines: Lines drawn on a map to indicate where on Earth's surface the magnetic declination is the same.

Isothermal lines: Lines drawn on a map indicating where on Earth's surface the atmospheric temperature is the same.

Latitude: Position on Earth's surface north or south of the equator.

Launch window: Because of thrust limitations in current launch vehicles and the fact that the planets are always in motion, spacecraft leaving Earth's orbit must be launched within very narrow windows of time. The lowest energy trajectory to Mars occurs once every twenty-six months and lasts about thirty days. Missing the launch window into that trajectory means waiting for the next window.

Libration point: The point at which the gravity fields of two large bodies balance. There are three such points for Earth and the sun (L-1, L-2, and L-3) and two for the Earth-moon system (L-4 and L-5). Libration points are also referred to as Lagrangian points.

Longitude: Position on Earth's surface east or west of Greenwich, England.

Lunar-Distance Method: The method of finding longitude by comparing the distance of the moon from either the sun or certain fixed stars and comparing them with the same distance as given in a printed ephemeredes.

Magnetic Crusade: The British endeavor to determine the variation of Earth's magnetic field through the establishment of a system of magnetic observatories throughout British possessions overseas.

Magnetic declination: The amount of angular deviation of the magnetic compass from true geographic north or south.

Manifest Destiny: First used by the New York journalist John O'Sullivan in 1845, the term refers to a nineteenth-century ideology that holds that it was America's destiny, divined by God, to march westward and develop the continent all the way to the Pacific Ocean.

Oceanology: Distinct from oceanography, which is the umbrella term that includes a wide range of ocean sciences, oceanology pertained to the practice of engineering technology capable of exploring and exploiting the ocean. The word was also used by Soviet scientists to refer to the ocean sciences in general.

Parallax: The apparent change in position of an object when viewed from two different points.

***Paramore*:** The vessel in which Edmund Halley surveyed the Atlantic Ocean on two separate voyages from 1699 through 1701 and in which he surveyed the English Channel in 1701.

Philosophes: Intellectuals who fostered the Enlightenment. Although far from a homogenous group, the philosophes championed economic and political reform, freedom of thought and religion, and the application of Newtonian principles to all aspects of civic life.

Plate tectonics: A dynamic theory that holds that continents consist of plates that ride over a fluid mantle. Some plates are being created such as the mid-Atlantic ridge, which is a location for seafloor spreading. Other plate boundaries scrape against each other or collide, causing one to ride over the other. These boundaries are often the sites of volcanic and seismic activity. Plate tectonics only became a matter of consensus in the 1960s.

***Principia*:** Isaac Newton's seminal text published in 1678 that introduced his theory of universal gravitation.

Quadrant: An instrument that mariners used to take measurements of altitudes of fixed stars, the sun, or the moon to determine latitude at sea.

Radioisotope thermal generator: Generally called RTGs, these generate electrical power from the natural decay of radioactive material. They have been used on all outer-planet missions because the weak sunlight beyond the orbit of Mars is insufficient to power a spacecraft given the low conversion efficiency of current solar arrays. An RTG provides about as much energy as several household light bulbs consume.

***Rattlesnake*:** The British vessel in which T. H. Huxley circumnavigated the globe.

Renaissance: The rebirth in learning in arts and letters from the fifteenth through the sixteenth centuries. It began in Italy and was coincident with the first great age of European discovery.

Royal Academy of Sciences in Paris: Founded in 1666 by King Louis XIV on the suggestion of Jean-Baptiste Colbert, France's finance minister, the Royal Academy was France's first national scientific society. Different from the Royal Society of London. Its membership was limited and those members were expected to work for the state.

Royal Society of London: Founded in the early 1660s and consciously modeled after Francis Bacon's Salomon's House in his *New Atlantis*, the Royal Society was England's first national scientific society, a central locale where natural philosophers could perform experiments, build instruments, and catalog observations from around the globe.

Sextant: An instrument that mariners used to take measurements of altitudes of heavenly bodies to determine latitude at sea. It resembles a quadrant except that it has a graduate arc equal to one-sixth of a circle.

Sondes: Automated balloons that collect data and radio it down to the ground. The two most common types are radiosondes, which are weather balloons equipped to make temperature, pressure, and humidity measurements, and rawinsondes, balloons that are tracked by radar to provide wind speeds.

Sonic depth finder (SDF): An electronic device mounted on a ship that emits and receives acoustic signals to detect the depth of oceans. SDFs quickly replaced the practice of plumbing the ocean depths by lowering a weighted line from a ship and have been widely used since the beginning of the twentieth century.

Synoptic data: Data accumulated from many different locations simultaneously, for instance, when twenty ships in the Gulf of Mexico take the surface temperature of the ocean at noon, 1:00, 2:00, etc. The technique was and is vital for atmospheric and oceanographic scientists interested in portraying the dynamic movements of the environments being explored.

Taxonomy: The science of classification, especially the classification of living organisms, that searches out general laws or principles.

Telemetry: A technology that allows unmanned remote-sensing instruments to wirelessly transmit environmental data to a central location. Telemetric instruments have allowed scientists to accumulate large amounts of environmental data without having to physically go into the field to collect it.

Terra firma: Latin for "firm ground." Environments that are distinct from marshes, lakes, rivers, and oceans. Explorers are fond of using such Latinate terminology. More common is terra incognita, land that is unknown to explorers but known to the people living there.

Terrestrial magnetism: The scientific study of Earth's changing magnetic field.

Texas Towers: Ocean-based structure, built in the 1950s, that was part of the U.S. early detection system for identifying hostile planes. The towers were named because of their resemblance to oil rigs. They quickly proved too dangerous and ineffective for military purposes, but several were used as stations for accumulating scientific data.

Theodolite: Surveying instrument with a telescope that measures both vertical and horizontal angles. Due to their unprecedented precision, they made possible the practice of conducting triangulation surveys.

Tidology: A word coined by William Whewell denoting the scientific study of the tides.

Topographical map: A map that represents the elevation and contours of a given geography. Still used today, topographical maps were invaluable for planning intercontinental railroads in the nineteenth century.

Transit of Venus: The eclipse of the sun by Venus, which occurs roughly in pairs every 112 years. Through accurate measurements of the transit from widely varying positions on Earth's surface, astronomers in the eighteenth century determined the distance between Earth and the sun, the astronomical unit (AU).

Triangulation survey: A precise surveying technique that consists of beginning with a fixed line segment of known distance and elevation. Using theodolites, other surveying equipment, and trigonometry, surveyors created a series of triangles over the landscape. The method was widely used in the nineteenth and early twentieth centuries.

Uniformitarianism: A theory of Earth generally attributed to James Hutton but made popular by Charles Lyell in the 1830s. The theory holds that geological change over history is slow and steady, caused by the same processes active in the present. Uniformitarianism was often contrasted with catastrophism, which held that geological changes were caused by infrequent cataclysmic events.

Upwellings: Oceanographic term for places where wind moves water in the surface of the ocean, causing often cold nutrient-rich water beneath it to rise to the surface. Upwellings are usually locations with very large fish populations.

Primary Source Documents

Chapter 1

Salomon's House, 289

Henry Oldenberg's Call for Observations on Sea Voyages, 291

Chapter 2

Captain James Cook's Voyage to Tahiti, 292

The Royal Academy's Instructions to Jean-Francois Galaup, Comte de La Pérouse, 294

Chapter 3

Alexander von Humboldt's Personal Narrative, 299

James Clark Ross and the Northern Magnetic Pole, 302

Chapter 4

Charles Darwin's Landing on Chatham Island, 305

Alfred Russel Wallace's Account of Living in Dobbo, 307

Joseph Dalton Hooker's Account of Preparations and Negotiations for Entering the Tibetan Passes, 309

Chapter 5

The Corps of Discovery and the Mandan Nation, 312

Dr. G. Suckley's Report on Pacific Salmon, 314

Chapter 6

William Scoresby on Sea Color in the Arctic, 316

Henry Bigelow's Report on Oceanography, 317

Matthew Fontaine Maury on Determining Accurate Seafloor Depth, 319

Chapter 8

John F. Kennedy's Special Message on Urgent National Needs, 321

Chapter 9

James M. Beggs's Letter on Budget Cuts and the Jet Propulsion Laboratory, 324

Daniel S. Goldin's May 1992 Remarks to the Jet Propulsion Laboratory, 327

Daniel S. Goldin's March 2000 Remarks to the Jet Propulsion Laboratory, 330

CHAPTER 1

Salomon's House

In the New Atlantis, *Sir Francis Bacon laid out his utopian vision of a society run by natural philosophers. In the passage below, Bacon recounts the workings of Salomon's House, a great institution of scientific and technological learning founded on the remote and uncharted island of Bensalem.*

Source: Francis Bacon, *The New Atlantis* (London: n.p., 1626).

God bless thee, my son; I will give thee the greatest jewel I have. For I will impart unto thee, for the love of God and men, a relation of the true state of Salomon's House. Son, to make you know the true state of Salomon's House, I will keep this order. First, I will set forth unto you the end of our foundation. Secondly, the preparations and instruments we have for our works. Thirdly, the several employments and functions whereto our fellows are assigned. And fourthly, the ordinances and rites which we observe.

The end of our foundation is the knowledge of causes, and secret motions of things; and the enlarging of the bounds of human empire, to the effecting of all things possible.

. . .

We have dispensatories or shops of medicines; wherein you may easily think, if we have such variety of plants, and living creatures, more than you have in Europe (for we know what you have), the simples, drugs, and ingredients of medicines, must likewise be in so much the greater variety. We have them likewise of divers ages, and long fermentations. And for their preparations, we have not only all manner of exquisite distillations, and separations, and especially by gentle heats, and percolations through divers strainers, yea, and substances; but also exact forms of composition, whereby they incorporate almost as they were natural simples.

We have also divers mechanical arts, which you have not; and stuffs made by them, as papers, linen, silks, tissues, dainty works of feathers of wonderful lustre, excellent dyes, and many others, and shops likewise as well for such as are not brought into vulgar use among us, as for those that are. For you must know, that of the things before recited, many of them are grown into use throughout the kingdom, but yet, if they did flow from our invention, we have of them also for patterns and principals.

. . .

We have also engine-houses, where are prepared engines and instruments for all sorts of motions. There we imitate and practise to make swifter motions than any you have, either out of your muskets or any engine that you have; and to make them and multiply them more easily and with small force, by wheels and other means, and to make them stronger and more violent than yours are, exceeding your greatest cannons and basilisks. We represent also ordnance and instruments of war and engines of all kinds; and likewise new mixtures and compositions of gunpowder, wild-fires burning in water and unquenchable, also fire-works of all variety, both for pleasure and use. We imitate also flights of birds; we have some degrees of flying in the air. We have ships and boats for going under water and brooking of seas, also swimming-girdles and support- ers. We have divers curious clocks and other like motions of return, and some perpetual motions. We imitate also motions of living creatures by images of men, beasts, birds, fishes, and serpents; we have also a great number of other various motions, strange for equality, fineness, and subtilty.

We have also a mathematical-house, where are represented all instru- ments, as well of geometry as astronomy, exquisitely made.

We have also houses of deceits of the senses, where we represent all man- ner of feats of juggling, false apparitions, impostures and illusions, and their fallacies. And surely you will easily believe that we, that have so many things truly natural which induce admiration, could in a world of particulars deceive the senses if we would disguise those things, and labor to make them more mi- raculous. But we do hate all impostures and lies, insomuch as we have severely forbidden it to all our fellows, under pain of ignominy and fines, that they do not show any natural work or thing adorned or swelling, but only pure as it is, and without all affectation of strangeness.

These are, my son, the riches of Salomon's House.

For the several employments and offices of our fellows, we have twelve that sail into foreign countries under the names of other nations (for our own we conceal), who bring us the books and abstracts, and patterns of experiments of all other parts. These we call merchants of light.

Henry Oldenberg's Call for Observations on Sea Voyages

Scientific societies such as the Royal Society of London and the Royal Academy of Sciences in Paris served as centers of calculation where observations from around the globe could be analyzed and compared. Excerpted below are the prefatory remarks made by Henry Oldenberg, the unofficial secretary of the Royal Society of London and editor of the prestigious Philosophical Transactions, *to his call for observations to be made by persons bound for far-off voyages of discovery.*

Source: [Henry Oldenberg], "Directions for Observations and Experiments to Be Made by Masters of Ships, Pilots, and Other Fit Persons in Their Sea-Voyages," *Philosophical Transactions* 2 (1666–1667): 433–434.

Though the Art of Navigation, one of the most useful in the world, be of late vastly improved, yet remain their [*sic*] many things to be known and done, the knowledge and performance whereof, would tend to the accomplishment of it: As the making of exact *Mapps* of all Coasts, Ports, Harbors, Bayes, Promontories, Islands, with their several Prospects and Bearings; Describing of Tydes, Depths, Currents, and other things considerable in the Seas; Turnings, Passages, Creeks, Sands, Shelves, Rocks, and other dangers; Nice Observations of the Variations and Dippings of the Needle, in different places, and in the same place, at different times; The Winds, Weather, and Tempers of the Seasons every where; The great Depths, Ground, and Vegetables at the bottom of the Sea; The various Degrees of Saltness of the Sea-water, in several places, and at several depths at the same place. If besides Astronomical things, to be hereafter lookt into, the following Experiments be carefully made, and Directions observed by as many Ingenious Persons, as have opportunity, it may fairly be hoped, that from multitudes of Experiments and Observations, such Rules may be framed, as may be of inestimable use for Seamen. To which purpose the *Royal Society*, having some years ago, ordered that Eminent Mathematician Master *Rooke*, one of their Fellows, and *Geometry* Professor of *Gresham College* (since deceased, to the great detriment of the Common-wealth of Learning) to draw up some Directions for Seamen, the better to capacitate them for making such Observations abroad, as might be pertinent and suitable to the purposes above-mentioned, such Directions were drawn up accordingly, and soon after printed in *Num*. 8. of these *Transactions*. But, further to incourage and facilitate the Work of those, that shall be engaged to put them into practice, it was thought fit, that what of this Kind was heretofore but barely proposed, should now be publisht with ample and particular Explanations, and considerable Additions; which done, a good number of such printed Copies is, by the Care, and at the Expences of the *R. Society*, to be lodged with the Master of *Trinity-house*, to be recommended to such, as are bound for far Sea-Voyages, and shall be judged fit for performance; who are also to be desired, to keep an exact Diary of such observations and Experiments, and deliver at their return a fair Copy thereof to the Lord High Admiral of *England*, his Royal Highness the *Duke of York*, and another to *Trinity-house*, to be perused by the said *R. Society*.

CHAPTER 2

Captain James Cook's Voyage to Tahiti

Captain James Cook made three voyages of discovery between 1768 and 1779 into the heart of the Pacific, Arctic, and Antarctic regions. Accompanying Cook's first voyage was Joseph Banks, future president of the Royal Society of London from 1780 to his death in 1820, and Banks's assistant, Daniel Solander, a pupil of Carl Linnaeus. In the account of Cook's first voyage excerpted below, Cook describes the crew's visit to Tahiti (Otaheite), including their relationship with the Tahitians, particularly Tupia, who left Tahiti with Cook.

Source: Captain James Cook, *An Account of a Voyage Round the World in 1768, 1769, 1770, and 1771*, "Book 1, Chapter XVI," in *The Voyages of Discovery of Captain James Cook, Complete in Two Volumes* (N.p.: Ward, Lock, Bowden and Co., 1888), 77–78.

Among the natives who were almost constantly with us, was Tupia, whose name has been often mentioned in this narrative. He had been, as I have before observed, the first minister of Oberea, when she was in the height of her power: he was also the chief Tahowa or priest of the island, consequently well acquainted with the religion of the country, as well with respect to its ceremonies as principles. He had also great experience and knowledge in navigation, and was particularly acquainted with the number and situation of the neighbouring islands. This man had often expressed a desire to go with us, and on the 12th in the morning, having with the other natives left us the day before, he came on board, with a boy about thirteen years of age, his servant, and urged us to let him proceed with us on our voyage. To have such a person on board was certainly desirable for many reasons; by learning his language, and teaching him ours, we should be able to acquire a much better knowledge of the customs, policy, and religion of the people, than our short stay among them could give us, I therefore gladly agreed to receive them on board. As we were prevented from sailing to-day, by having found it necessary to make new stocks to our small and best bower anchors, the old ones having been totally destroyed by the worms, Tupia said, he would go once more on shore, and make a signal for the boat to fetch him off in the evening. He went accordingly, and took with him a miniature picture of Mr. Banks's, to show his friends, and several little things to give them as parting presents.

After dinner, Mr. Banks being desirous to procure a drawing of the Morai belonging to Tootahah at Eparré, I attended him thither, accompanied by Dr. Solander, in the pinnace. As soon as we landed, many of our friends came to meet us, though some absented themselves in resentment of what had happened the day before. We immediately proceeded to Tootahah's house, where we were joined by Oberea, with several others who had not come out to meet us, and a perfect reconciliation was soon brought about; in consequence of which they promised to visit us early the next day, to take a last farewel of us, as we told them we should certainly set sail in the afternoon. At this place also we found Tupia, who returned with us, and slept this night on board the ship for the first time.

On the next morning, Thursday the 13th of July, the ship was very early crowded with our friends, and surrounded by a multitude of canoes, which were filled with the natives of an inferior class. Between eleven and twelve we weighed anchor, and as soon as the ship was under sail, the Indians on board took their leaves, and wept, with a decent and silent sorrow, in which there was something very striking and tender: the people in the canoes, on the contrary, seemed to vie with each other in the loudness of their lamentations, which we considered rather as affectation than grief. Tupia sustained himself in this scene with a firmness and resolution truly admirable: he wept indeed, but the effort that he made to conceal his tears, concurred, with them, to do him honour. He sent his last present, a shirt, by Otheothea, to Potomai, Tootahah's favourite mistress, and then went with Mr. Banks to the mast-head, waving to the canoes as long as they continued in sight.

Thus we took leave of Otaheite, and its inhabitants, after a stay of just three months; for much the greater part of the time we lived together in the most cordial friendship, and a perpetual reciprocation of good offices. The accidental differences which now and then happened, could not be more sincerely regretted on their part than they were on ours: the principal causes were such as necessarily resulted from our situation and circumstances, in conjunction with the infirmities of human nature, from our not being able perfectly to understand each other, and from the disposition of the inhabitants to theft, which we could not at all times bear with or prevent. They had not, however, except in one instance, been attended with any fatal consequence; and to that accident were owing the measures that I took to prevent others of the same kind. I hoped, indeed, to have availed myself of the impression which had been made upon them by the lives that had been sacrificed in their contest with the Dolphin, so as that the intercourse between us should have been carried on wholly without bloodshed; and by this hope all my measures were directed during the whole of my continuance at the island, and I sincerely wish, that whoever shall next visit it, may be still more fortunate. Our traffic here was carried on with as much order as in the best regulated market in Europe. It was managed principally by Mr. Banks, who was indefatigable in procuring provisions and refreshments while they were to be had; but during the latter part of our time they became scarce, partly by the increased consumption at the fort and ship, and partly by the coming on of the season in which cocoa-nuts and bread-fruit fail. All kind of fruit we purchased for beads and nails, but no nails less than forty penny were current: after a very short time we could never get a pig of more than ten or twelve pounds, for less than a hatchet; because, though these people set a high value upon spike nails, yet these being an article with which many people in the ship were provided, the women found a much more easy way of procuring them than by bringing down provisions.

The best articles for traffic here are axes, hatchets, spikes, large nails, looking-glasses, knives, and beads, for some of which, every thing that the natives have may be procured. They are indeed fond of fine linen cloth, both white and printed; but an ax worth half a crown, will fetch more than a piece of cloth worth twenty shillings.

The Royal Academy's Instructions to Jean-Francois Galaup, Comte de La Pérouse

The French mounted several voyages of discovery into the Pacific in the late eighteenth century, partly in response to the successful voyages of Captain James Cook. Rear Admiral Jean-Francois Galaup, comte de La Pérouse, was chosen to lead one such voyage and, like Cook, combined natural history with geopolitical and commercial motives. The Royal Academy of Sciences in Paris crafted detailed instructions for the voyage, excerpted

*below, that served as a template for French voyages of exploration
throughout the late eighteenth and early nineteenth centuries.*

Source: Jean-Francios de Galaup de la Pérouse, "Instructions—Part the Third,"
in *Voyage Round the World Performed in the Year 1785, 1786, 1787, and
1788 by the Boussole and Astrolabe* (London: n.p., 1799), 33–38.

Operations Respecting Astronomy, Geography, Navigation, Physics, and the Different Branches of Natural History

1. His majesty having appointed two astronomers, to be employed under the
orders of the sieur de la Perouse, in the expedition entrusted to his charge, and
his two frigates being provided with all the instruments for the purpose of astronomy and navigation, of which they can make use either by sea or by land; he
will take care, that neither neglects any opportunity in the course of the voyage,
of making every astronomical observation, which he may deem useful.

The most important object for the safety of navigation is, to determine
with accuracy the latitudes and longitudes of the places at which he may touch,
or in sight of which he may sail. With this view he will recommend to the astronomer employed on board each frigate to attend with the greatest exactness
to the time-keepers, and to neglect no favourable opportunity of ascertaining
on shore, whether they have kept the regularity of their going at sea, and of determining by observations any change that may take place in their mean rate, in
order to take account of such change, for the purpose of fixing with more precision the longitude of the islands, capes, or other remarkable objects, which he
may have descried and laid down in the interval between two verifications.

As often as the state of the heavens will permit, he will be careful that
the distance of the moon from the sun or stars be taken with the proper instruments, to find out from it the longitude of the vessel, and compare it with that
given by the time-keepers at the same instant, and for the same spot: and he
will cause the observations, of every kind, to be repeated, in order that the
greater precision may be obtained from the mean of different operations. When
he passes in sight of any land or isle, at which he does not intend to touch, he
will take care to keep himself as closely as possible on the parallel of the object, while the observation of the meridian altitude of the sun, or of some other
star, is taking, to determine the latitude of the vessel; and he will keep in the
meridian of the same object, during the observations by which its longitude is
to be ascertained. Thus he will avoid every mistake of position and estimation
of distance, which might diminish the accuracy of the determination.

Every day, when the weather will permit, he will cause the variation and dip of the magnetic needle to be observed.

On his arrival at any port, he will choose some convenient situation, for the erection of the tents and portable observatory with which he is furnished, and will set a guard over them.

Independently of the observations for determining latitudes and longitudes, for which he will cause every known practicable method to be employed, and of those for ascertaining the variation and dip of the needle, he will take care, that every celestial phenomenon, capable of being seen, be observed; and on all occasions he will procure the two astronomers every convenience and assistance, that can ensure the success of the operations.

His majesty is persuaded, that the superior and petty officers employed on board the two frigates will be eager of themselves, to make, in concert with the astronomers, every observations, that can be of advantage to navigation; and that the astronomers, on their parts, will be ready to impart to those officers the fruits of their study, and all the theoretical knowledge, that may contribute to the improvement of the nautical art.

The sieur de la Perouse will direct a double register to be kept, on board each frigate, in which shall be entered daily, both at sea and on shore, the astronomical observations, those relating to the use of the time keepers, and all others. These observations will be entered simply in the register, that is to say, the number of degrees, minutes, &c. given by the instrument at the moment of observation, will be set down without any calculation, but merely noting the known error of the instrument employed, if such error have been ascertained by the usual proofs.

Each of the astronomers will keep one of these registers to himself, and the other will remain in the hands of the captains commanding the vessels.

The astronomer will likewise keep a second register, in which he will insert, daily, all the observations he may make, adding, to each operation, all the calculations, by means of which the result is to be deduced.

At the end of the voyage, the sieur de Ia Perouse will cause the two registers kept by the astronomers to be delivered to him, after they have certified them to be the true ones, and signed them.

2. When the sieur de la Perouse shall touch at any port, which it may be of importance to examine in a military view, he will direct the country to be reconnoitred by the engineer in chief, who will deliver to him a circumstantial report of all the remarks he shall have made, and such plans as he may be able to take.

The sieur de la Perouse will cause accurate charts of all the coasts and islands he shall visit to be drawn; and, if they be already known, he will examine the accuracy of the descriptions and charts given by other navigators.

For this purpose, as he sails along coasts, and in sight of islands, he will cause their bearings to be accurately taken, by means of the reflecting circle, or the azimuth compass; and he will remember, that the bearings, which may be employed with most security for the construction of charts, are those, in which the situation of one cape, or any other remarkable object, can be rectified by those of another.

He will employ the officers of the two frigates, and the geographical engineer, in carefully taking plans of the coasts, bays, harbours, and anchoring places, which he may have an opportunity of visiting and examining: and to every plan he will add instructions containing every thing that can be of use for distinguishing and approaching the land, sailing in and out of the harbours, the manner of coming to an anchor and mooring ship, and the best place for watering; the soundings, bottom, dangers, rocks, and shoals; the prevailing winds, sea and land breezes, monsoons, time of their duration, and periods of their change; in short, every nautical information, that can be of use to navigators.

Of all the plans of countries, coasts, and harbours, two copies will be made; one will be kept by each of the captains of the vessels; and at the termination of the voyage, the sieur de la Perouse will cause all the charts, plans, and instructions relative to them, to be delivered to himself.

His majesty leaves it to the sieur de la Perouse to determine the period at which he will cause the decked boats, put on board each frigate, to be set up: a business, however, which, he will no doubt reserve for his stay at Otaheitee. These boats may be very usefully employed in the service of the frigates, either to visit the archipelagoes of the grand equatorial ocean; to explore minutely parts of the coast, and sound bays, harbours, and passages; or to facilitate every research, which requires a vessel drawing little water, and capable of carrying a few days provision for its crew.

3. The natural philosophers and naturalists, intended to make observations in their respective sciences during the course of the voyage, will be employed each in that branch of natural history or physics, to which his studies have been particularly turned.

Accordingly, the sieur de la Perouse will point out to them the researches they will have to make, and will distribute to them the proper instruments.

In appointing their different occupations, he will take care to give to each a single subject, that his knowledge and zeal may thereby have their full effect in promoting the general success of the expedition.

He will communicate to them the paper drawn up by the academy of sciences, in which the academy points out the particular observations, to which it wishes the natural philosophers and naturalists to turn their attention during the voyage; and he will direct them to concur, each in his particular department, and as circumstances shall offer; in fulfilling the objects indicated in this paper.

In like manner he will communicate to the surgeon of each frigate the paper drawn up by the society of medicine, in order that both may employ themselves in such observations, as will tend to accomplish the wish of the society.

Throughout the course of the voyage, and during his stay in port, the sieur de la Perouse will cause a journal to be kept on board each of the vessels, containing observations, from day to day, of every thing that relates to the state of the heavens and of the sea, the winds, currents, variations of the atmosphere, and whatever pertains to the science of meteorology.

During the stay he shall make in any harbour, he will observe the genius, character, manners, customs, bodily constitution, language, government, and number of the inhabitants.

He will cause the nature of the soil and the productions of the different countries to be examined, and every thing relating to the natural history of the globe.

He will direct natural curiosities, both of land and sea, to be collected to be arranged in order; and a descriptive catalogue of each kind to be drawn up, in which shall be mentioned the places where they were found, the uses to which they are applied by the natives of the country, and, if they be plants, the virtues ascribed to them.

In like manner he will order the garments, arms, ornaments, utensils, tools, musical instruments and every thing used by the different people he shall visit, to be collected and classed; and each article to be ticketed, and marked with a number corresponding to that assigned it in the catalogue.

He will direct the draughtsmen embarked on board the frigates, to take views of all remarkable places and countries, portraits of the natives of different parts, their dresses, ceremonies, games, buildings, boats and vessels, and all the productions of the sea and land, in each of the three kingdoms of nature, if he shall think that drawings of them will render the descriptions more intelligible.

All the drawings made during the course of the voyage, all the boxes containing natural curiosities, together with the descriptions given of them, and all the collections of astronomical observations, shall be delivered to the sieur de la Perouse at the end of it; and no man of science, no artist, shall reserve for himself or others, any specimen of natural history, or other object, which the sieur de la Perouse shall deem worthy to be included in the collection intended for his majesty.

4. Before he enters the port of Brest, at the termination of the voyage or before arriving at the Cape of Good Hope, if he should deem it proper to put in there, the sieur de la Perouse will demand all the journals kept [a]board the two frigates by the superior and petty officers, by the astronomers, men of science, and artists, by the pilots and all other persons, to be delivered into his hands. He will enjoin them, to observe the strictest silence respecting the object of the voyage, and the discoveries that may have been made, and for this he will require them to pledge their words.

CHAPTER 3

Alexander von Humboldt's Personal Narrative

Alexander von Humboldt and his traveling companion Aimé Bonpland spent five years, from 1799 to 1804, traveling through Venezuela, Colombia, Ecuador, Peru, Cuba, and Mexico. They collected massive amounts of dried plants and accumulated information on plant geography, zoology, geology, mineralogy, and climatology and a host of writings on the political economy, archaeology, and ethnography of South America. His published material, including his widely read Personal Narrative, *excerpted below, served as a foundation text for scientific travelers throughout the nineteenth century.*

Source: Alexander von Humboldt, *Personal Narrative of Travels to the Equinoctial Regions of America during the Years 1799–804*, vol. 1, translated and edited by Thomasina Ross (London: George Bell and Sons, 1907), Introduction.

Many years have elapsed since I quitted Europe, to explore the interior of the New Continent. Devoted from my earliest youth to the study of nature, feeling with enthusiasm the wild beauties of a country guarded by mountains and shaded by ancient forests, I experienced in my travels, enjoyments which have amply compensated for the privations inseparable from a laborious and often agitated life. These enjoyments, which I endeavoured to impart to my readers in my 'Remarks upon the Steppes,' and in the 'Essay on the Physiognomy of Plants,' were not the only fruits I reaped from an undertaking formed with the design of contributing to the progress of natural philosophy. I had long prepared myself for the observations which were the principal object of my journey to the torrid zone. I was provided with instruments of easy and convenient use, constructed by the ablest makers, and I enjoyed the special protection of a government which, far from presenting obstacles to my investigations, constantly honoured me with every mark of regard and confidence. I was aided by a courageous and enlightened friend, and it was singularly propitious to the success of our participated labour, that the zeal and equanimity of that friend never failed, amidst the fatigues and dangers to which we were sometimes exposed.

Under these favourable circumstances, traversing regions which for ages have remained almost unknown to most of the nations of Europe, I might add even to Spain, M. Bonpland and myself collected a considerable number of materials, the publication of which may throw some light on the history of nations, and advance the study of nature.

I had in view a two-fold purpose in the travels of which I now publish the historical narrative. I wished to make known the countries I had visited; and to collect such facts as are fitted to elucidate a science of which we as yet possess scarcely the outline, and which has been vaguely denominated Natural History of the World, Theory of the Earth, or Physical Geography. The last of these two objects seemed to me the most important. I was passionately devoted to botany and certain parts of zoology, and I flattered myself that our investigations might add some new species to those already known, both in the animal and vegetable kingdoms; but preferring the connection of facts which have been long observed, to the knowledge of insulated facts, although new, the discovery of an unknown genus seemed to me far less interesting than an observation on the geographical relations of the vegetable world, on the migrations of the social plants, and the limit of the height which their different tribes attain on the flanks of the Cordilleras.

The natural sciences are connected by the same ties which link together all the phenomena of nature. The classification of the species, which must be considered as the fundamental part of botany, and the study of which is rendered attractive and easy by the introduction of natural methods, is to the geography of plants what descriptive mineralogy is to the indication of the rocks constituting the exterior crust of the globe. To comprehend the laws observed in the position of these rocks, to determine the age of their successive formations, and their identity in the most distant regions, the geologist should be previously acquainted with the simple fossils which compose the mass of mountains, and of which the names and character are the object of oryctognostical knowledge. It is the same with that part of the natural history of the globe which treats of the relations plants have to each other, to the soil whence they spring, or to the air which they inhale and modify. The progress of the geography of plants depends in a great measure on that of descriptive botany; and it would be injurious to the advancement of science, to attempt rising to general ideas, whilst neglecting the knowledge of particular facts.

I have been guided by these considerations in the course of my inquiries; they were always present to my mind during the period of my preparatory studies. When I began to read the numerous narratives of travels, which compose so interesting a part of modern literature, I regretted that travellers, the most enlightened in the insulated branches of natural history, were seldom possessed of sufficient variety of knowledge to avail themselves of every advantage arising from their position. It appeared to me, that the importance of the results hitherto obtained did not keep pace with the immense progress which, at the end of the eighteenth century, had been made in several departments of science, particularly geology, the history of the modifications of the atmosphere, and the physiology of animals and plants. I saw with regret, (and all scientific men have shared this feeling) that whilst the number of accurate instruments was daily increasing, we were still ignorant of the height of many mountains and elevated plains; of the periodical oscillations of the aerial ocean; of the limit of perpetual snow within the polar circle and on the borders of the torrid zone; of the variable intensity of the magnetic forces, and of many other phenomena equally important.

Maritime expeditions and circumnavigatory voyages have conferred just celebrity on the names of the naturalists and astronomers who have been appointed by various governments to share the dangers of those undertakings; but though these eminent men have given us precise notions of the external configuration of countries, of the natural history of the ocean, and of the productions of islands and coasts, it must be admitted that maritime expeditions are less fitted to advance the progress of geology and other parts of physical science, than travels into the interior of a continent. The advancement of the natural sciences has been subordinate to that of geography and nautical astronomy. During a voyage of several years, the land but seldom presents itself to the observation of the mariner, and when, after lengthened expectation, it is descried, he often finds it stripped of its most beautiful productions. Sometimes, beyond a barren coast, he perceives a ridge of mountains covered with verdure, but its distance forbids examination, and the view serves only to excite regret.

Journeys by land are attended with considerable difficulties in the conveyance of instruments and collections, but these difficulties are compensated by advantages which it is unnecessary to enumerate. It is not by sailing along a coast that we can discover the direction of chains of mountains, and their geological constitution, the climate of each zone, and its influence on the forms and habits of organized beings. In proportion to the extent of continents, the greater on the surface of the soil are the riches of animal and vegetable productions; the more distant the central chain of mountains from the sea-shore, the greater is the variety in the bosom of the earth, of those stony strata, the regular succession of which unfolds the history of our planet. As every being considered apart is impressed with a particular type, so, in like manner, we find the same distinctive impression in the arrangement of brute matter organized in rocks, and also in the distribution and mutual relations of plants and animals. The great problem of the physical description of the globe, is the determination of the form of these types, the laws of their relations with each other, and the eternal ties which link the phenomena of life, and those of inanimate nature.

James Clark Ross and the Northern Magnetic Pole

James Clark Ross was the premier polar explorer in the Victorian era, traveling to both the Arctic and Antarctic regions, including a privately funded voyage between 1829 and 1833 with his uncle, John Ross, also a seasoned Arctic explorer. In the passage below, Ross describes his triumph of standing on the northern magnetic pole, the first European in history to do so.

Source: John Ross, "Chapter XLII," in *Narrative of a Second Voyage in Search of a North-West Passage and of a Residence in the Arctic Region during the Years 1829, 1830, 1831, 1832, Including the Reports of Commander, Now Captain, J. C. Ross and the Discovery of the Northern Magnetic Pole* (London: n.p., 1835), 549–557.

Having given to the Royal Society a paper on the subject of the North Magnetic Pole, which they have done me the honour to print, I need not here repeat the preliminary or other general remarks which it contains, but confine this narrative, as I have done my former ones, to the facts and reflections which occurred during our voyage and our travelling: thus conforming to the journal character of the volume in which I have borne the share assigned to me. If there are scientific readers who desire to see what I have written on this subject since my return, they will find it in the Philosophical Transactions for 1834.

It must be known to many more readers than those, that the subject here in question had engaged the attention of our predecessors, Parry and Franklin, during their several voyages and travels in these regions for those purposes of geographical discovery which are now so familiar to every one. If all general

praise of these conspicuous men is now superfluous, I must here however re-
mark, that the numerous and accurate observations on the subject of magnetism, made by them and the officers under their command, have proved of great
value towards the advancement of magnetic science in general, if more particularly to the assignment of the laws by which that of the globe, as it regards the
needle, is regulated.

The geographical restrictions, however, to which these discoveries had
been subjected, were such as to prevent them from extending their observations over so large a space as was to be desired. They had at different times
made nearer approximations to the expected place of the North magnetic pole
than had ever before been effected, but the spot where it ought to exist had
been a sealed place to them: more than once tantalizing with hopes which, it
was destined, were not then to be fulfilled. Observations were still wanting
at other and nearer points to this desired and almost mysterious spot; that its
place might be at least assigned with still more security and precision than it
had been from those already made, that, if possible, the observer might even
assure himself that he had reached it, had placed his needle where no deviation
from the perpendicular was assignable, and had so set his foot that it now lay
between him and the centre of the earth.

These hopes were at length held out to us; we had long been drawing near
to this point of so many desires and so many anxieties, we had conjectured and
calculated, once more, its place, from many observations and from nearer approaches than had ever yet been made, and with our now acquired knowledge
of the land on which we stood, together with the power of travelling held out
to us, it at last seemed certain that this problem was reserved for us, that we
should triumph over all difficulties, and plant the standard of England on the
North magnetic pole, on the keystone of all these labours and observations.

Under the determinations of the navigators who had preceded us the place
of this important spot had been calculated, and with a degree of precision, as
it afterwards proved, far greater than could have been expected. At the time
of our departure from England, it was presumed to be situated in 70° of north
latitude, and in 98° 30' of west longitude. Thus it appeared, that in the course of
my land journey to the westward in the preceding year (1830), I had been within
ten miles of this assigned place, when near Cape Felix: but, as I was not then
provided with the necessary instruments, I could do nothing towards verifying
the fact, and had the mortification of being obliged to return, when thus, as I
believed, on the point of accomplishing this long wished-for object.

We had now, however, been compelled to pass another winter in our ship,
not far from the place which we had occupied in the former year, and I thus
hoped that I should be able to investigate this spot more effectually in the

coming spring. With this view I carried on a series of magnetic observations during the winter, and thus at length succeeded in assigning a place for this magnetic pole which I believed to be much more accurate than the one which had previously been supposed. The dip of the needle at the place of observation exceeded 89°; and it was thus a much nearer approximation in distance than had yet been attained.

These observations were continued till within a few hours of our departure from the ship, on a journey which was undertaken for this sole purpose, and we set out on our expedition on the 27th of May, accompanied by Captain Ross and a party under his direction, as far as the shores of the western ocean, when they separated from us for the purpose of returning to the ship by the way of Neitchillee.

Unfortunately, however, the weather became so very unfavourable that I could no longer continue these magnetic observations: and this vexatious state of things attended us during nearly the whole of our journey across the country. We were, nevertheless, obliged to persist, as it was impossible to wait for better weather when our time was always so much contracted by the state of our supplies. At three in the afternoon of the same day, therefore, we crossed to the opposite shore of the inlet into which the Stanley river flows, and travelled along the land towards the west until eight in the morning of the twenty-eighth, when we were compelled to halt, in consequence of the ophthalmia, which, from the usual cause, had severely affected four of our party. We had gained but ten miles, and our encampment was made in latitude 69° 34' 45", and longitude 94° 54' 23" west.

. . .

We were, however, fortunate in here finding some huts of Esquimaux, that had not long been abandoned. Unconscious of the value which not only we, but all the civilized world, attached to this place, it would have been a vain attempt on our part to account to them for our delight, had they been present. It was better for us that they were not; since we thus took possession of their works, and were thence enabled to establish our observations with the greater ease; encamping at six in the evening on a point of land about half a mile to the westward of those abandoned snow houses.

The necessary observations were immediately commenced, and they were continued throughout this and the greater part of the following day. Of these, the details for the purposes of science have been since communicated to the Royal Society; as a paper containing all that philosophers require on the subject has now also been printed in their Transactions. I need not therefore repeat them here, even had it not been the plan of the whole of this volume to refer every scientific matter which had occurred to Captain Ross and myself, to a separate work, under the name of an appendix.

But it will gratify general curiosity to state the most conspicuous results in a simple and popular manner. The place of the observatory was as near to the magnetic pole as the limited means which I possessed enabled me to determine. The amount of the dip, as indicated by my dipping needle, was 89° 59', being thus within one minute of the vertical; while the proximity at least of this pole, if not its actual existence where we stood, was further confirmed by the action, or rather by the total inaction of the several horizontal needles then in my possession. These were suspended in the most delicate manner possible, but there was not one which showed the slightest effort to move from the position in which it was placed: a fact, which even the most moderately informed of readers must now know to be one which proves that the centre of attraction lies at a very small horizontal distance, if at any.

As soon as I had satisfied my own mind on this subject, I made known to the party this gratifying result of all our joint labours; and it was then, that amidst mutual congratulations, we fixed the British flag on the spot, and took possession of the North Magnetic Pole and its adjoining territory, in the name of Great Britain and King William the Fourth. We had abundance of materials for building, in the fragments of limestone that covered the beach; and we therefore erected a cairn of some magnitude, under which we buried a canister, containing a record of the interesting fact: only regretting that we had not the means of constructing a pyramid of more importance, and of strength sufficient to withstand the assaults of time and of the Esquimaux.

CHAPTER 4

Charles Darwin's Landing on Chatham Island

Charles Darwin believed that his circumnavigation of the globe on HMS Beagle was the most important event in his scientific life. During his five-year voyage, he studied the geographical distribution of species, research that would later form the foundation of his evolutionary theory. Of particular importance were the finches and tortoises on the Galápagos Islands. In the passage excerpted below, Darwin describes his experiences on first landing on Chatham Island in the Galápagos Islands.

Source: Charles Darwin, *The Voyage of the Beagle* (New York: P. F. Collier and Son, 1909), 376–379.

SEPTEMBER 15th.—This archipelago consists of ten principal islands, of which five exceed the others in size. They are situated under the Equator, and between five and six hundred miles westward of the coast of America. They are all formed of volcanic rocks; a few fragments of granite curiously glazed and altered by the heat, can hardly be considered as an exception. Some of the craters, surmounting the larger islands, are of immense size, and they rise to a height of between three and four thousand feet. Their flanks are studded by innumerable smaller orifices. I scarcely hesitate to affirm, that there must be in the whole archipelago at least two thousand craters. These consist either of lava or scoriae, or of finely stratified, sandstone-like tuff. Most of the latter are beautifully symmetrical; they owe their origin to eruptions of volcanic mud without any lava: it is a remarkable circumstance that every one of the twenty-eight tuff-craters which were examined, had their southern sides either much lower than the other sides, or quite broken down and removed. As all these craters apparently have been formed when standing in the sea, and as the waves from the trade wind and the swell from the open Pacific here unite their forces on the southern coasts of all the islands, this singular uniformity in the broken state of the craters, composed of the soft and yielding tuff, is easily explained.

Considering that these islands are placed directly under the equator, the climate is far from being excessively hot; this seems chiefly caused by the singularly low temperature of the surrounding water, brought here by the great southern Polar current. Excepting during one short season, very little rain falls, and even then it is irregular; but the clouds generally hang low. Hence, whilst the lower parts of the islands are very sterile, the upper parts, at a height of a thousand feet and upwards, possess a damp climate and a tolerably luxuriant vegetation. This is especially the case on the windward sides of the islands, which first receive and condense the moisture from the atmosphere.

In the morning (17th) we landed on Chatham Island, which, like the others, rises with a tame and rounded outline, broken here and there by scattered hillocks, the remains of former craters. Nothing could be less inviting than the first appearance. A broken field of black basaltic lava, thrown into the most rugged waves, and crossed by great fissures, is everywhere covered by stunted, sunburnt brushwood, which shows little signs of life. The dry and parched surface, being heated by the noon-day sun, gave to the air a close and sultry feeling, like that from a stove: we fancied even that the bushes smelt unpleasantly. Although I diligently tried to collect as many plants as possible, I succeeded in getting very few; and such wretched-looking little weeds would have better become an arctic than an equatorial Flora. The brushwood appears, from a short distance, as leafless as our trees during winter; and it was some time before I discovered that not only almost every plant was now in full leaf, but that the greater number were in flower. The commonest bush is one of the Euphorbiaceae: an acacia and a great odd-looking cactus are the only trees which afford any shade. After the season of heavy rains, the islands are said to appear for a short time partially green. The volcanic island of Fernando Noronha, placed in many respects under nearly similar conditions, is the only other country where I have seen a vegetation at all like this of the Galapagos Islands.

The Beagle sailed round Chatham Island, and anchored in several bays. One night I slept on shore on a part of the island, where black truncated cones were extraordinarily numerous: from one small eminence I counted sixty of them, all surmounted by craters more or less perfect. The greater number consisted merely of a ring of red scoriae or slags, cemented together: and their height above the plain of lava was not more than from fifty to a hundred feet; none had been very lately active. The entire surface of this part of the island seems to have been permeated, like a sieve, by the subterranean vapours: here and there the lava, whilst soft, has been blown into great bubbles; and in other parts, the tops of caverns similarly formed have fallen in, leaving circular pits with steep sides. From the regular form of the many craters, they gave to the country an artificial appearance, which vividly reminded me of those parts of Staffordshire, where the great iron-foundries are most numerous. The day was glowing hot, and the scrambling over the rough surface and through the intricate thickets, was very fatiguing; but I was well repaid by the strange Cyclopean scene. As I was walking along I met two large tortoises, each of which must have weighed at least two hundred pounds: one was eating a piece of cactus, and as I approached, it stared at me and slowly walked away; the other gave a deep hiss, and drew in its head. These huge reptiles, surrounded by the black lava, the leafless shrubs, and large cacti, seemed to my fancy like some antediluvian animals. The few dull-coloured birds cared no more for me than they did for the great tortoises.

Alfred Russel Wallace's Account of Living in Dobbo

Alfred Russel Wallace, the co-discoverer of evolution through natural selection, traveled widely as a natural history collector throughout the Amazon and the Malay Archipelago (Indonesia and Malaysia). His prize possessions included a unique type of bird of paradise, which he obtained during his visit to the Aru Islands, off the coast of New Guinea. In the passage below, Wallace recounts his experiences with insects, birds of paradise, and the natives while living in Dobbo, a trading village in the Aru Islands.

Source: Alfred Russel Wallace, *The Malay Archipelago: The Land of the Orang-Utan, and the Bird of Paradise* (New York: Harper and Brothers, 1869), 465–467.

Ever since leaving Dobbo I had suffered terribly from insects, who seemed here bent upon revenging my long-continued persecution of their race. At our first stopping-place sand-flies were very abundant at night, penetrating to every part of the body, and producing a more lasting irritation than mosquitoes. My feet and ankles especially suffered, and were completely covered with little red swollen specks, which tormented me horribly. On arriving here we were delighted to find the house free from sand-flies or mosquitoes, but in the plantations where my daily walks led me, the day-biting mosquitoes swarmed, and seemed especially to delight in attacking my poor feet. After a month's incessant punishment, those useful members rebelled against such treatment and broke into open insurrection, throwing out numerous inflamed ulcers, which were very painful, and stopped me from walking. So I found myself confined to the house, and with no immediate prospect of leaving it. Wounds or sores in the feet are especially difficult to heal in hot climates, and I therefore dreaded them more than any other illness. The confinement was very annoying, as the fine hot weather was excellent for insects, of which I had every promise of obtaining a fine collection; and it is only by daily and unremitting search that the smaller kinds, and the rarer and more interesting specimens, can be obtained. When I crawled down to the river-side to bathe, I often saw the blue-winged Papilio ulysses, or some other equally rare and beautiful insect; but there was nothing for it but patience, and to return quietly to my bird-skinning, or whatever other work I had indoors. The stings and bites and ceaseless irritation caused by these pests of the tropical forests, would be borne uncomplainingly; but to be kept prisoner by them in so rich and unexplored a country where rare and beautiful creatures are to be met with in every forest ramble—a country reached by such a long and tedious voyage, and which might not in the present century be again visited for the same purpose—is a punishment too severe for a naturalist to pass over in silence.

I had, however, some consolation in the birds my boys brought home daily, more especially the Paradiseas, which they at length obtained in full plumage. It was quite a relief to my mind to get these, for I could hardly have torn myself away from Aru had I not obtained specimens.

But what I valued almost as much as the birds themselves was the knowledge of their habits, which I was daily obtaining both from the accounts of my hunters, and from the conversation of the natives. The birds had now commenced what the people here call their "sacaleli," or dancing-parties, in certain trees in the forest, which are not fruit trees as I at first imagined, but which have an immense tread of spreading branches and large but scattered leaves, giving a clear space for the birds to play and exhibit their plumes. On one of these trees a dozen or twenty full-plumaged male birds assemble together, raise up their wings, stretch out their necks, and elevate their exquisite plumes, keeping them in a continual vibration. Between whiles they fly across from branch to branch in great excitement, so that the whole tree is filled with waving plumes in every variety of attitude and motion. . . . The bird itself is nearly as large as a crow, and is of a rich coffee brown colour. The head and neck is of a pure straw yellow above and rich metallic green beneath. The long plumy tufts of golden orange feathers spring from the sides beneath each wing, and when the bird is in repose are partly concealed by them. At the time of its excitement, however, the wings are raised vertically over its back, the head is bent down and stretched out, and the long plumes are raised up and expanded till they form two magnificent golden fans, striped with deep red at the base, and fading off into the pale brown tint of the finely divided and softly waving points. The whole bird is then overshadowed by them, the crouching body, yellow head, and emerald green throat forming but the foundation and setting to the golden glory which waves above. When seen in this attitude, the Bird of Paradise really deserves its name, and must be ranked as one of the most beautiful and most wonderful of living things. I continued also to get specimens of the lovely little king-bird occasionally, as well as numbers of brilliant pigeons, sweet little parroquets, and many curious small birds, most nearly resembling those of Australia and New Guinea.

Joseph Dalton Hooker's Account of Preparations and Negotiations for Entering the Tibetan Passes

Joseph Dalton Hooker traveled first to the Antarctic region with James Clark Ross and then into the Himalayas, traversing the states of Sikkim and eastern Nepal. Hooker was the first naturalist in history, including Darwin himself, to explore a new world armed with Darwin's theory of evolution through natural selection. In the passage below, Hooker describes his preparations and the negotiations required to enter the Tibetan passes near Kinchinjunga, then thought to be the highest mountain in the world.

Source: J. D. Hooker, *Himalayan Journals; Or, Notes of a Naturalist in Bengal, the Sikkim and Nepal Himalayas, the Khasia Mountains &c.* (London: Ward, Lock, Bowden and Co., 1891), 123–126.

Owing to the unsatisfactory nature of our relations with the Sikkim authorities, to which I have elsewhere alluded, my endeavours to procure leave to penetrate further beyond the Dorjiling territory than Tonglo, were attended with some trouble and delay.

In the autumn of 1848, the Governor-General communicated with the Rajah, desiring him to grant me honourable and safe escort through his dominions; but this was at once met by a decided refusal, apparently admitting of no compromise. Pending further negotiations, which Dr. Campbell felt sure would terminate satisfactorily, though perhaps too late for my purpose, he applied to the Nepal Rajah for permission for me to visit the Tibetan passes, west of Kinchinjunga; proposing in the meanwhile to arrange for my return through Sikkim. Through the kindness of Col. Thoresby, the Resident at that Court, and the influence of Jung Bahadoor, this request was promptly acceded to, and a guard of six Nepalese soldiers and two officers was sent to Dorjiling to conduct me to any part of the eastern districts of Nepal which I might select. I decided upon following up the Tambur, a branch of the Arun river, and exploring the two easternmost of the Nepalese passes into Tibet (Wallanchoon and Kanglachem), which would bring me as near to the central mass and loftiest part of the eastern flank of Kinchinjunga as possible.

For this expedition (which occupied three months), all the arrangements were undertaken for me by Dr. Campbell, who afforded me every facility which in his government position he could command, besides personally superintending the equipment and provisioning of my party. Taking horses or loaded animals of any kind was not expedient: the whole journey was to be performed on foot, and everything carried on men's backs. As we were to march through wholly unexplored countries, where food was only procurable at uncertain intervals, it was necessary to engage a large body of porters, some of whom should carry bags of rice for the coolies and themselves too. The difficulty of selecting these carriers, of whom thirty were required, was very great. The Lepchas, the best and most tractable, and over whom Dr. Campbell had the most direct influence, disliked employment out of Sikkim, especially in so warlike a country as Nepal: and they were besides thought unfit for the snowy regions. The Nepalese, of whom there were many residing as British subjects in Dorjiling, were mostly run-aways from their own country, and afraid of being claimed, should they return to it, by the lords of the soil. To employ Limboos, Moormis, Hindoos, or other natives of low elevations, was out of the question;

and no course appeared advisable but to engage some of the Bhotan run-aways domiciled in Dorjiling, who are accustomed to travel at all elevations, and fear nothing but a return to the country which they have abandoned as slaves, or as culprits: they are immensely powerful, and though intractable to the last degree, are generally glad to work and behave well for money. The choice, as will hereafter be seen, was unfortunate, though at the time unanimously approved.

My party mustered fifty-six persons. These consisted of myself, and one personal servant, a Portuguese half-caste, who undertook all offices, and spared me the usual train of Hindoo and Mahometan servants. My tent and equipments (for which I was greatly indebted to Mr. Hodgson), instruments, bed, box of clothes, books and papers, required a man for each. Seven more carried my papers for drying plants, and other scientific stores. The Nepalese guard had two coolies of their own. My interpreter, the coolie Sirdar (or headman), and my chief plant collector (a Lepcha), had a man each. Mr. Hodgson's bird and animal shooter, collector, and stuffer, with their ammunition and indispensables, had four more; there were besides, three Lepcha lads to climb trees and change the plant-papers, who had long been in my service in that capacity; and the party was completed by fourteen Bhotan coolies laden with food, consisting chiefly of rice with ghee, oil, capsicums, salt, and flour.

I carried myself a small barometer, a large knife and digger for plants, note-book, telescope, compass, and other instruments; whilst two or three Lepcha lads who accompanied me as satellites, carried a botanising box, thermometers, sextant and artificial horizon, measuring-tape, azimuth compass and stand, geo-logical hammer, bottles and boxes for insects, sketch-book, etc., arranged in compartments of strong canvass bags. The Nepal officer (of the rank of serjeant, I believe) always kept near me with one of his men, rendering innumerable little services. Other sepoys were distributed amongst the remainder of the party; one went ahead to prepare camping-ground, and one brought up the rear.

The course generally pursued by Himalayan travellers is to march early in the morning, and arrive at the camping-ground before or by noon, breakfast-ing before starting, or *en route.* I never followed this plan, because it sacrificed the mornings, which were otherwise profitably spent in collecting about camp; whereas, if I set off early, I was generally too tired with the day's march to em-ploy in any active pursuit the rest of the daylight, which in November only lasted till 6 P.M. The men breakfasted early in the morning, I somewhat later, and all had started by 10 A.M., arriving between 4 and 6 P.M. at the next camping-ground. My tent was formed of blankets, spread over cross pieces of wood and a ridge-pole, enclosing an area of 6 to 8 feet by 4 to 6 feet. The bedstead, table, and chair were always made by my Lepchas, as described in the Tonglo excursion. The evenings

I employed in writing up notes and journals, plotting maps, and ticketing the plants collected during the day's march.

I left Dorjiling at noon, on the 27th October, accompanied by Dr. Campbell, who saw me fairly off, the coolies having preceded me. Our direct route would have been over Tonglo, but the threats of the Sikkim authorities rendered it advisable to make for Nepal at once; we therefore kept west along the Goong ridge, a western prolongation of Sinchul.

On overtaking the coolies, I proceeded for six or seven miles along a zig-zag road, at about 7,500 feet elevation, through dense forests, and halted at a little hut within sight of Dorjiling. Rain and mist came on at nightfall, and though several parties of my servants arrived, none of the Bhotan coolies made their appearance, and I spent the night without food or bed, the weather being much too foggy and dark to send back to meet the missing men. They joined me late on the following day, complaining unreasonably of their loads, and without their Sirdar, who, after starting his crew, had returned to take leave of his wife and family. On the following day he appeared, and after due admonishment we started, but four miles further on were again obliged to halt for the Bhotan coolies, who were equally deaf to threats and entreaties. As they did not come up till dusk, we were obliged to encamp here (alt. 7,400 feet) at the common source of the Balasun, which flows to the plains, and the Little Rungeet, whose course is north.

Chapter 5

The Corps of Discovery and the Mandan Nation

The annals of Lewis and Clark's transcontinental expedition demonstrate how the practice of exploratory science worked hand in glove with nation-building. The following entry deals with an early contact experience between the members of the Corps of Discovery and the Mandan nation.

Source: History of the Expedition under the Command of Captains Lewis & Clark, Reporting of the 1814 Biddle Edition (New York: Allerton Book Co., 1922), 35–36, 78–79.

On the acquisition of Louisiana, in the year 1803, the attention of the Government of the United States, was early directed toward exploring and improving the new territory. Accordingly, in the summer of the same year, an expedition was planned by the President for the purpose of discovering the courses and sources of the Missouri, and the most convenient water communication thence to the Pacific ocean. His private secretary captain Meriwether Lewis, and captain William Clark, both officers of the Army of the United States, were associated in the command of this enterprise. After receiving the requisite instructions, captain Lewis left the seat of government, and being joined by captain Clark at Louisville, in Kentucky, proceeded to St. Louis, where they arrived in the month of December. Their original intention was to pass the winter at La Charrette, the highest settlement on the Missouri. But the Spanish commandant of the province, not having received an official account of its transfer to the United States, was obliged by the general policy of his government, to prevent strangers from passing through the Spanish territory. They therefore camped at the mouth of Wood river, on the eastern side of the Mississippi, out of his jurisdiction, where they passed the winter in disciplining the men, and making the necessary preparations for setting out early in the Spring, before which the cession was officially announced.

The party consisted of nine young men from Kentucky, fourteen soldiers of the United States army who volunteered their services, two French watermen—an interpreter and hunter—and a black servant belonging to captain Clark—All these, except the last, were enlisted to serve as privates during the expedition, and three sergeants appointed from amonst them by the captains. In addition to these were engaged a corporal and six soldiers, and nine watermen to accompany the expedition as far as the Mandan nation, in order to assist in carrying the stores, or repelling an attack which was most to be apprehended between Wood river and that tribe.

August 3 The next morning the Indians, with their six chiefs, were all assembled under an awning, formed with the mainsail, in presence of all our party, paraded for the occasion. A speech was then made, announcing to them the change in the government, our promises of protection, and advice as to their future conduct. All the six chiefs replied to our speech, each in his turn, according to rank: they expressed their joy at the change in the government; their hopes that we would recommend them to their Great Father (the President), that they might obtain trade and necessaries; they wanted arms as well for hunting as for defense, and asked our mediation between them and the Mahas, with whom they are now at war. We promised to do so, and wished some of them to accompany us to that nation, which they declined, for fear of being killed by them. We then proceeded to distribute our presents. The grand chief of the nation not being of the party, we sent him a flag, a medal, and some ornaments for clothing. To the six chiefs who were present, we gave a medal of the second grade to one Ottoe chief, and one Missouri chief; a medal of the third grade to two inferior chiefs of each nation: the customary mode of recognizing a chief, being to place a medal round his neck, which is considered among his tribe as a proof of his consideration abroad. Each of these medals was accompanied by a present of paint, garters, and cloth ornaments of dress; and to this we added a cannister of powder, a bottle of whisky, and a few presents to the whole, which appeared to make them perfectly satisfied. The air-gun too was fired, and astonished them greatly. The absent grand chief was an Ottoe, named Weahrushhah, which in English, degenerates into Little Thief. The two principal chieftains present were, Shongotongo, or Big Horse; and Wethea, or Hospitality; also Shosguscan, or White Horse, an Ottoe; the first an Ottoe, the second a Missouri. The incident just related, induced us to give to this place the name of the Council-bluff; the situation of it is exceedingly favourable for a fort and trading factory, as the soil is well calculated for bricks, and there is an abundance of wood in the neighbourhood, and the air is pure and healthy. It is also central to the chief resorts of the Indians: one day's journey to the Ottoes; one and a half to the great Pawnees; two days from the Mahas; two and a quarter from the Pawnees Loups village; convenient to the hunting grounds of the Sioux; and twenty-five days journey to Santa Fee.

The ceremonies of the council being concluded, we set sail in the afternoon, and camped at the distance of five miles, on the south side, where we found the mosquitoes very troublesome.

Dr. G. Suckley's Report on Pacific Salmon

Before the actual construction of transcontinental railroads in the 1860s and 1870s, the federal government sponsored various exploratory expeditions to recommend possible routes. Other than geography, these expeditions were also general natural history surveys that, as Dr. G. Suckley's report on Pacific salmon demonstrates, cataloged natural species and assessed their potential economic value.

Source: Dr. G. Suckley, "Chapter I: Report upon the Salmonidae," in "Report upon the Fishes Collected on the Survey," in *Reports of Explorations and Surveys, to Ascertain the Most Practicable and Economical Route for a Railroad from the Mississippi River to the Pacific Ocean*, vol. 12, bk. 2, no. 5 (Washington, DC: Thomas H. Ford, Printer, 1860), 307–308, 310–311.

As the *salmon* family holds the strongest position in economical importance among the fishes of the northwest, the first portion of this report is devoted to the consideration of that group. . . . The fisheries of Washington Territory will, at an early day, be considered of great importance to our commerce. The various kinds of salmon form the bulk of the valuable fishes there found, but there are, in addition, many others which, although not so numerous, are yet abundant and of fair relative commercial value. Among these are the *cod*, found in moderate quantity in Puget Sound, and said to be very abundant on a deep bar or bank off the mouth of the Straits of Fuca; the *halibut*, found in the same situations; the *eulachon*, a very delicious fish, in some years coming in great shoals in the bays in the lower part of Puget Sound, and along the coast near the mouth of Frazer's river; the *herring* arriving in vast quantities in the same waters at regular periods, besides a vast number of good table fish, such as sole, flounders, the so-called "rock-rod," viviparous perch, &c., &c., which, although not valuable for trade, are useful additions to the fare of the inhabitants greater than eight or ten. This size is very convenient for packing. . . .

Several points on the Columbia river are most excellent locations for the taking of salmon and the establishment of "packing" houses. These are generally at the greater falls and rapids. The best fish are there taken in the spring and early summer months. Salmon of different kinds are taken at other seasons. The species of salmon which is principally used for salting in Puget Sound is the *Skowitz*, an autumnal visitor. Of these, Messrs. Riley & Swan, proprietors of the salmon packing establishment at the mouth of the Puyallup river, have taken 3,000 at one haul of the seine! The average weight of the species cannot be said to exceed twelve pounds, and is perhaps not greater than eight or ten. This size is very convenient for packing. . . .

Puget Sound *proper* has scarcely *any* rock bottom, and but two or three reefs. Near Bellingham bay, and along the north side of the straits, many rock islands occur. Along the shores are many sandspits [*sic*] partially surrounding shallow bays, in which vast numbers of young salmonidae feed and live, and where for a short time before the season of entering the rivers the adult individuals of each kind may be found. It is in these situations that most of the good salmon taken by the Indians during the cold months are caught. Although salmon have been as yet unknown to take bait or the *fly* after entering the rivers of that region, they nevertheless are caught in the *salt bays* in large numbers by the natives. The following plan is pursued. A small herring four or five inches long, is tied to a hook. Some six or eight feet from the bait a small round stone is fastened to the line. The stone acts as a "sinker," keeping the bait sunk some six or eight feet below the surface while being "trolled." The Indian in a light canoe paddles about slowly and noiselessly trolling the line with a jerking motion, and not unfrequently taking in the course of a couple of hours several handsome fish, weighing from ten to thirty pounds each. The time chosen for this business is generally the two hours succeeding day break and an hour or two towards evening.

CHAPTER 6

William Scoresby on Sea Color in the Arctic

Oceanography as a field of study was built on the natural observations of ship surgeons, ocean-traveling naturalists, and captains such as William Scoresby. Scoresby relates here some of his knowledge on the color of the sea that was gained through his extensive exploration as a whaler in the Arctic.

Source: William Scoresby, *An Account of the Arctic Regions, with a History and Description of the Northern Whale-Fishery,* vol. 1 (Edinburgh: Archibald Constable and Co., 1820), 172–175.

In the Spitzbergen quarter, the hydrography of which I have most particularly to consider, the sea is different in colour, transparency, saltness, and temperature, from what it generally is in the Atlantic Ocean.

The water of the main ocean is well known to be as transparent and as colourless as that of the most pure springs; and it is only when seen in very deep seas, that any certain and unchangeable colour appears. This colour is commonly ultramarine blue, differing but a shade from the colour of the atmosphere, when free from the obscurity of cloud or haze. Where this ultramarine blue occurs, the rays of light seem to be absorbed in the water, without being reflected from the bottom; the blue rays only being intercepted. But, where the depth is not considerable, the colour of the water is affected by the quality of the bottom. Thus, fine white sand, in very shallow water, affords a greenish grey, or apple-green colour, becoming of a deeper shade as the depth increases, or as the degree of light decreases; yellow sand, in soundings, produces a dark green colour in the water; dark sand a blackish green; rocks a brownish or a blackish colour; and loose sand or mud, in a tide-way, a greyish colour. From this effect of the bottom, the names of the White Sea, the Black Sea, and the Red Sea, have doubtless been derived. Near the mouths of large rivers, the sea is often of a brownish colour, owing to the admixture of mud and other substances held in suspension, together with vegetable or mineral dyes, brought down with the fresh water from the land. But, in the main ocean, in deep water, the prevailing colour is blue, or greenish blue. It may be observed, that there is a good deal of deception in the colour of the sea, owing to the effect of the sun, and the colour of clouds; and, its true tinge can only be observed, with accuracy, by looking downward through a long tube, reaching nearly to its surface, so as to intercept the lateral rays of light, which, by their reflection, produce the deception, and thus obtain a clear view of the interior of the sea. The trunk of the rudder answers this purpose tolerably well. When thus examined, the colour of the sea is not materially affected, either by sun or clouds. But, if examined superficially, from an exposed situation, the sea, in all places, will be found to vary in appearance with every change in the state of the atmosphere. Hence the surface generally partakes of the colour of the clouds; and, when the sky is chiefly clear, a small cloud partially intercepting the sun's rays, casts a deep brown or blackish shadow over the surface, and sometimes gives the appearance of shallow water, or rocks, and thus occasions, in the navigator, unnecessary alarm. It is not, therefore, the varying aspect of the surface of the water that is meant by the colour of the sea; but the appearance of the interior of a body of waters, when looked into through a perpendicular tube. The only effect then produced by a change in the aspect of the sky, is to give the water a lighter or darker shade; but it has little effect on its real colour. For, observed in this way, the same colours may be recognised in storm, or calm, in fine weather or foul, clear or cloudy, fair or showery, being always nearly the same.

Henry Bigelow's Report on Oceanography

Oceanography was a costly field of scientific exploration that often required government patronage. European countries pioneered this relationship at the end of the nineteenth and beginning of the twentieth centuries. The following 1931 report by oceanographer Henry Bigelow called on American scientists and politicians to appreciate the vital economic and military rationale for a robust U.S. program in oceanographic exploration.

Source: Henry B. Bigelow, *Oceanography: Its Scope, Problems, and Economic Importance* (Boston and New York: Houghton Mifflin, 1931), 3–4, 8–11.

Oceanography has been aptly defined as the study of the world below the surface of the sea: it should include the contact zone between sea and atmosphere. According to present-day acceptance it has to do with all the characteristics of the bottom and margins of the sea, of the sea water, and of the inhabitants of the latter. Thus widely combining geophysics, geochemistry, and biology, it is inclusive, as is, of course, characteristic of any 'young' science: and modern oceanography is in its youth. But in this case it is not so much immaturity that is responsible for the fact that these several sub-sciences are still grouped together, but rather the realization that the physics, chemistry, and biology of the sea water are not only important *per se*, but that in most of the basic problems of the sea all three of these subdivisions have a part. And with every advance in our knowledge of the sea making this interdependence more and more apparent, it is not likely that we shall soon see any general abandonment of this concept of oceanography as a mother science, the branches of which, though necessarily attacked by different disciplines, are intertwined too closely to be torn apart. Every oceanic biologist should, therefore, be grounded in the principles of geophysics and geochemistry; every chemical or physical oceanographer in some of the oceanic aspects of biology.

. . .

While this regional-descriptive era of oceanography will never definitely close so long as the science of the sea is pursued, there came a change, toward the end of the century just past, when persistence in the old discursive methods, determined by established habits of thought, no longer yielded new and wonderful discoveries at the rate that had been the order of the day when no one knew what was to be found at the bottom of the sea. Thenceforth, with increasing frequency, continued exploration along these preliminary lines yielded results more corroborative than novel. And a period of general oceanographic stagnation might then have succeeded to the preceding peak of activity (this did, in fact, happen in America), had there not arisen new schools, centering their attention on the biologic economy of the inhabitants of the ocean as related to their physical-chemical environment, on mathematical analysis of the internal dynamics of the sea water, and on the geologic bearing of submarine topography and sedimentation, rather than on areal [*sic*] surveys of one or another feature of the sea.

This conscious alteration of viewpoint, from the descriptive to the analytic, is one of two chief factors that gives to oceanography its present tone: the other is the growth of an economic demand that oceanography afford practical assistance to the sea fisheries. This demand developed first in northwestern Europe, where, as it chanced, the fisheries were so rapidly expanding, and increasing in intensity through the adoption of more effective methods of fishing, that dread of depletion began to loom in the offing, just when oceanography was approaching the end of its nineteenth-century boom; i.e., just when it needed a fresh stimulus.

The immediate, practical result was a concentration of attention on limited coastwise areas (sites of important fisheries) as contrasted with the broad oceans, and the development of an international and official organization—the Conseil International pour l'Exploration de la Mer—with power to coordinate the scientific efforts of the Fisheries Bureaus of the several nations fronting on these areas in northwestern Europe.

. . .

The foregoing remarks are introductory to the thesis that a discussion of certain of the underlying problems that seem most clearly to illustrate the general fields of research falling within the province of the oceanographer, and that are now most to the fore, is integral in any rational exposition of the scope and present status of this inclusive branch of science. This, and no more, is attempted in the following chapters. To list all the problems that await the oceanographer will never be possible so long as science lives, for new ones will constantly unfold, as the boundaries of knowledge are rolled back.

Matthew Fontaine Maury on Determining Accurate Seafloor Depth

Matthew Fontaine Maury's The Physical Geography of the Sea and Its Meteorology *(1855) is one of the founding texts in the history of oceanography. Here, Maury discusses the problems of disclosing the accurate depth of the seafloor and evaluates some of the new exploratory technologies developed to improve accuracy.*

Source: Matthew Fontaine Maury, *The Physical Geography of the Sea and Its Meteorology* (Boston: Belknap, 1963), 280–282, 286–287.

Chapter XIII: The Depths of the Ocean

561. *Ignorance concerning the depth of "blue water."* Until the commencement of the plan of deep-sea soundings, as they have been conducted in the American navy, the bottom of what the sailors call "blue water" was as unknown to us as is the interior of any of the planets of our system. Ross and Dupetit Thouars, with other officers of the English, French, and Dutch navies, had attempted to fathom the deep sea, some with silk threads, some with spun-yarn (coarse

hemp threads twisted together), and some with the common lead and line of navigation. All of these attempts were made upon the supposition that when the lead reached the bottom, either a shock would be felt, or the line, becoming slack, would cease to run out.

562. *Early attempts at deep-sea soundings—unworthy of reliance.* The series of systematic experiments recently made upon this subject shows that there is no reliance to be placed on such a supposition, for the shock caused by striking bottom can not be communicated through very great depths. Furthermore, the lights of experience show that, as a general rule, the under currents of the deep sea have force enough to take the line out long after the plummet has ceased to do so. Consequently, there is but little reliance to be placed upon deep-sea soundings of former methods, when the depths reported exceeded eight or ten thousand feet.

. . .

571. *Method of making a deep-sea sounding.* In making these deep-sea soundings, the practice is to time the hundred fathom marks as they successively go out; and by always using a line of the same size and "make," and a sinker of the same shape and weight, we at last established the law of descent. Thus the mean of our experiments gave us, for the sinker and twine used,

2 m. 21 s. as the average time of descent from 400 to 500 fathoms.
3 m. 26 s.   "   "   "   1000 to 1100
4 m. 29 s.   "   "   "   1800 to 1900

572. *The law of the plummet's descent.* Now, by aid of the law here indicated, we could tell very nearly when the ball ceased to carry the line out, and when, of course, it began to go out in obedience to the current and drift alone; for currents would sweep the line out at a uniform rate, while the cannon ball would drag it out at a decreasing rate. The development of this law was certainly an achievement, for it enabled us to show that the depth of the sea at the places named was not as great as reports made it. These researches were interesting; the problem in hand was important, and it deserved every effort that ingenuity could suggest for reducing it to a satisfactory solution.

. . .

574. *The deepest part of the Atlantic Ocean.* The greatest depths at which the bottom of the sea has been reached with the plummet are in the North Atlantic

Ocean, and the places, where it has been fathomed do not show it to be deeper than twenty-five thousand feet. The deepest place in this ocean is probably between the parallels of 35<dg> and 40<dg> north latitude, and immediately to the southward of the Grand Banks of Newfoundland. The first specimens have been received from the coral sea of the Indian Archipelago and from the North Pacific. They were collected by the surveying expedition employed in those seas. A few soundings have been made in the South Atlantic, but not enough to justify deduction as to its depths or the precise shape of its floor.

CHAPTER 8

John F. Kennedy's Special Message on Urgent National Needs

Delivered before a joint session of Congress, President John F. Kennedy's Special Message on Urgent National Needs launched the race to the moon. Kennedy also sought in this speech additional funds for the war in Vietnam, civil defense, military assistance to European allies, domestic concerns, and construction of a weather satellite system, framing all of these requests around the need to promote freedom at home as well as abroad.

Source: John F. Kennedy, Special Message to the Congress on Urgent National Needs, 25 May 1961, Public Papers of the Presidents, John F. Kennedy, 1961, Document 205, 396–405.

Mr. Speaker, Mr. Vice President, my co-partners in Government, gentlemen—and ladies:

The Constitution imposes upon me the obligation to "from time to time give to the Congress information of the State of the Union." While this has traditionally been interpreted as an annual affair, this tradition has been broken in extraordinary times.

These are extraordinary times. And we face an extraordinary challenge. Our strength as well as our convictions have imposed upon this nation the role of leader in freedom's cause.

No role in history could be more difficult or more important. We stand for freedom. That is our conviction for ourselves—that is our only commitment to others. No friend, no neutral and no adversary should think otherwise. We are not against any man—or any nation—or any system—except as it is hostile to freedom. Nor am I here to present a new military doctrine, bearing any one name or aimed at any one area. I am here to promote the freedom doctrine.

. . .

[Section IX. Space]

Finally, if we are to win the battle that is now going on around the world between freedom and tyranny, the dramatic achievements in space which occurred in recent weeks should have made clear to us all, as did the Sputnik in 1957, the impact of this adventure on the minds of men everywhere, who are attempting to make a determination of which road they should take. Since early in my term, our efforts in space have been under review. With the advice of the Vice President, who is Chairman of the National Space Council, we have examined where we are strong and where we are not, where we may succeed and where we may not. Now it is time to take longer strides—time for a great new American enterprise—time for this nation to take a clearly leading role in space achievement, which in many ways may hold the key to our future on earth.

I believe we possess all the resources and talents necessary. But the facts of the matter are that we have never made the national decisions or marshalled the national resources required for such leadership. We have never specified long-range goals on an urgent time schedule, or managed our resources and our time so as to insure their fulfillment.

Recognizing the head start obtained by the Soviets with their large rocket engines, which gives them many months of lead-time, and recognizing the likelihood that they will exploit this lead for some time to come in still more impressive successes, we nevertheless are required to make new efforts on our own. For while we cannot guarantee that we shall one day be first, we can guarantee that any failure to make this effort will make us last. We take an additional risk by making it in full view of the world, but as shown by the feat of astronaut Shepard, this very risk enhances our stature when we are successful. But this is not merely a race. Space is open to us now; and our eagerness to share its meaning is not governed by the efforts of others. We go into space because whatever mankind must undertake, free men must fully share.

I therefore ask the Congress, above and beyond the increases I have earlier requested for space activities, to provide the funds which are needed to meet the following national goals:

First, I believe that this nation should commit itself to achieving the goal, before this decade is out, of landing a man on the moon and returning him safely to the earth. No single space project in this period will be more impressive to mankind, or more important for the long-range exploration of space; and none will be so difficult or expensive to accomplish. We propose to accelerate the development of the appropriate lunar space craft. We propose to develop alternate liquid and solid fuel boosters, much larger than any now being developed, until certain which is superior. We propose additional funds for other engine development and for unmanned explorations—explorations which are particularly important for one purpose which this nation will never overlook: the survival of the man who first makes this daring flight. But in a very real sense, it will not be one man going to the moon—if we make this judgment affirmatively, it will be an entire nation. For all of us must work to put him there.

Secondly, an additional 23 million dollars, together with 7 million dollars already available, will accelerate development of the Rover nuclear rocket. This gives promise of some day providing a means for even more exciting and ambitious exploration of space, perhaps beyond the moon, perhaps to the very end of the solar system itself.

Third, an additional 50 million dollars will make the most of our present leadership, by accelerating the use of space satellites for world-wide communications.

Fourth, an additional 75 million dollars—of which 53 million dollars is for the Weather Bureau—will help give us at the earliest possible time a satellite system for world-wide weather observation.

Let it be clear—and this is a judgment which the Members of the Congress must finally make—let it be clear that I am asking the Congress and the country to accept a firm commitment to a new course of action—a course which will last for many years and will carry very heavy costs: 531 million dollars in fiscal '62—an estimated seven to nine billion dollars additional over the next five years. If we are to go only half way, or reduce our sights in the face of difficulty, in my judgment it would be better not to go at all.

Now this is a choice which this country must make, and I am confident that under the leadership of the Space Committees of the Congress, and the Appropriating Committees, that you will consider the matter carefully.

It is a most important decision that we make as a nation. But all of you have lived through the last four years and have seen the significance of space and the adventures in space, and no one can predict with certainty what the ultimate meaning will be of mastery of space.

I believe that we should go to the moon. But I think every citizen of this country as well as the Members of the Congress should consider the matter carefully in making their judgment, to which we have given attention over many weeks and months, because it is a heavy burden, and there is no sense in agreeing or desiring that the United States take an affirmative position in outer space, unless we are prepared to do the work and bear the burdens to make it successful.

This decision demands a major national commitment of scientific and technical manpower, material and facilities, and the possibility of their diversion from other important activities where they are already thinly spread. It means a degree of dedication, organization and discipline which have not always characterized our research and development efforts. It means we cannot afford undue work stoppages, inflated costs of material or talent, wasteful interagency rivalries, or a high turnover of key personnel.

New objectives and new money cannot solve these problems. They could in fact, aggravate them further—unless every scientist, every engineer, every serviceman, every technician, contractor, and civil servant gives his personal pledge that this nation will move forward, with the full speed of freedom, in the exciting adventure of space.

CHAPTER 9

James M. Beggs's Letter on Budget Cuts and the Jet Propulsion Laboratory

When President Ronald Reagan's budget office started cutting NASA's budget, the agency did not have an administrator. Reagan finally appointed James M. Beggs, who quickly set out to contest the cuts. In the letter below, he threatens to close the Jet Propulsion Laboratory, near Reagan's home in California, and end the nation's planetary science program if the cuts continued. In addition to being the president's home state, California also had the largest congressional delegation. Beggs's letter served as a reminder that there would be political repercussions from cuts of this magnitude.

Source: James M. Beggs to Honorable David A. Stockman, [stamped "SEP 29 1981"], Document II-31, "Exploring the Unknown, vol. 5: Exploring the Cosmos," Washington, DC, NASA SP-2001-4407, 433–437.

Honorable David A. Stockman
Director
Office of Management and Budget
Washington, DC 20503

Dear Mr. Stockman:

We have reviewed the guidelines you provided for the NASA budget outlays for FY 1983 at $6,041 million and FY 1984 at $5,687 million. In examining the NASA program to determine what must be done to reach these ceiling levels, I have come to the firm conclusion that they take us past the point at which we can simply take percentage reductions out of each of our programs to meet the proposed guidelines. Rather, we are now at the point at which it becomes necessary, in order to maintain viable programs in some areas, to close down other major programs that NASA has operated since its inception.

The proposed guidelines represent real reductions over the current spending level in FY 1981 of 10% in FY 1983 and 20% in 1984 when the effects of inflation are properly taken into account. The conclusion I have reached is inescapable when added to the fact that the NASA budget has already been reduced by over 20% in real terms, in the 10 years since the Shuttle decision was taken and it was agreed between NASA and OMB (Cap Weinberger) that NASA's budget would remain constant in real terms at the $3.4 billion level in 1971 dollars.

What I am compelled to do to meet these guidelines is to delete the planetary exploration program to reach an intermediate $6.5 billion level, and then to wipe out the space applications program and to make significant reductions in the Space Shuttle program to meet the $6.0 billion level.

It is important to understand the reasons behind the decisions I have outlined. The planetary exploration program is one of the most successful and viable NASA programs. However, it is our judgment that in terms of scientific priority it ranks below space astronomy and astro-physics. Planetary exploration is much more highly dependent on launch vehicles, and it is our opinion that the most important missions that can reasonably be done within the current launch vehicle capability have, more or less, been done. The next step in planetary exploration is to do such things as landing missions and sample return missions, and these require full development of the capability of the

Shuttle and the ability to assemble elements in earth orbit before sending the assembled spacecraft on its way. In our judgment, it is ultimately better for future planetary exploration to concentrate on developing the Shuttle capabilities rather than to attempt to run a "sub-critical" planetary program given the current financial restrictions we face. Of course, elimination of the planetary exploration program will make the Jet Propulsion Laboratory in California surplus to our needs.

In the case of the space applications program our intention is to attempt to retain the scientific aspects of the program but to delete the future efforts in weather, remote sensing, and communications oriented toward commercialization and demonstration. We would make this decision recognizing that applications have been politically popular and that these are the areas that are being emphasized in the European and Japanese Space efforts. Finally, to reach the $6 billion level, we would be forced to reduce the flight rate of the Shuttle below the 32 flight level through 1985 recommended in the FY 1983 budget.

We believe that, while painful and unpopular, the above approach will provide a program which marginally meets the objectives of the basic NASA mission. Making cuts in all programs, with the result that none are truly viable, will not advance the interests of the United States.

Therefore the budget guidelines you have proposed go beyond simple cut backs and require policy decisions from the appropriate officials in the White House. A few weeks ago, I met with Admiral Garrick of Mr. Meese's staff and informed him that any NASA budget for FY 1983 much below the level of $7.0 billion would require such policy decisions. I am now requesting a meeting with you and the President's Counselor, Mr. Meese, to discuss the policy questions raised by the budget guidelines you have proposed. It is most important, in my opinion, that all involved clearly understand the political implications and the international complications that result from these decisions.

Sincerely yours,
[signature]
James M. Beggs
Administrator

Concurrence: "Hans Mark" [hand-signed]/AD
A:JMBeggs:A24716:tm:53918:9/29/81

Daniel S. Goldin's May 1992 Remarks to the Jet Propulsion Laboratory

Daniel S. Goldin was appointed by President George H. W. Bush to be an agent of change in NASA. The Bush administration was frustrated with the increasing cost and slowing pace of space missions, a product of reliance on the Space Shuttle and of NASA management practices. Goldin, who had worked as a manager of TRW corporation's efforts supporting the Strategic Defense Initiative, was an advocate of new management methods and advanced technologies that could dramatically reduce the cost of space science. In a visit to the Jet Propulsion Laboratory on 28 May 1992, he challenged the lab's staff to adopt his "faster, better, cheaper" ways.

Source: Remarks to Jet Propulsion Lab Workers, Thursday, 28 May 1992, Speeches of Daniel G. Goldin, CD-ROM compiled by the NASA History Office, January 2001, Document e000020.pdf, edited for length.

[marginal notation: Goldin's copy]

Thank you, Ed Stone. It's a delight to be here. I've been looking forward to coming here for quite some time. It's simply inspiring being in the presence of such scientific talent. (You can almost feel the brain waves bouncing off the walls in here. It's amazing!)

JPL and Cal Tech have [among] the best people in the space business. In field after field—electronics, astrophysics, planetary science—and I could go on—you are the tops.

What we as a civilization know about our past—where we come from—and about our neighbors in the solar systems—comes [in most part] from you. To do this took the ingenuity, talent and innovation of the NASA team—and our partners in business and academia.

But for all the glories of the past, it's time we face the fact that we have real problems in the way we're managing our space programs—including robotic missions. Planetary missions can take a decade to plan, another decade to build, and another decade to get to their destination. [Who will be here to read the data that comes back!]

We're capable of so much more. We've got technology sitting in [your] labs that people drool over elsewhere. But it's taking too long to incorporate into spacecraft that will unlock the mysteries of the solar system. It's taking too long to transfer the technology to our industry so that we're more competitive world-wide creating Jobs. [*sic*]

That we're not measuring up to our potential is not the fault of the employees. It's the system's fault, and we must work together to bring needed change.

The new NASA is committed to restore the technical and managerial excellence that made NASA the hallmark of quality in the 1960s and 1970s. We simply must do some serious self-evaluation to see how we can recapture the pioneering spirit and excellence of those days.

Now, there must be as many rumors here at JPL as there are in Washington about the new NASA—so let me tell you what we are doing.

Over the coming months, the NASA senior management team will work with other NASA employees and JPL to come up with a shared vision—a roadmap to define the goals and objectives of the space and aeronautics program—and then tell us how we can implement our missions in a more cost effective manner. How can we do everything better, faster, cheaper, without compromising safety.

Why have we initiated this activity? Well, when I arrived at NASA seven weeks ago, the agency had set up [budget] expectations that were unrealistic and that could not be supported by the President and Congress in these tough fiscal times.

Yes, we might have been able to get through FY 1993—if Congress provided the President's budget request—but the budget requirements for the outyears were out of sight.

To correct that disaster-in-the-making, we have established a series of Blue and Red Teams where the best and brightest minds challenge the content, timing, technology and schedule of a program.

These teams were finalized this week. To start, these teams will review our major programs—but as we progress—the teams will look at our other programs, as well as our institution. We must find ways to cut operating costs so that we can generate enough savings to invest in new projects. The teams have been asked to find ways to [significantly] cut [development costs and to reduce operating] costs by a factor of two to four—and even orders of magnitude.

Is that possible? You bet. For all the talent in this room—for all the human and technological *brilliance* of JPL—we must do much more with the dollars we spend. Projects are too big, schedules are too long, costs are too high, and in the end we still have problems on some of our projects like the Galileo antenna and Hubble mirror.

I'm here today to challenge you to do more—*much* more. Building spacecraft didn't used to take so long. Alan Shepard's [Mercury] capsule was delivered from the contractor 22 months after a contract was signed and Shepard flew in it a few months later. The Surveyor program [developed here at JPL] went from contract award to putting its first robot on the moon in five years—and seven Surveyor spacecraft were flown within [the next] two years. It was a fantastic accomplishment, paving the way for Apollo, even though two Surveyors failed. The voyager spacecraft—JPL's pride and joy—were ordered and launched in only five years.

But now, years [later], with all our experience and advanced technology, we measure programs in decades. Plot a chart on how much money was spent and how many probes were launched in the 60s, 70s, and 80s. Now ask yourselves: are we today getting all we should? We can do better. We *must* do better.

We need to stretch ourselves. Be bold—take risks. [A] project that's 20 for 20 isn't successful. It's proof that we're playing it too safe. If the gain is great, risk is warranted. Failure is OK, as long as it's on a project that's pushing the frontiers of technology.

. . .

This civilization's future prosperity lies in the exploration of space. We owe it to future generations to continue the human journey of exploration.

Two years ago, little Voyager 2—one of the most *priceless* hunks of metal ever assembled—left our solar system carrying a copper disk—a cosmic *message*-in-a-bottle from Planet Earth. From the very *heart* of all *humanity*, it carries *this* message: "We step out of our solar system into the universe seeking only peace and friendship—to *teach* if we are called upon, to be *taught* if we are *fortunate.* We know full well that [we] are but a *small* part of the *immense* universe—and it is with *humility* and *hope* that we *take* this *step.*"

Ladies and gentlemen, that *step* was part of an *unstoppable* march, begun 500 years ago and stretching 500 years hence. The Balboas and Magellans of the *Space* Age of Exploration are seated here today. The Lewis and Clarks will come after us, until that *inevitable* day when we venture out to the stars.

Will we *do* it, or will it remain just a fantasy? It's really up to *us*. *Join* me on this most *noble* of endeavors. Thank you very much.

Daniel S. Goldin's March 2000 Remarks to the Jet Propulsion Laboratory

After the Jet Propulsion Laboratory's loss of the Mars Climate Orbiter and Mars Polar Lander in 1998 and after the publication of the relevant investigation reports, Daniel S. Goldin returned to address the laboratory's staff. He congratulated the lab for its successes during the faster, better, cheaper era and apologized for pushing Mars '98 to the point of failure. After this, Goldin backed off, and mission costs rose. The faster, better, cheaper era was over.

Source: "When The Best Must Do Even Better," Remarks by NASA Administrator Daniel S. Goldin at the Jet Propulsion Laboratory, Pasadena, California, 29 March 2000, Speeches of Daniel G. Goldin, CD-ROM compiled by the NASA History Office, January 2001, document number e000861.pdf, edited for length.

I'd like to thank David Baltimore, the head of Cal Tech. David actually canceled plans to be here today. I think that is just a small expression of his strong support to people here and the work they do. Thank you, David.

I'd also like to acknowledge Admiral Inman, head of the JPL Oversight Committee at Cal Tech. He couldn't be here today, but I talked to him by phone. His commitment to the team here is also unwavering. And I thank him for that.

As I was flying out to California last night, I thought back to one of my early visits to JPL.

It was about eight years ago—just two months after I was fortunate enough to be named Administrator of NASA, by the President of the United States. The head of the best organization in the world, bar none.

Some of you may remember that visit—I was the quiet and unassuming one, Mr Congeniality.

Kidding aside, for a lot of us, it wasn't an easy day.

How do you tell the group that has no equal when it comes to exploring the outer limits . . .

That it was time to explore the inner-workings?

That it was time to do things differently?

That if we wanted to continue along the path of discovery . . .

We all needed a new roadmap to get there?

How do you say all that? Well, I just said it straight out.

I'm quiet . . . unassuming . . . and don't forget . . . I'm very subtle.

I remember one of the meetings I had in particular. I like to call it the "Catch 22" meeting. The abridged version goes something like this.

I started with the pitch.

I said to that particular group the same thing I said to all of you in a larger meeting.

Tell us how we can implement our missions in a more cost-effective manner.

Tell us how we can improve everything we do without compromising safety.

Tell us how NASA can be "Faster, Better, Cheaper."

When I finished, a few of the engineers stood up and said: "Mr. Goldin, we hear what you're saying. We even have some great ideas. But there's more to it."

"If we take a chance . . . if we use new technologies . . . if we try new managerial approaches . . . if we develop that exciting new design . . . the powers-that-be will ask: Is it risk-free? Has it flown before?" "We will say, 'No, it hasn't flown. It's new and different. It won't fly unless and until *you* [senior management] fly it.'"

"They will say: OK. Fine. Sounds good."

"Bring it back to me after it has flown.'"

It's hard to change . . . even when you have little or no choice. That's why I made a pledge to all of you. I said that we wouldn't be overly prescriptive because you know this place better than anyone else does. I said we wouldn't be micromanagers—authority and responsibility would belong totally to you.

But most of all, I said never forget that you are the best and brightest. I empower you to take some risks. Don't be afraid to push the envelope. Press the boundaries. Do things that have not been done. Fly things that haven't been flown. There will be some failure, but when it occurs . . . we will face the problem together . . . and fix the problem together. So, by all means, have the confidence to take risk. Your courage will be rewarded.

Today, before I say anything else, let me say thank you to each and every one of you for accepting that charge. And let me recognize one group in particular that I am incredibly proud of. I had dinner with the leaders of the Mars 1998 Team last night at Monty's here in downtown Pasadena.

I was very candid with them. I told them that in my effort to empower people, I pushed too hard . . . and in so doing, stretched the system too thin.

It wasn't intentional. It wasn't malicious. I believed in the vision . . . but it may have made failure inevitable. I wanted to demonstrate to the world that we could do things much better than anyone else.

And you delivered—you delivered with Mars Pathfinder . . . With Mars Global Surveyor . . . With Deep Space 1. We pushed the boundaries like never before . . . And had not yet reached what we thought was the limit.

Not until Mars 98.

I salute that team's courage and conviction. And make no mistake: they need not apologize to anyone. They did not fail alone. As the head of NASA, I accept the responsibility. If anything, the system failed them. And for that reason, even the best must do better. And just as we cannot let a mission failure diminish our success . . . we certainly cannot allow it to detract us from our goals . . . cloak us in the shroud of the timid . . . or worse, paralyze the agency in a straight jacket of indifference and inactivity.

If there is one thing that I have learned as NASA Administrator, it is that our nation's space program is strong, it is relevant and it is vital to every American . . . not because how we react to success, of which there have been many . . . but how we learn from failure, of which there have been few.

When we suffer a setback, we must learn from mistakes and, yes, move on— continuing to push back the boundaries of the unknown . . . continuing to imagine what might be possible . . . And by God, continuing our effort to land on Mars and one day send astronauts there. You will do it.

. . .

So I'll leave you today with a challenge similar to the one you responded to so well 8 years ago.

You are still the best and brightest—nothing in any report tells me otherwise. Assess yourselves and this Laboratory with the same vigor—the same courage, conviction and confidence—the same heart and soul—that have always set you apart from everyone in the world. And as you do, remember this:

A problem can be fixed. A spacecraft is hardware that can be rebuilt. The science has not been denied . . . it has only been delayed. But the exuberance, pride, and spirit that each of you bring here day-in and day-out is absolutely irreplaceable.

I'll say it again: irreplaceable.

Just outside my office at headquarters, there is [a] quote from Teddy Roosevelt—something I find important in helping me carry out my job, something he said at the dawn of the 20th century. "Far better to dare mighty things, to win glorious triumphs, even those checkered with failure, than to rank with those poor spirits who neither enjoy much nor suffer much, because they live in the gray twilight that knows not victory or defeat."

On the dawn of this new century and this new millennium, we know both victory and defeat. And because we do, let our spirit be both rich and ready. Let's do as we have always done—dare the mightiest of things. And then, let's do what the American people have come to expect from us . . . have the confidence and take the risks that can bring those glorious triumphs . . . on Earth . . . on Mars . . . and wherever America's dreams take us.

Thank you very much.

Notes

CHAPTER 1

1. King James Bible, Daniel 12:4.
2. [Henry Oldenberg], "Epistle Dedicatory," *Philosophical Transactions of the Royal Society of London* 1 (1665): n.p.

CHAPTER 2

1. Isaac Greenwood, "New Method for Composing a Natural History of Meteors," *Philosophical Transactions* (1727): 399.
2. Charles Marie de la Condamine, *Succinct Abridgment of a Voyage Made within the Inland Parts of South-America* (London: n.p., 1747), 24.
3. Ibid., 37.
4. Ibid., 26.
5. Lisbet Koerner, *Linnaeus: Nature and Nation* (Cambridge, MA: Harvard University Press, 1999), 2.
6. Alan Frost, *The Voyage of the Endeavour: Captain Cook and the Discovery of the Pacific* (St. Leonards, New South Wales: Allen and Unwin, 1998), 60.

CHAPTER 3

1. Alexander Humboldt, *Personal Narrative of a Journey to the Equinoctial Regions of the New Continent*, translated with an introduction by Jason Wilson and a "Historical Introduction" by Malcolm Nicolson (London: Penguin, 1995), ix.
2. Michael Dettelbach, "Humboldtian Science," in *Cultures of Natural History*, ed. Nicholas Jardine, Emma Spary, and J. A. Secord (Cambridge: Cambridge University Press, 1996), 298.
3. John Herschel, ed., *Admiralty Manual of Scientific Enquiry: Prepared for the Use of Officers in Her Majesty's Navy; and Travellers in General* (London: HMSO, 1849), 280.
4. The Beaufort quotation in the text comes from a letter from Beaufort to the Committee of the Royal Geographical Society, printed in "Communications

on a North-West Passage, and Further Survey of the Northern Coast of America," *Journal of the Royal Geographical Society* 6 (1836): 46.

5. Ann Savours, *The Search for the North West Passage* (New York: St. Martin's, 1999), 273.

CHAPTER 4

1. Francis Darwin, ed., *The Autobiography of Charles Darwin* (1882; reprint, Amherst, NY: Prometheus, 2000), 28.
2. Frederick Burckhardt and Syndey Smith, eds., *The Correspondence of Charles Darwin* (Cambridge: Cambridge University Press, 1985–2003), 1:237.
3. Charles Darwin, *On the Origin of Species* (1859; reprint, London: Penguin, 1982), 65.
4. Burckhardt and Smith, *The Correspondence of Charles Darwin*, 3:2.
5. Ibid., 5:178.
6. J. D. Hooker, *Himalayan Journals: Notes of a Naturalist in Bengal, the Sikkim and Nepal Himalayas* (London: John Murray, 1854), v.
7. Ibid., ix.
8. Peter Raby, *Bright Paradise: Victorian Scientific Travellers* (Princeton, NJ: Princeton University Press, 1996), 138.
9. Alfred Russel Wallace, *My Life: A Record of Events and Opinions*, 2 vols. (London: Chapman and Hall, 1905), 1:256–257.
10. Ibid., 1:363.
11. James Marchant, ed., *Alfred Russel Wallace: Letters and Reminiscences*, 2 vols. (London: Cassell, 1916), 1:241–242.
12. Adrian Desmond, *Huxley: From Devil's Disciple to Evolution's High Priest* (1994; reprint, Boston, MA: Addison-Wesley, 1997), 117.
13. Darwin, *Origin of the Species*, 458.
14. Alfred Russel Wallace, "The Origin of Human Races and the Antiquity of Man Deduced from the Theory of 'Natural Selection,'" paper presented at a meeting of the Anthropological Society of London on 1 March 1864; reprinted in *Journal of the Anthropological Society of London*, vol. 2 (1864), clviii–clxx.
15. Wallace, *My Life*, 1:366.

CHAPTER 5

1. Hugh Richard Slotten, *Patronage, Practice, and the Culture of American Science: Alexander Dallas Bache and the U.S. Coast Survey* (Cambridge: Cambridge University Press, 1994), 43–44.
2. Ibid., 59.

3. Ibid., 82.

4. Ibid., 106.

5. "John Charles Fremont—Phrenological Character and Biography," *New York Times*, 7 August 1856, 2.

6. Michael L. Smith, *Pacific Visions: California Scientists and the Environment, 1850–1915* (New Haven, CT: Yale University Press, 1987), 32.

7. John Muir, *The Cruise of the Corwin: Journal of the Arctic Expedition of 1881 in Search of De Long and the Jeannette* (Boston: Houghton Mifflin, 1917), 191–192.

8. Charles Haskins Townsend, "Lower California: Access to the Gulf of California Versus the Acquisition of the Peninsula," *New York Times*, 26 January 1919, 34.

CHAPTER 6

1. Chester Hearn, *Tracks in the Sea: Matthew Fontaine Maury and the Mapping of the Oceans* (Camden, ME: International Marine/Ragged Mountain Press, 2003), 208.

2. Ibid., 159.

3. William Scoresby, *An Account of the Arctic Regions, with a History and Description of the Northern Whale-Fishery*, vol. 1 (Edinburgh: Archibald Constable and Co., 1820), 177.

4. Susan Schlee, *The Edge of an Unfamiliar World: A History of Oceanography* (New York: E. P. Dutton, 1973), 128.

CHAPTER 7

1. Cindy Lee Van Dover, *Deep-Ocean Journeys: Discovering New Life at the Bottom of the Sea* (Reading, MA: Addison-Wesley, 1996), 41.

2. Van Dover, *Deep-Ocean Journeys*, 17–18.

3. Ibid., 171.

4. Roy Waldo Miner, "Miracles in Coral and Mud," *World's Work* 59 (November 1939): 43.

5. William Beebe, *The Arcturus Adventure: An Account of the New York Zoological Society's First Oceanographic Expedition* (New York: Putnam, 1926), 341–342.

6. Henry Chester Tracy, *American Naturists* (New York: Dutton, 1930), 222–223; and Alexander Petrunkevitch, "An Explorer of Nature," *Yale Review* 16 (January 1927): 404–406.

7. Willard Bascom and Roger Revelle, "Free-Diving: A New Exploratory Tool," *American Scientist* 41 (October 1953): 624–625.

8. John E. Rexine, Review of *Archaeology under Water* by George Bass, *American Journal of Archaeology* 72 (January 1968): 85.

9. James Dugan, "Portrait of Homo Aquaticus," *New York Times*, 21 April 1963, SM20.

10. Ibid.

11. Jacques Cousteau, "Ocean-Bottom Homes for Skin Divers," *Popular Mechanics* 120 (July 1963): 98.

12. George F. Bond, *Papa Topside: The Sealab Chronicles of Capt. George F. Bond, USN* (Annapolis, MD: Naval Institute Press, 1993), 81–82.

13. Joseph N. Bell, "Chilly, Wet 'Walk': Window under Water," *Christian Science Monitor*, 27 October 1965, 1.

14. "Man against the Deep," *New York Times*, 24 February 1969, 36.

15. Jocelyn Arundel, "Undersea Lab: Aquanauts Chart Spiny-Lobster Habits," *Christian Science Monitor*, 6 July 1970, 3.

16. Sylvia Earle, "All-Girl Team Tests the Habitat," *National Geographic Magazine* 140 (April 1971): 296.

17. Sylvia Earle and Al Giddings, *Exploring the Deep Frontier: The Adventure of Man in the Sea* (Washington, DC: National Geographic Society, 1980), 281.

18. William Beebe, "Beebe and Aide Descend 1,426 Feet into Sea in Steel Sphere, Phoning Observations to Tug," *New York Times*, 13 June 1930, sec. 9, p. 4; "Dr. Beebe Returns, Dived to 1,426 Feet," *New York Times*, 31 October 1930, 20.

19. "Beebe Out to Set Sea Descent Mark," *New York Times*, 15 April 1934, N1.

20. Arnold, Henry, "Manned Submersibles for Research," *Science* 158 (6 October 1967): 84–90, 95.

21. Kenneth G. Slocum, "Exploiting the Deep: Tapping Sea's Wealth Sparks Investor Interest, but Payoff Is Remote," *Wall Street Journal*, 16 September 1968, 1, 24.

22. John Travis, "Deep-Sea Debate Pits *Alvin* against *Jason*," *Science* 259 (12 March 1993): 1534–1536.

23. Bruce Robison, "Submersibles in Oceanographic Research," in *Oceanographic History: The Pacific and Beyond*, ed. Keith Benson and Philip Rehbock (Seattle: University of Washington Press, 2002), 387–390.

Bibliography

Chapter 1

Andrewes, William J. H., ed. *The Quest for Longitude.* Cambridge, MA: Harvard Collection of Historical Scientific Instruments, 1996.

Bennett, Jim. "The Challenge of Practical Mathematics." Pp. 176–190 in *Science, Culture and Popular Belief in Renaissance Europe*, ed. Stephen Pumfrey et al. Manchester, UK: Manchester University Press, 1991.

———. "Projection and the Ubiquitous Virtue of Geometry in the Renaissance." Pp. 27–38 in *Making Space for Science: Territorial Themes in the Shaping of Knowledge*, ed. Crosbie Smith and Jon Agar. London: Macmillan, 1998.

Butterfield, H. *The Origins of Modern Science, 1300–1800.* London: G. Bell, 1957.

Cook, Alan H. *Edmond Halley: Charting the Heavens and the Seas.* Oxford, UK: Clarendon, 1998.

Dear, Peter. "'*Totius in verba*': Rhetoric and Authority in the Early Royal Society." *Isis* 76 (1985): 145–161.

Dictionary of Scientific Biography. 18 vols., including index and supplements. Edited by Charles Coulston Gillespie. New York: Scribner, 1970–1990.

Flaum, Eric. *Discovery: Exploration through the Centuries.* New York: Gallery Books, 1992.

Forbes, Eric G. *Greenwich Observatory*, vol. 1, *Origins and Early History (1675–1835).* London: Taylor and Francis, 1975.

Frost, Orcutt. *Bering: The Russian Discovery of America.* New Haven, CT: Yale University Press, 2003.

Goodman, David, and Colin A. Russell, eds. *The Rise of Scientific Europe, 1500–1800.* London: Hodder and Stoughton, 1991.

Greenberg, Leonard. *The Problem of the Shape of the Earth from Newton to Clairaut.* Cambridge: Cambridge University Press, 1995.

Hahn, Roger. *The Anatomy of a Scientific Institution: The Paris Academy of Sciences, 1666–1803.* Berkeley: University of California Press, 1971.

Hall, A. R. *The Revolution in Science, 1500–1750*. London: Longman, 1983.

Halley, Edmund. "An Historical Account of the Trade Winds, and Monsoons, Observable in the Seas between and near the Tropicks, with an Attempt to Assign the Physical Cause of the Said Winds." *Philosophical Transactions* 16 (1686–1692): 153–168.

———. "The True Theory of the Tides, Extracted from That Admired Treatise of Mr. Isaac Newton." *Philosophical Transactions* 19 (1696): 445–457.

Hooykaas, R. "The Rise of Modern Science: When and Why?" *British Journal for the History of Science* 20 (1987): 453–473.

Konvitz, Josef W. *Cartography in France, 1660–1848: Science, Engineering, and Statecraft*. Chicago, IL: University of Chicago Press, 1987.

Levathes, Louise. *When China Ruled the Seas: The Treasure Fleet of the Dragon Throne, 1405–1433*. Oxford: Oxford University Press, 1997.

MacPike, Eugene Fairfield, ed. *Correspondence and Papers of Edmond Halley*. 1932. Reprint, Oxford, UK: Clarendon, 1975.

Maupertuis, Pierre-Louis Moreau de. *Oeuvres*. 4 vols. Lyons: Jean Marie Bruyset, 1756.

Menzies, Gavin. *1421: The Year China Discovered America*. New York: Harper Perennial, 2004.

Needham, Joseph. *Science and Civilization in China*. 7 vols. Cambridge: Cambridge University Press, 1954–2004.

[Oldenberg, Henry]. "Epistle Dedicatory." *Philosophical Transactions of the Royal Society of London* 1 (1665): n.p.

Ornstein, M. *The Role of Scientific Societies in the Seventeenth Century*. New York: Arno, 1975.

Pagden, Anthony. *European Encounters with the New World: From Renaissance to Romanticism*. New Haven, CT: Yale University Press, 1994.

Sargent, Rose-Mary. "Bacon As an Advocate for Cooperative Scientific Research." Pp. 146–171 in *The Cambridge Companion to Bacon*, ed. Markku Peltonen. Cambridge: Cambridge University Press, 1996.

Shaffer, Lynda. *Maritime Southeast Asia to 1500*. Armonk, NY: M. E. Sharpe, 1996.

Shapin, Steven. "The House of Experiment in Seventeenth-Century England." *Isis* 79 (1988): 373–404.

Sobel, Dava. *Longitude: The True Story of a Lone Genius Who Solved the Greatest Scientific Problem of His Time*. New York: Penguin, 1996.

Spedding, J., et al., eds. *The Works of Francis Bacon*. 14 vols. London: Longmans, 1857–1874.

Teresi, Dick. *Lost Discoveries: The Ancient Roots of Modern Science, from the Babylonians to the Maya*. New York: Simon and Schuster, 2002.

Terrall, Mary. *The Man Who Flattened the Earth: Maupertuis and the Sciences in the Enlightenment.* Chicago, IL: University of Chicago Press, 2002.

Van Deen, Abbeele Georges. *Travel As Metaphor: From Montaigne to Rousseau.* Minneapolis: University of Minnesota Press, 1992.

Waters, D. W. "Science and the Techniques of Navigation in the Renaissance." Pp. 189–237 in *Art, Science and History in the Renaissance,* ed. C. Singleton. Baltimore, MD: Johns Hopkins University Press, 1967.

Zagorin, Perez. *Francis Bacon.* Princeton, NJ: Princeton University Press, 1999.

Chapter 2

Badger, G. M., ed. *Captain Cook: Navigator and Scientist.* London: C. Hurst, 1970.

Blunt, Wilfred. *The Compleat Naturalist: A Life of Linnaeus.* New York: Viking, 1971.

Bougainville, Louis de. *A Voyage Round the World.* 1772. Reprint, New York: Da Capo, 1967.

Boxer, C. R. *Two Pioneers of Tropical Medicine.* London: Wellcome Historical Medical Library, 1963.

Brockway, Lucille. *Science and Colonial Expansion: The Role of the British Royal Botanical Gardens.* New York: Academic Press, 1979.

Chappe d'Auteroche, Jean-Baptiste. *A Voyage to California, to Observe the Transit of Venus by Mons. Chappe d'Auteroche; with an Historical Description of the Author's Route through Mexico, and the Natural History of That Province.* London: n.p., 1778.

Clark, William, Jan Golinski, and Simon Schaffer, eds. *The Sciences in Enlightened Europe.* Chicago, IL: University of Chicago Press, 1999.

Condamine, Charles Marie de la. *Succinct Abridgment of a Voyage Made within the Inland Parts of South-America.* London: n.p., 1747.

Dick, Steven J. *Sky and Ocean Joined: U.S. Naval Observatory, 1830–2000.* Cambridge: Cambridge University Press, 2003.

Dictionary of Scientific Biography. 18 vols., including index and supplements. Edited by Charles Coulston Gillespie. New York: Scribner, 1970–1990.

Drayton, Richard. *Nature's Government: Science, Imperial Britain, and the "Improvement" of the World.* New Haven, CT: Yale University Press, 2000.

Dunmore, John, trans. and ed. *The Journal of Jean-Francois de Galaup de la Perouse, 1785–7.* London: Hakluyt Society, 1994.

———. *French Explorers in the Pacific, I: The Eighteenth Century.* Oxford, UK: Clarendon, 1965.

Duyker, Edward. *Nature's Argonaut: Daniel Solander, 1733–1782*. Melbourne: Miegunyah Press, 1998.

Engstrand, Iris Wilson. *Spanish Scientists in the New World: The Eighteenth-Century Expeditions*. Seattle: University of Washington Press, 1981.

Farber, Paul. *Finding Order in Nature: The Naturalist Tradition from Linnaeus to E. O. Wilson*. Baltimore, MD: Johns Hopkins University Press, 2000.

Findlen, Paula. "Courting Nature." Pp. 57–74 in *Cultures of Natural History*, ed. Nicholas Jordine, Emma Spray, and J. A. Secord, Cambridge: Cambridge University Press, 1994.

Forster, Georg. *A Voyage Round the World*. 2 vols. Edited by Nicholas Thomas and Oliver Berghof, assisted by Jennifer Newell. Honolulu: University of Hawaii Press, 2000.

Forster, Johann Reinhold. *Observations Made during a Voyage Round the World*. Edited by Nicholas Thomas, Harriet Guest, and Michael Dettelbach, with a linguistics appendix by Karl H. Rensch. Honolulu: University of Hawaii Press, 1996.

Frost, Alan. "Science for Political Purposes: European Exploration of the Pacific Ocean, 1764–1806." Pp. 27–44 in *Scientific Aspects of European Expansion*, ed. William K. Storey. London: Variorum, 1996.

———. *The Voyage of the Endeavour: Captain Cook and the Discovery of the Pacific*. St. Leonards, New South Wales: Allen and Unwin, 1998.

Gascoigne, John. "Joseph Banks and the Expansion of Empire." Pp. 39–51 in *Science and Exploration: European Voyages in the Southern Oceans in the 18th Century*, ed. Margarette Lincoln. Woodbridge, Suffolk, UK: Boydel, 1998.

Goodman, David, and Colin A. Russell, eds. *The Rise of Scientific Europe, 1500–1800*. London: Hodder and Stoughton, 1991.

Greenwood, Isaac. "New Method for Composing a Natural History of Meteors." *Philosophical Transactions* (1727): 399.

Haycox, Stephen, James K. Barnett, and Caedmon A. Liburd, eds. *Enlightenment and Exploration in the North Pacific, 1741–1805*. Seattle: University of Washington Press, 1997.

Hoare, Michael E., ed. *The Resolution Journal of Johann Reinhold Forster, 1772–1775*. London: Hakluyt Society, 1982.

Jardine, Nicholas, Emma Spary, and J. A. Secord, eds. *Cultures of Natural History*. Cambridge: Cambridge University Press, 1996.

Koerner, Lisbet. *Linnaeus: Nature and Nation*. Cambridge, MA: Harvard University Press, 1999.

———. "Linnaeus' Floral Transplants." *Representations* 47 (1994): 144–169.

Lincoln, Margarette, ed. *Science and the Exploration of the Pacific: European Voyages to the Southern Oceans in the 18th Century.* Woodbridge, Suffolk, UK: Boydell, 1998.

Linnaeus, Carl. *A General System of Nature.* London: Lackington, Allen, 1802–1806.

———. *Lachesis Lapponica, or a Tour in Lapland, in Two Volumes.* Edited by James Edward Smith. London: White and Cochrane, 1811.

McConnell, Anita. "La Condamine's Scientific Journey down the River Amazon, 1743–4." *Annals of Science* 48 (1991): 1–19.

O'Brian, Patrick. *Joseph Banks: A Life.* Chicago, IL: University of Chicago Press, 1987.

Pratt, Mary Louise. *Imperial Eyes: Travel Writing and Transculturation.* London: Routledge, 1992.

Rousseau, G. S., and Roy Porter, eds. *The Ferment of Knowledge: Studies in the Historiography of Eighteenth Century Science.* Cambridge: Cambridge University Press, 1980.

Schiebinger, Londa. *Nature's Body: Gender in the Making of Modern Science.* Boston, MA: Beacon, 1993.

Stafleu, Frans A. *Linnaeus and the Linnaeans: The Spreading of Their Ideas in Systematic Botany, 1735–1789.* Utrecht: A. Oosthoek's Uitgeversmaatschapij, 1971.

Stroup, Alice. *A Company of Scientists: Botany, Patronage, and Community at the Seventeenth Century Parisian Royal Academy of Sciences.* Berkeley: University of California Press, 1990.

Tournefort, Joseph Pitton de. *A Voyage into the Levant . . . Perform'd by Command of the Late French King.* London: D. Browne, 1741.

Voyages and Adventures of La Pérouse. Translated from the French by Julius S. Gassner. Honolulu: University of Hawaii Press, 1969.

Whitaker, Katie. "The Culture of Curiosity," Pp. 75–90 in *Cultures of Natural History* [ibid.].

Williams, Roger L. *French Botany in the Enlightenment: The Ill Fated Voyages of La Perouse and His Rescuers.* Dordrecht: Kluwer Academic, 2003.

Withey, Lynne. *Voyages of Discovery: Captain Cook and the Exploration of the Pacific.* New York: William Morrow, 1987.

Woolf, Harry. *The Transits of Venus: A Study of Eighteenth Century Science.* Princeton, NJ: Princeton University Press, 1959.

Chapter 3

Anderson, Katharine. *Predicting the Weather: Victorians and the Science of Meteorology.* Chicago, IL: University of Chicago Press, 2005.

————. "The Weather Prophets: Science and Reputation in Victorian Meteorology." *History of Science* 37 (1999): 179–216.

Brock, W. H. "Humboldt and the British: A Note on the Character of British Science." *Annals of Science* 50 (1993): 365–372.

Browne, Janet. "Biogeography and Empire." Pp. 305–321 in *Cultures of Natural History*, ed. Nicholas Jardine, Emma Spary, and J. A. Secord. Cambridge: Cambridge University Press, 1996.

Cannon, Susan Faye. *Science in Culture: The Early Victorian Period.* New York: Science History Publications, 1978.

Cawood, John. "The Magnetic Crusade." *Isis* 70 (1979): 493–518.

————. "Terrestrial Magnetism and the Development of International Collaboration in the Early Nineteenth Century." *Annals of Science* 34 (1977): 551–587.

"Communications on a North-West Passage, and Further Survey of the Northern Coast of America." *Journal of the Royal Geographical Society* 6 (1836): 34–50.

Courtney, Nicholas. *Gale Force 10: The Life and Legacy of Admiral Beaufort.* London: Headline Books, 2003.

Crawford, Elisabeth. *Nationalism and Internationalism in Science, 1880–1939.* Cambridge: Cambridge University Press, 2002.

Day, Archibald. *The Admiralty Hydrographic Service, 1795–1919.* London: HMSO, 1976.

Deacon, Margaret. *Scientists and the Sea, 1650–1900: A Study of Marine Science.* London: Academic Press, 1971.

Deacon, Margaret. *Scientists and the Sea, 1650–1900: A Study of Marine Science.* 2nd ed. Brookfield, USA: Ashgate, 1977.

Deacon, Margaret, Tony Rice, and Colin Summerhayes (eds.). *Understanding the Oceans.* London: VCL Press, 2001.

Dettelbach, Michael. "Humboldtian Science." Pp. 287–304 in *Cultures of Natural History*, ed. Nicholas Jardine, Emma Spary, and J. A. Secord. Cambridge: Cambridge University Press, 1996.

Fara, Patricia. *Sympathetic Attractions: Magnetic Practices, Beliefs and Symbolism in 18th Century England.* Cambridge: Cambridge University Press, 1994.

Fleming, James Rodger. *Meteorology in America, 1800–1870.* Baltimore, MD: Johns Hopkins University Press, 1990.

Friendly, Alfred. *Beaufort of the Admiralty: The Life of Sir Francis Beaufort, 1774–1857.* New York: Random House, 1977.

Guberlet, Muriel L. *Explorers of the Sea: Famous Oceanographic Expeditions.* New York: Ronald Press Company, 1964.

Hagen, Victor Wolfgang von. *South America Called Them.* New York: Knopf, 1945.

Hall, Marie Boas. *All Scientists Now: The Royal Society in the Nineteenth Century.* Cambridge: Cambridge University Press, 1984.

Herschel, John, ed. *Admiralty Manual of Scientific Enquiry: Prepared for the Use of Officers in Her Majesty's Navy; and Travellers in General.* London: HMSO, 1849.

Hill, J. R., ed. *The Oxford Illustrated History of the Royal Navy.* Oxford: Oxford University Press, 1995.

Home, R. W. "Humboldtian Science Revisited: An Australian Case Study." *History of Science* 33 (1995): 1–22.

Home, R. W., and Sally Gregory Kohlstedt, eds. *International Science and National Scientific Identity: Australia between Britain and America.* Boston: Kluwer Academic Publishers, 1991.

Humboldt, Alexander. *Cosmos: A Sketch of a Physical Description of the Universe.* Translated by E. C. Otté with an introduction by Nicolas A. Rupke. Baltimore, MD: Johns Hopkins University Press, 1997.

———. *Personal Narrative of a Journey to the Equinoctial Regions of the New Continent.* Translated with an introduction by Jason Wilson and a "Historical Introduction" by Malcolm Nicolson. London: Penguin, 1995.

Jankovic, Vladimir. "Ideological Crests versus Empirical Troughs: John Herschel's and William Radcliffe Birt's Research on Atmospheric Waves, 1843–50." *British Journal for the History of Science* 31 (1998): 21–40.

———. *Reading the Skies: A Cultural History of English Weather, 1650–1820.* Manchester, UK: Manchester University Press, 2000.

Kellner, L. *Alexander von Humboldt.* Oxford: Oxford University Press, 1963.

Kennedy, Paul M. *The Rise and Fall of British Naval Mastery.* New York: Scribner, 1976.

Lambert, Andrew D. "Preparing for the Long Peace: The Reconstruction of the Royal Navy, 1815–1830." *Mariner's Mirror* 82 (1996): 41–54.

Levere, Trevor H. *Science and the Canadian Arctic: A Century of Exploration, 1818–1918.* Cambridge: Cambridge University Press, 1993.

Lewis, Michael. *The Navy in Transition, 1814–1864: A Social History.* London: Hodder and Stoughton, 1965.

Livingstone, David N. *The Geographical Tradition: Episodes in the History of a Contested Enterprise.* Oxford, UK: Blackwell, 1992.

MacLeod, Roy, and Peter Collins, eds. *Parliament of Science: The British Association for the Advancement of Science, 1831–1981.* Northwood, Midx.: Science Reviews, 1981.

Manning, Thomas G. *U.S. Coast Survey vs Naval Hydrographic Office: A 19th-Century Rivalry in Science and Politics.* Tuscaloosa: University of Alabama Press, 1988.

Miller, David Philip. "The Revival of the Physical Sciences in Britain, 1815–1840." *Osiris* 2 (1986): 107–134.

Morrell, Jack, and Arnold Thackray. *Gentlemen of Science: Early Years of the British Association for the Advancement of Science.* Oxford, UK: Clarendon, 1981.

Morris, Roger. "200 Years of Admiralty Charts and Surveys." *Mariner's Mirror* 83 (1996): 420–435.

Neatby, Leslie H. *The Search for Franklin.* London: Barker, 1970.

Nicolson, Malcolm. "Alexander Von Humboldt, Humboldtian Science and the Origins of the Study of Vegetation." *History of Science* 25 (1987): 167–194.

———. "Humboldtian Plant Geography after Humboldt: The Link to Ecology." *British Journal for the History of Science* 29 (1996): 289–310.

Pratt, Mary Louise. *Imperial Eyes: Travel Writing and Transculturation.* London: Routledge, 1992.

Richards, Joan. "Observing Science in Early Victorian England: Recent Scholarship on William Whewell." *Perspectives on Science* 4 (1996): 231–247.

Riffendburgh, Beau. *The Myth of the Explorer: The Press, Sensationalism, and Geographical Discovery.* Oxford: Oxford University Press, 1994.

Ross, John. *A Voyage of Discovery, Made under the Orders of the Admiralty, in His Majesty's Ships Isabella and Alexander, for the Purpose of Exploring Baffin's Bay, and Inquiring into the Probability of a Northwest Passage.* London: John Murray, 1819.

Rozwadowski, Helen. *Fathoming the Ocean: The Discovery and Exploration of the Deep Sea.* Cambridge, MA: Belknap Press of Harvard University Press, 2005.

Savours, Ann. *The Search for the North West Passage.* New York: St. Martin's, 1999.

Schlee, Susan, *Edge of an Unfamiliar World: A History of Oceanography.* New York: Dutton, 1973.

Slotten, Hugh Richard. *Patronage, Practice, and the Culture of American Science: Alexander Dallas Bache and the U.S. Coast Survey.* Cambridge: Cambridge University Press, 1994.

Smith, Anthony. *Explorers of the Amazon.* Chicago, IL: University of Chicago Press, 1990.

Terra, Helmut de. *Humboldt: The Life and Times of Alexander von Humboldt, 1769–1859.* New York: Knopf, 1955.

Todhunter, Isaac. *William Whewell, D.D., Master of Trinity College, Cambridge: An Account of His Writings with Selections from His Literary and Scientific Correspondence.* 2 vols. 1876. Reprint, Westmead, Farnborough, Hampshire, UK: Gregg Publishers, 1970.

Winter, Alison. "'Compasses All Awry': The Iron Ship and the Ambiguities of Cultural Authority in Victorian Britain." *Victorian Studies* 38 (1994): 69–98.

Yeo, Richard. *Defining Science: William Whewell, Natural Knowledge, and Public Debate in Early Victorian Britain.* Cambridge: Cambridge University Press, 1993.

Chapter 4

Adas, Michael. *Machines As the Measure of Men.* Ithaca, NY: Cornell University Press, 1990.

Allan, Mea. *The Hookers of Kew.* London: M. Joseph, 1967.

Allen, David Elliston. *The Naturalist in Britain: A Social History.* Princeton, NJ: Princeton University Press, 1976.

Barlow, Nora, ed. *The Autobiography of Charles Darwin.* 1882. Reprint, New York: Norton, 1993.

Barr, Alan P., ed. *Thomas Henry Huxley's Place in Science and Letters.* Athens: University of Georgia Press, 1997.

Bowler, Peter J. *Evolution: The History of an Idea.* 1984. Revised and expanded ed., Berkeley: University of California Press, 2003.

Brackman, Arnold C. *A Delicate Arrangement.* New York: Times Books, 1980.

Brockway, Lucille. *Science and Colonial Expansion: The Role of the British Royal Botanical Gardens.* New York: Academic Press, 1979.

Brooke, John Hedley. *Science and Religion: Some Historical Perspectives.* Cambridge: Cambridge University Press, 1991.

Browne, Janet. *Charles Darwin: The Power of Place.* New York: Knopf, 2002.

———. *Charles Darwin: Voyaging.* 1995. Reprint, Princeton, NJ: Princeton University Press, 1996.

Burckhardt, Frederick, and Syndey Smith, eds. *The Correspondence of Charles Darwin.* Cambridge: Cambridge University Press, 1985–2003.

Camerini, Jane. *The Alfred Russel Wallace Reader: A Selection of Writings from the Field.* Baltimore, MD: Johns Hopkins University Press, 2002.

———. "Evolution, Biogeography, and Maps: An Early History of Wallace's Line." Pp. 70–109 in *Darwin's Laboratory: Evolutionary Theory and Natural History in the Pacific,* ed. Roy MacLeod and Philip F. Rehbock. Honolulu: University of Hawaii Press, 1994.

————. "Wallace in the Field." Pp. 44–65 in *Osiris, Volume 11: Science in the Field*, 2nd series, ed. Henrika Kuklick and Robert Kohler. Chicago, IL: University of Chicago Press, 1996.

Darwin, Charles. *The Descent of Man.* London: John Murray, 1871.

————. *On the Origin of Species.* 1859. Reprint, London: Penguin, 1982.

————. *Voyage of the Beagle.* London: Penguin, 1989.

Darwin, Francis, ed. *The Autobiography of Charles Darwin.* 1882. Reprint, Amherst, NY: Prometheus, 2000.

————, ed. *Life and Letters of Charles Darwin.* London: John Murray, 1887.

Dentith, Simon. *Society and Cultural Forms in Nineteenth Century England.* Basingstoke, UK: Palgrave Macmillan, 1999.

Desmond, Adrian. *Huxley: From Devil's Disciple to Evolution's High Priest.* 1994. Reprint, Boston, MA: Addison-Wesley, 1997.

Desmond, Adrian, and James Moore. *Darwin: The Life of a Tormented Evolutionist.* New York: Warner Books, 1991.

Drayton, Richard. *Nature's Government: Science, Imperial Britain, and the "Improvement" of the World.* New Haven, CT: Yale University Press, 2000.

Farber, Paul. *The Temptations of Evolutionary Ethics.* 1994. Reprint, Berkeley: University of California Press, 1998.

Gould, Stephen J. *Ever Since Darwin: Reflections on Natural History.* New York: Norton, 1977.

Greene, John C. *The Death of Adam: Evolution and Its Impact on Western Thought.* 1959. Revised ed., Ames: Iowa State University Press, 1996.

Hagen, Victor Wolfgang von. *South America Called Them.* New York: Knopf, 1945.

Hooker, J. D. *Himalayan Journals: Notes of a Naturalist in Bengal, the Sikkim and Nepal Himalayas.* London: John Murray, 1854.

Hopkins, Robert S. *Darwin's South America.* New York: John Day, 1969.

Huxley, Julian, ed. *Thomas Henry Huxley's Diary of the Voyage of the H.M.S. Rattlesnake.* 1935. Reprint, New York: Kraus, 1972.

Huxley, Leonard, ed. *Life and Letters of Sir Joseph Dalton Hooker.* 2 vols. London: John Murray, 1918.

————, ed. *Life and Letters of Thomas Huxley.* 2 vols. London: Macmillan, 1900.

Irvine, William. *Apes, Angels, and Victorians: Darwin, Huxley, and Evolution.* New York: McGraw-Hill, 1955.

Lansing, Alfred. *Endurance: Shackleton's Incredible Voyage.* 1959. Reprint, New York: Carroll and Graf, 1999.

Largent, Mark A., ed. *Sourcebook on the History of Evolution.* Dubuque, IA: Kendall/Hunt, 2002.

MacLeod, Roy, and Philip F. Rehbock, eds. *Darwin's Laboratory: Evolutionary Theory and Natural History in the Pacific*. Honolulu: University of Hawaii Press, 1994.

———, eds. *Nature in Its Greatest Extent: Western Science in the Pacific*. Honolulu: University of Hawaii Press, 1988.

Marchant, James, ed. *Alfred Russel Wallace: Letters and Reminiscences*. 2 vols. London: Cassell, 1916.

Mix, Michael C., et al., eds. *Biology: The Network of Life*. New York: Harper-Collins, 1992.

Moorehead, Alan. *Darwin and the Beagle*. New York: Harper and Row, 1969.

Paradis, James G. *T. H. Huxley: Man's Place in Nature*. Lincoln: University of Nebraska Press, 1978.

Paradis, James, and George Christopher Williams, eds. *Evolution and Ethics: T. H. Huxley's Evolution and Ethics with New Essays on Its Victorian and Sociobiological Context*. Princeton, NJ: Princeton University Press, 1989.

Pratt, Mary Louise. *Imperial Eyes: Travel Writing and Transculturation*. London: Routledge, 1992.

Quammen, David. *The Reluctant Mr. Darwin: An Intimate Portrait of Charles Darwin and the Making of His Theory of Evolution*. New York: Norton, 2006.

Raby, Peter. *Alfred Russel Wallace: A Life*. Princeton, NJ: Princeton University Press, 2001.

———. *Bright Paradise: Victorian Scientific Travellers*. Princeton, NJ: Princeton University Press, 1996.

Reidy, Michael S. *Tides of History: Organizing the Oceans and Creating the Scientist*. Chicago, IL: University of Chicago Press, forthcoming.

Turrill, W. B. *Joseph Dalton Hooker: Botanist, Explorer, and Administrator*. London: Nelson, 1963.

Wallace, Alfred Russel. *The Malay Archipelago: The Land of the Orang-Utan and the Bird of Paradise; A Narrative of Travel with Studies of Man and Nature*. New York: Harper and Brothers, 1869.

———. *My Life: A Record of Events and Opinions*. 2 vols. London: Chapman and Hall, 1905.

———. "The Origin of Human Races and the Antiquity of Man Deduced from the Theory of 'Natural Selection.'" Paper presented at a meeting of the Anthropological Society of London on 1 March 1864. Reprinted in *Journal of the Anthropological Society of London*, vol. 2 (1864), clviii–clxx.

———. *Travels on the Amazon and Rio Negro*. London: Reeve, 1853.

Chapter 5

Ambrose, Stephen. *Undaunted Courage: Meriwether Lewis, Thomas Jefferson, and the Opening of the American West.* New York: Simon and Schuster, 1996.

Bartlett, Richard A. *Great Surveys of the American West.* Norman: University of Oklahoma Press, 1962.

Berkelman, Robert. "Clarence King: Scientific Pioneer." *American Quarterly* 5 (Winter 1953): 301–324.

Bruce, Robert V. *The Launching of Modern American Science, 1846–1876.* New York: Knopf, 1987.

Camerini, Jane. "Wallace in the Field." Pp. 44–65 in *Osiris, Volume 11: Science in the Field,* 2nd series, ed. Henrika Kuklick and Robert Kohler. Chicago, IL: University of Chicago Press, 1996.

Coues, Elliot, ed. *The History of the Lewis and Clark Expedition.* 3 vols. New York: Dover, 1893.

Daniels, George H., ed. *Nineteenth-Century American Science: A Reappraisal.* Evanston, IL: Northwestern University Press, 1972.

DeVoto, Bernard Augustine. *Course of Empire.* Boston: Houghton Mifflin, 1952.

Dorsey, Kurkpatrick. *The Dawn of Conservation Diplomacy: U.S.-Canadian Wildlife Protection Treaties in the Progressive Era.* Seattle: University of Washington Press, 1998.

Dupree, A. Hunter. *Science in the Federal Government: A History of Policies and Activities.* 1957. Reprint, Baltimore, MD: Johns Hopkins University Press, 1986.

Egan, Ferol. *Frémont: Explorer for a Restless Nation.* Garden City, NY: Doubleday, 1977.

Fox, Stephen. *John Muir and His Legacy: The American Conservation Movement.* Boston: Little, Brown, 1981.

Furtwanglers, Albert. *Acts of Discovery: Visions of America in the Lewis and Clark Journals.* Urbana: University of Illinois Press, 1993.

Goetzmann, William. *Army Exploration in the American West, 1803–1863.* New Haven, CT: Yale University Press, 1959.

———. *Exploration and Empire: The Explorer and the Scientist in the Winning of the American West.* New York: Knopf, 1966.

———. *New Lands, New Men: America and the Second Great Age of Discovery.* New York: Viking, 1986.

Hays, Samuel P. *Conservation and the Gospel of Efficiency.* Cambridge. MA: Harvard University Press, 1959.

Hevley, Bruce. "The Heroic Science of Glacier Motion." Pp. 66–86 in *Osiris, Volume 11: Science in the Field*, 2nd series, ed. Henrika Kuklick and Robert Kohler. Chicago, IL: University of Chicago Press, 1996.

"John Charles Fremont—Phrenological Character and Biography." *New York Times*, 7 August 1856, 2.

Kirsch, Scott. "Regions of Government Science: John Wesley Powell in Washington and the American West." *Endeavour* 223 (1999): 155–158.

Muir, John. *The Cruise of the Corwin: Journal of the Arctic Expedition of 1881 in Search of De Long and the Jeannette*. Boston: Houghton Mifflin, 1917.

Nash, Gerald D. "The Conflict between Pure and Applied Science in Nineteenth-Century Public Policy: The California State Geological Survey, 1860–1874." *Isis* 54 (1963): 217–228.

Nash, Roderick. *Wilderness in the American Mind*. New Haven, CT: Yale University Press, 1967.

Pauly, Philip. *Biologists and the Promise of American Life: From Meriwether Lewis to Alfred Kinsey*. Princeton, NJ: Princeton University Press, 2000.

Porter, Charlotte. *The Eagle's Nest: Natural History and American Ideas, 1812–1842*. Tuscaloosa: University of Alabama Press, 1986.

Reingold, Nathan. *Science in America since 1820*. New York: Science History Publications, 1976.

Rindge, Debora. "Science and Art Meet in the Parlor: The Role of Popular Magazine Illustration in the Pictorial Record of the 'Great Surveys.'" Pp. 173–196 in *Surveying the Record: North American Scientific Exploration to 1930*, ed. Edward Carter. Philadelphia: American Philosophical Society, 1999.

Rolle, John. *Charles Frémont: Character as Destiny*. Norman: University of Oklahoma Press, 1991.

Ronda, James. *Finding the West: Explorations with Lewis and Clark*. Albuquerque: University of New Mexico Press, 2001.

———. *Lewis and Clark among the Indians*. Lincoln: University of Nebraska Press, 1984.

Short, John Rennie. "A New Mode of Thinking: Creating a National Geography in the Early Republic." Pp. 19–50 in *Surveying the Record: North American Scientific Exploration to 1930*, ed. Edward Carter. Philadelphia: American Philosophical Society, 1999.

Simpson, John W. *Visions of Paradise: Glimpses of Our Landscape's Legacy*. Berkeley: University of California Press, 1999.

Slaughter, Thomas. *Exploring Lewis and Clark: Reflections on Men and Wilderness*. New York: Knopf, 2003.

Slotten, Hugh Richard. *Patronage, Practice, and the Culture of American Science: Alexander Dallas Bache and the U.S. Coast Survey*. Cambridge: Cambridge University Press, 1994.

Smith, Henry Nash. *Virgin Land: The American West As Symbol and Myth*. Cambridge, MA: Harvard University Press, 1950.

Smith, Michael L. *Pacific Visions: California Scientists and the Environment, 1850–1915*. New Haven, CT: Yale University Press, 1987.

Stegner, Wallace. *Beyond the Hundredth Meridian: John Wesley Powell and the Second Opening of the West*. Boston: Houghton Mifflin, 1953.

Taylor, Joseph. *Making Salmon: An Environmental History of the Northwest Fisheries Crisis*. Seattle: University of Washington Press, 1999.

Terrall, Mary. "Heroic Narratives of Quest and Discovery." *Configurations* 6 (Spring 1998): 223–242.

Townsend, Charles Haskins. "Lower California: Access to the Gulf of California Versus the Acquisition of the Peninsula." *New York Times*, 26 January 1919, 34.

Tucker, Jennifer. "Voyages of Discovery on Oceans of Air: Scientific Observation and the Image of Science in an Age of 'Balloonacy.'" Pp. 144–176 in *Osiris, Volume 11: Science in the Field*, 2nd series, ed. Henrika Kuklick and Robert Kohler. Chicago, IL: University of Chicago Press, 1996.

White, Richard. *The Organic Machine*. New York: Hill and Wang, 1995.

Worster, Donald. *A River Running West: The Life of John Wesley Powell*. New York: Oxford University Press, 2002.

Wrobel, David. *The End of American Exceptionalism: Frontier Anxiety from the Old West to the New Deal*. Lawrence: University of Kansas Press, 1993.

Chapter 6

Benson, Keith, and Philip Rehbock, eds. *Oceanographic History: The Pacific and Beyond*. Seattle: University of Washington Press, 2002.

Brosco, Jeffrey. "Henry Bryant Bigelow, the US Bureau of Fisheries and Intensive Area Study." *Social Studies of Science* 19 (May 1989): 239–264.

Corfield, Richard. *The Silent Landscape: The Scientific Voyage of HMS Challenger*. Washington, DC: Joseph Henry, 2003.

Deacon, Margaret. *Scientists and the Sea, 1650–1900: A Study of Marine Science*. London: Academic Press, 1971.

Greene, Mott. "Oceanography's Double Life." *Earth Sciences History* 12 (1993): 48–53.

Hamblin, Jacob Darwin. "Environmental Diplomacy in the Cold War: The Disposal of Radioactive Waste at Sea during the 1960s." *International History Review* 24 (June 2002): 348–375.

———. "Science in Isolation: American Marine Geophysics Research, 1950–1968." *Physics in Perspective* 2 (2000): 293–312.

———. "Visions of International Scientific Cooperation: The Case of Oceanic Science, 1920–1955." *Minerva* 38 (2000): 393–423.

Hearn, Chester. *Tracks in the Sea: Matthew Fontaine Maury and the Mapping of the Oceans.* Camden, ME: International Marine/Ragged Mountain Press, 2003.

Landauer, Lyndall Baker. *From Scoresby to Scammon: Nineteenth Century Whalers in the Foundations of Cetology.* N.p., n.d.

Mills, Eric. *Biological Oceanography: An Early History, 1870–1960.* Ithaca, NY: Cornell University Press, 1989.

———. "The Historian of Science and Oceanography after Twenty Years." *Earth Sciences History* 12 (1993): 5–18.

Mukerji, Chandra. *A Fragile Power: Science and the State.* Princeton, NJ: Princeton University Press, 1989.

Oreskes, Naomi. "Laissez-tomber: Military Patronage and Women's Work in Mid-Twentieth Century Oceanography." *Historical Studies in the Physical and Biological Sciences* 30 (2000): 373–392.

———. "Objectivity or Heroism? On the Invisibility of Women in Science." Pp. 87–113 in *Osiris, Volume 11: Science in the Field,* 2nd series, ed. Henrika Kuklick and Robert Kohler. Chicago, IL: University of Chicago Press, 1996.

Oreskes, Naomi, and Ronald Rainger. "Science and Security before the Atomic Bomb: The Loyalty Case of Harald U. Sverdrup." *Studies in the History of Modern Physics* 31 (2000): 309–369.

Pauly, Philip. *Biologists and the Promise of American Life: From Meriwether Lewis to Alfred Kinsey.* Princeton, NJ: Princeton University Press, 2000.

Philbrick, Nathaniel. *Sea of Glory: America's Voyage of Discovery—the U.S. Exploring Expedition, 1838–1842.* New York: Viking, 2003.

Rainger, Ronald. "Adaptation and the Importance of Local Culture: Creating a Research School at the Scripps Institution of Oceanography." *Journal of the History of Biology* 36 (2003): 461–500.

———. "Constructing a Landscape for Postwar Science: Roger Revelle, the Scripps Institution and the University of California, San Diego." *Minerva* 39 (2001): 327–353.

Rozwadowski, Helen. *Fathoming the Ocean: The Discovery and Exploration of the Deep Sea.* Cambridge, MA: Belknap Press of Harvard University Press, 2005.

———. *The Sea Knows No Boundaries: A Century of Marine Science.* Copenhagen: International Council for the Exploration of the Sea in association with University of Washington Press, Seattle, 2002.

————. "Small World: Forging a Scientific Maritime Culture for Oceanography." *Isis* 87 (September 1996): 409–429.

Schlee, Susan. *The Edge of an Unfamiliar World: A History of Oceanography.* New York: Dutton, 1973.

————. *On Almost Any Wind: The Saga of the Oceanographic Vessel "Atlantis."* Ithaca, NY: Cornell University Press, 1978.

Scoresby, William. *An Account of the Arctic Regions, with a History and Description of the Northern Whale-Fishery.* Vol. 1. Edinburgh: Archibald Constable and Co., 1820.

Sears, Mary, and Daniel Merriman, eds. *Oceanography: The Past.* New York: Springer-Verlag, 1980.

Shor, Elizabeth Noble. *Scripps Institution of Oceanography: Probing the Oceans, 1936–1976.* San Diego, CA: Tofua, 1978.

Stanton, William Roger. *The Great United States Exploring Expedition of 1838–1842.* Berkeley: University of California Press, 1975.

Vanderpool, Christopher. "Marine Science and the Law of the Sea." *Social Studies of Science* 13 (February 1983): 107–129.

Weir, Gary E. *An Ocean in Common: American Naval Officers, Scientists, and the Ocean Environment.* College Station: Texas A&M University Press, 2001.

Weisgall, Johnathan M. *Operation Crossroads: The Atomic Tests at Bikini Atoll.* Annapolis, MD: Naval Institute Press, 1994.

Winsor, Mary. *Reading the Shape of Nature: Comparative Zoology and the Agassiz Museum.* Chicago, IL: University of Chicago Press, 1991.

Chapter 7

Arnold, Henry. "Manned Submersibles for Research." *Science* 158 (6 October 1967): 84–90, 95.

Arundel, Jocelyn. "Undersea Lab: Aquanauts Chart Spiny-Lobster Habits." *Christian Science Monitor,* 6 July 1970.

Ballard, Robert, with Will Hively. *The Eternal Darkness: A Personal History of Deep-Sea Exploration.* Princeton, NJ: Princeton University Press, 2000.

Barton, Otis. *The World beneath the Sea.* New York: Crowell, 1953.

Bascom, Willard, and Roger Revelle. "Free-Diving: A New Exploratory Tool." *American Scientist* 41 (October 1953): 624–627.

Beebe, William. *The Arcturus Adventure: An Account of the New York Zoological Society's First Oceanographic Expedition.* New York: Putnam, 1926.

————. *Beneath Tropical Seas.* New York: Halcyon House, 1926.

———. *Half Mile Down*. 1934. Reprint, New York: Duell, Sloane and Pearce, 1951.

"Beebe and Aide Descend 1,426 Feet into Sea in Steel Sphere, Phoning Observations to Tug." *New York Times*, 13 June 1930, sec. 9, p. 4.

"Beebe Out to Set Sea Descent Mark." *New York Times*, 15 April 1934, N1.

Bond, George F. *Papa Topside: The Sealab Chronicles of Capt. George F. Bond, USN*. Annapolis, MD: Naval Institute Press, 1993.

Clark, Eugenie. *Lady with a Spear*. New York: Harper, 1953.

———. *The Lady and the Sharks*. New York: Harper and Row, 1969.

Cousteau, Jacques. *The Living Sea*. New York: Harper and Row, 1963.

———. "Ocean-Bottom Homes for Skin Divers." *Popular Mechanics* 120 (July 1963): 98.

———. *Silent World*. New York: Harper, 1953.

Crane, Kathleen. *Sea Legs: Tales of a Woman Oceanographer*. Boulder, CO: Westview, 2003.

Craven, John Piña. *The Silent War: The Cold War Battle beneath the Sea*. New York: Simon and Schuster, 2001.

Dugan, James. "Portrait of Homo Aquaticus." *New York Times*, 21 April 1963, SM20, 58.

Earle, Sylvia. "All-Girl Team Tests the Habitat," *National Geographic Magazine* 140 (April 1971): 296.

———. *Sea Change: A Message of the Oceans*. New York: Fawcett Columbine, 1995.

Earle, Sylvia, and Al Giddings. *Exploring the Deep Frontier: The Adventure of Man in the Sea*. Washington, DC: National Geographic Society, 1980.

Heirtzler, J. R., and J. F. Grassle. "Deep-Sea Research by Manned Submersibles." *Science*, n.s., 194 (15 October 1976): 294–299.

Kaharl, Victoria. *Water Baby: The Story of Alvin*. New York: Oxford University Press, 1990.

Madsen, Axel. *Cousteau: An Unauthorized Biography*. New York: Beaufort Books, 1986.

"Man against the Deep." Editorial, *New York Times*, 24 February 1969, 36.

Miner, Roy Waldo. "Coral Castle Builders in Tropical Seas." *National Geographic* 65 (June 1934): 703–728.

———. "Diving in Coral Gardens." *Natural History Magazine* (September–October, 1933). Reprinted in Guide Leaflet Series No. 80, New York: American Museum of Natural History, 1933.

———. "Forty Tons of Coral." *Natural History Magazine* (July–August 1931). Reprinted in Guide Leaflet Series No. 78, New York: American Museum of Natural History, 1933.

————. "Miracles in Coral and Mud." *World's Work* 59 (November 1939).

Moffett, C. "Taking Moving Pictures at the Bottom of the Sea." *American Magazine* 79 (January 1915): 11–16.

Oreskes, Naomi. "A Context of Motivation: US Navy Oceanographic Research and the Discovery of Sea-Floor Hydrothermal Vents." *Social Studies of Science* 33 (October 2003): 679–742.

Piccard, Jacques. *Seven Miles Down: The Story of the Bathyscaph Trieste.* New York: Putnam, 1961.

Pitts, John A. *The Human Factor: Biomedicine in the Manned Space Program to 1980.* NASA SP-4213. Washington, DC: U.S. Government Printing Office, 1985.

Powell, David. *A Fascination for Fish: Adventures of an Underwater Pioneer.* Berkeley: University of California Press, 2001.

Rainey, Froelich, and Elizabeth K. Ralph. "Archeology and Its New Technology." *Science,* n.s., 1153 (23 September 1966): 1481–1491.

Randall, John. "Grazing Effect on Sea Grasses by Herbivorous Reef Fishes in the West Indies." *Ecology* 46 (May 1965): 255–260.

————. "Overgrazing of Algae by Herbivorous Marine Fishes." *Ecology* 42 (October 1961): 812.

Rexine, John E. Review of *Archaeology under Water* by George Bass. *American Journal of Archaeology* 72 (January 1968): 85.

Robison, Bruce. "Submersibles in Oceanographic Research." Pp. 387–390 in *Oceanographic History: The Pacific and Beyond,* ed. Keith Benson and Philip Rehbock. Seattle: University of Washington Press, 2002.

Slocum, Kenneth G. "Exploiting the Deep: Tapping Sea's Wealth Sparks Investor Interest, but Payoff is Remote." *Wall Street Journal,* 16 September 1968, 1, 24.

Taves, Brian. "With Williamson Beneath the Sea." *Journal of Film Preservation* 25 (April 1996): 54–61.

Tracy, Henry Chester. *American Naturists.* New York: Dutton, 1930.

Travis, John. "Deep-Sea Debate Pits *Alvin* against *Jason.*" *Science* 259 (12 March 1993): 1534–1536.

Van Dover, Cindy Lee. *Deep-Ocean Journeys: Discovering New Life at the Bottom of the Sea.* Reading, MA: Addison-Wesley, 1996.

Van Hoek, Susan, and Marion Clayton Link. *From Sky to Sea: A Story of Edwin A. Link.* Flagstaff, AZ: Best Publishing, 2003.

Weir, Gary E. *An Ocean in Common: American Naval Officers, Scientists, and the Ocean Environment.* College Station: Texas A&M University Press, 2001.

Welker, Robert Henry. *Natural Man: The Life of William Beebe.* Bloomington: Indiana University Press, 1975.

Williamson, Ernest. *Twenty Years under the Sea.* New York: Junior Literary Guild and Hale, Chushman and Flint, 1936.

Chapter 8

Bilstein, Roger. *Stages to Saturn: A Technological History of the Apollo/Saturn Launch Vehicles.* 1980. Reprint, Gainesville: University Press of Florida, 2003.

Byers, Bruce. *Destination Moon: A History of the Lunar Orbiter Program.* NASA TM-3487. Washington, DC: U.S. Government Printing Office, 1977.

Columbia Accident Investigation Board Report. Washington, DC: U.S. Government Printing Office, August 2003.

Compton, W. David. *Where No Man Has Gone Before: A History of Apollo Lunar Exploration Missions.* NASA SP-4214. Washington, DC: U.S. Government Printing Office, 1989.

Compton, W. David, and Charles D. Benson. *Living and Working in Space: A History of Skylab.* NASA SP-4208. Washington, DC: U.S. Government Printing Office, 1983.

DeVorkin, David. *Race to the Stratosphere: Manned Scientific Ballooning in America.* New York: Springer-Verlag. 1989.

Hacker, Barton C., and James M. Grimwood. *On the Shoulders of Titans: A History of Project Gemini,* NASA SP-4203. Washington, DC: U.S. Government Printing Office, 1977.

Hall, R. Cargill. *Lunar Impact: A History of Project Ranger.* NASA SP-4210. Washington, DC: U.S. Government Printing Office, 1977.

———. "SAMOS to the Moon: The Clandestine Transfer of Reconnaissance Technology between Federal Agencies." Washington, DC: National Reconnaissance Office, 2001.

Heppenheimer, T. A. *The Space Shuttle Decision.* NASA SP-4221. Washington, DC: U.S. Government Printing Office, 1999.

Jenkins, Dennis. *The History of Developing the National Space Transportation System: The Beginning through STS-50.* Melbourne Beach, FL: Broadfield, 1992.

Koppes, Clayton. *JPL and the American Space Program.* New Haven, CT: Yale University Press. 1982.

Link, Mae Mills. *Space Medicine in Project Mercury.* NASA SP-4003. Washington, DC: U.S. Government Printing Office, 1965.

Logsdon, John M. *Together in Orbit: The Origins of International Participation in the Space Station.* Washington, DC: NASA, 1998.

McCurdy, Howard. *Space and the American Imagination.* Washington, DC: Smithsonian Institution Press, 1997.

———. *The Space Station Decision: Incremental Politics and Technological Choice.* Baltimore, MD: Johns Hopkins University Press, 1990.

McDougall, Walter. *From the Heavens to the Earth: A Political History of the Space Age.* New York: Basic Books, 1985.

Neufeld, Michael. *The Rocket and the Reich.* Cambridge, MA: Harvard University Press, 1995.

Pielke, Roger, Jr. "A Reappraisal of the Space Shuttle Programme." *Space Policy* (May 1993): 133–157.

Pitts, John A. *The Human Factor: Biomedicine in the Manned Space Program to 1980.* NASA SP-4213. Washington, DC: U.S. Government Printing Office, 1985.

Siddiqi, Asif A. *Challenge to Apollo: The Soviet Union and the Space Race, 1945–1974.* NASA SP-4408. Washington, DC: U.S. Government Printing Office, 2000.

———. *Deep Space Chronicle: A Chronology of Deep Space and Planetary Probes, 1958–2000.* NASA SP-4524. Washington, DC: U.S. Government Printing Office, 2002.

Swenson, Loyd S., et al. *This New Ocean: A History of Project Mercury.* NASA SP-4201. Washington, DC: U.S. Government Printing Office, 1966.

Weitekamp, Margaret. *Right Stuff, Wrong Sex: America's First Women in Space Program.* Baltimore, MD: Johns Hopkins University Press, 2004.

Winter, Frank. *Prelude to the Space Age: The Rocket Societies, 1924–1940.* Washington, DC: Smithsonian Institution Press, 1983.

Chapter 9

Boston, P. J., M. V. Ivanov, and C. P. McKay. "On the Possibility of Chemosynthetic Ecosystems in Subsurface Habitats on Mars." *Icarus* 95(2) (1992): 300–308.

Burrows, William E. *Exploring Space: Voyages in the Solar System and Beyond.* New York: Random House, 1990.

Committee on Planetary and Lunar Exploration. *A Science Strategy for the Exploration of Europa.* Washington, DC: National Academy Press, 1999.

Cooper, Henry. *The Evening Star: Venus Observed.* New York: Farrar, Straus and Giroux, 1993.

Dethloff, Henry C., and Ronald A. Schorn. *Voyager's Grand Tour.* Washington, DC: Smithsonian Books, 2002.

Dick, Steven J., and James E. Strick. *The Living Universe: NASA and the Development of Astrobiology.* New Brunswick, NJ: Rutgers University Press, 2004.

Ezell, Edward, and Linda Ezell. *On Mars: Exploration of the Red Planet, 1958–1978.* NASA SP-4212. Washington, DC: U.S. Government Printing Office, 1984.

Fimmel, Richard O., et al. *Pioneering Venus: A Planet Unveiled.* NASA SP-518. Washington, DC: U.S. Government Printing Office, 1995.

Hartmann, William K., and Odell Raper. *The New Mars.* NASA SP-337 Washington, DC: U.S. Government Printing Office, 1974.

Horowitz, Norman. *To Utopia and Back: The Search for Life in the Solar System.* New York: W. H. Freeman, 1986.

Kargel, Jeffrey. *Mars: A Warmer, Wetter Planet.* New York: Springer-Praxis, 2004.

Koppes, Clayton. *JPL and the American Space Program.* New Haven, CT: Yale University Press. 1982.

Marov, Mikhail Ya, and David Grinspoon. *The Planet Venus.* New Haven, CT: Yale University Press. 1998.

McCurdy, Howard. *Faster, Better, Cheaper: Low-Cost Innovation in the U.S. Space Program.* Baltimore, MD: Johns Hopkins University Press, 2001.

Mishkin, Andrew. *Sojourner.* New York: Berkley Books, 2003.

Morton, Oliver. *Mapping Mars: Science, Imagination and the Birth of a World.* New York: Picador, 2002.

"On Mars, a Second Chance for Life." *Science* (17 December 2004): 2010–2012.

Sheehan, William. *The Planet Mars.* Tucson: University of Arizona Press, 1996.

Shirley, Donna. *Managing Martians.* New York: Broadway Books, 1998.

Siddiqi, Asif A. *Deep Space Chronicle: A Chronology of Deep Space and Planetary Probes, 1958–2000.* NASA SP-4524. Washington, DC: U.S. Government Printing Office, 2002.

Squyres, S. W., et al. "The Spirit Rover's Athena Science Investigation at Gusev Crater, Mars." *Science* (6 August 2004): 794–799.

Index

Abbot, Charles Greeley, 221

Accademia del Cimento (Academy of
    Experiments), 14

Account of the Arctic Regions
    (Scoresby), 166, 167

Adams, Henry, 147

Advanced Composition Explorer
    (ACE), 264

Advancement of Learning, The
    (Bacon), 2

Agassiz, Alexander, 171

Agassiz, Louis, 96, 148

Agnew, Spiro, 239

Albert I (prince), 172

Aldrin, Edwin ("Buzz"), 235–236, 236
    (photo), 239

Allen, E. J., 209

American Association for the
    Advancement of Science, 138

American Coast and Geodetic Survey,
    177

American Museum of Natural History,
    Hall of Ocean Life, 193, 193 (photo)

American Philosophical Society, 136, 145

Amphitrite, 211

Anders, William, 235

Andros Island, 192

Anthropology, 131

Arago, François, 87

Archaeology under Water (Bass), 200

Arcturus Adventure (Beebe), 194–195

Armstrong, Neil, 235–236, 237

Army Ballistic Missile Agency, 226

Army Corps of Topographical
    Engineers, 142–146

Aru Islands, 119

Astronomia nova (Kepler), 10

Astronomy, 2–3, 9–10
    observational astronomy, 34, 35

Atomic Energy Commission, 181

Babbage, Charles, 73

Bache, Alexander Dallas, 90–91,
    136–138

Bacon, Francis, 1, 2, 11–12, 34, 289
    as the father of induction, 11–12

Bailey, William Whitman, 148

Baird, Spencer, 145, 146, 148

Ballard, Robert, 214

Ballooning, 219–220, 221–223

Banks, Joseph, 51, 58, 59, 63, 64 (figure),
    65, 76, 111, 112, 145, 167, 292

Baret, Jean, 53

Barton, Otis, 208, 208 (photo)

Bass, George, 199–200

Bates, Henry Walter, 117, 118

Bathythermograph (BT), 178

Beagle, voyage of, 104–111

Beale, Thomas, 167

Bean, Alan LaVern, 238

Beaufort, Francis, 82, 84–85, 89, 90, 92,
    106

Beebe, William, 193–195, 208–210, 208
    (photo)

Beggs, James M., 324

Benton, Thomas Hart, 143, 145

Bering, Vitus Jonassen, 21–22

Bering Fur Seal Commission, 157

Berson, Arthur, 220

Bigelow, Henry, 177, 178

Biogeography, 79

Biology, 131

Birt, William Radcliff, 90

Boerhaave, Hermann, 40

Bond, George, 201–202

Bonpland, Aimé, 75, 77, 299

Borman, Frank, 235

Bort, Teisserenc de, 221

Botanical gardens, 43–44. *See also*
    Royal Academy of Sciences in
    Paris, and the Jardin du Roi; Royal
    Botanic Gardens (Kew)

Botany Bay, 60

Botany of the Antarctic Voyage of
    HMS Discovery Ships Erebus and
    Terror, The (J. Hooker), 112

Bougainville, Louis-Antoine de, 41,
    52–54, 67, 71

Bouger, Pierre, 45
Boyle, Robert, 15
Brahe, Tycho, 10
Braun, Werner von, 223, 229, 233, 242, 246, 255
Breeder, Charles, 198
Brewer, William, 147
Britain, 11, 63, 65, 131
   Humboldt's influence on, 80–82
   industrialization in, 34
   "informal" empire of, 131
   investment of in oceanography, 97
   nautical instrument industry in, 31–32
   as nineteenth-century naval power, 71
   as technological and scientific leader, 103
   and the telegraph, 95
British Admiralty, scientific branch of, 82
   and Humboldtian initiatives, 83–91
British Association for the Advancement of Science (BA), 72–74, 84, 87–88, 89, 90, 170
British Meteorological Department, 91
Brooke, John, 164–165
Brush, George Jarvis, 148
Buchan, David, 92
Buffon. *See* Leclerc, Georges Louis
Bureau of Fisheries, 156, 157, 177
Bureau of Lighthouses, 178
Bureau of Ships, 181
Bush, George W., 242

Cabinets of curiosities. *See* Collections
California Geological Survey, 152–155
Callizo, Jean, 221–222
Campe, J. H., 74
Candide (Voltaire), 41
Cape Haze Marine Laboratory, 197, 198
Carlos III (king), 65
Carpenter, Scott, 202
Carr, Archie, 199
Carson, Rachel, 195
Cartography, 6
Casa de la Contratación (House of Trade), 8
Cassini, Dominique, 15, 19, 24, 25, 26, 27
Cassini, Jacques, 22
Cassini/Huygens, 268–26
Celsius, Olaf, 41

Century of Progress, 222
Chaffee, Roger, 233
Challenger Expedition, 96–97, 98, 169–171
Chappe d'Auteroche, Jean-Baptiste, 56
Charles I (king), 7
Charles II (king), 14, 26
Charles IV (king), 75, 76
Charles, Jacques Alexandre César, 220, 220 (figure)
Chatelet, Marquise de, 24
China, 3–5
Chronometer, 31–34, 58
Clark, Chris, 182
Clark, Eugenie, 197, 198
Clark, William, x, 74, 139–142, 141 (painting)
Clementine, 263
Clifford, George, 40, 48
Cluster mission, 264
Colbert, Jean-Baptiste, 19, 20 (painting)
Cold War, the, 223, 243. *See also* Space race, the
Collections, 43, 139
Collins, Michael, 235, 237
Columbus, Christopher, 7
Commerson, Philibert, 53
Compass, 1, 12, 34
   and magnetic declination, 8
Comte, Auguste, 127
Condamine, Charles-Marie de la, 23, 41, 46–47, 71
Conrad, Pete, 238
Conshelf I, 201
Conshelf II, 201
Conshelf III, 201
Cook, James, 18, 33, 41, 51, 52, 54, 58, 60 (figure), 63, 67, 74, 140, 292
   three voyages of, 57–62
Cooper, L. Gordon, Jr., 203
Copernicus, Nicolaus, 9
Coral Reefs (Darwin), 105
Corps of Discovery, 135, 139–142
Cosmology, 241
Cosmos (Humboldt), 75
Cotter, Richard, 153
Cousteau, Jacques, 196, 199, 200–201, 211
Coxwell, Henry, 220
Craven, John P., 203

Crocker, Templeton, 192
Crookham, George, 150
Cunningham, R. Walter, 233
Cuvier, Georges, 124

D'Alembert, Jean, 42
Dana, James Dwight, 148
Darwin, Charles, 33, 62, 91, 95, 104–111,
    112–113, 115, 116–117, 119–120, 130,
    131–132, 305
Davenport, Charles, 17
Davis, Jefferson, 145
De magnete (Gilbert), 11
De revolutionibus (Copernicus), 9
Dee, John, 11
Deep Frontier (Earle), 206
Deep Ocean Engineering, 206
Deep-Ocean Journeys (Van Dover), 191
Deep Ocean Technologies, 206
Deep-sea submersibles, 205, 208–214
    Alvin, 190–191, 212–214
    bathyscaphs, 210–211
    bathysphere, 208–210, 208 (photo)
    photosphere, 192
    remote operating vehicles (ROVs), 214
    Trieste, 210–211
Deep Submergence Systems Review
    Group, 211
Delisle, Joseph, 55, 56
D'Entrecasteaux, Antoine Raymond
    Jesopeh Bruni, 53, 66
Dépôt des Chartes, Plans, Journaux et
    Mémoires Relatifs à la Navigation,
    166
Depot of Charts and Instruments, 162
Desaguilier, John T., 31
Descartes, René, 12, 19
Descent of Man, The (Darwin), 105
*Dialogues on the Simples, Drugs, and
    Materia Medica of India* (d'Orta),
    42–43
Dias, Bartolomeu, 7
Diderot, Denis, 42, 47
Diving suits and helmets, 191, 195
Dixon, Jeremiah, 56
D'Orta, Garcia, 42–43
Drake, Francis, 11
Dugan, James, 200–201
Dumas, Frederick, 196
Dyna-Soar, 224

Earle, Sylvia, 205, 206–207, 207 (photo)
Earth Resources Technology Satellite
    (ERTS), 182
East India Company, 88
*Ecology of Deep-Sea Hydrothermal
    Vents, The* (Van Dover), 191
Economic botany, 52, 67
Edgerton, H. E., 200
Edison, Thomas, 176
Eisele, Donn, 233
Eisenhower, Dwight, 225–226, 228
Ekman, Gustav, 173
Elemens de botanique (Tournefort), 44
Elizabeth I (queen), 11
Ellicott, Andrew, 140
Ellis, John, 59
Emerson, Ralph Waldo, 115, 155
Emile (Rousseau), 41
Emmons, Samuel Franklin, 148
Enlightenment, the, 42, 65, 67–68, 129
Essay on the Principle of Population
    (Malthus), 110–111, 117, 119
Ethnography, 131
Euler, Leonard, 30, 31, 33
European Space Agency (ESA), 258, 259
Evidence As to Man's Place in Nature
    (Huxley), 123, 126 (figure)
Evolution, 116–117, 127–128; Darwin's
    theory of, 95, 104, 111–112, 116–117,
    119–120, 170, 127–128, 126; Huxley's
    theory of, 125–129; Spencer's theory
    of, 127–128; Wallace's theory of, 119,
    128
*Explanations and Sailing Directions
    to Accompany the Wind and
    Current Charts* (Maury), 165
Explorer probes, 248, 263
Explorer 1, 247 (photo), 248, 255
Explorer 2, 222
Explorer 3, 248

Falck, Pehr, 50
Ferdinand (king), 7
Field, Cyrus W., 165
Fitzroy, Robert, 91, 106, 110, 111
Flamsteed, John, 26
Flechsig, Arthur, 203
Fleming, Richard, 179
Flora lapponica (Linnaeus), 40
FNRS, 222

Fobos probes
    Fobos 1, 260
    Fobos 2, 260
Forbes, Edward, 122, 170
Forbes, James, 89
Forster, Georg, 61, 62, 75, 76
Forster, Johann Reinhold, 50, 61–62
Fram, 173, 174
France, 24, 65
Franklin, Benjamin, 56
Franklin, John, 66, 92, 93–94
Frederick II (king), 10, 75
Frémont, John C., 143–145, 144 (figure)
Fundamenta botanica (Linnaeus), 40
Furneaux, Tobias, 61

Gagarin, Yuri, 229, 230
Galápagos Islands, 109
Galaup, Jean-Francoise de. *See* La
    Pérouse
Galileo, 10, 12, 26, 27
Galileo, 260–261
Gama, Vasco de, 7
Gardiner, James, 147, 148
Gassendi, Pierre, 19
Gauss, Carl Friedrich, 86, 87, 98
Gay-Lussac, Joseph Louis, 220
*General and Rare Memorials
    Pertayning to the Perfect Arte of
    Navigation* (Dee), 11
Genesis Project, 201
Geographia (Ptolemy), 6
Geographical Distribution of Animals
    (Wallace), 118
*Geological Observations on South
    America* (Darwin), 105
Geotail, 264
German Rocket Society, 223–224
Gilbert, William, 11
Giotto, 259
Glasier, James, 220
Glenn, John, 229
Goethe, Johann, 76, 79
Goldin, Daniel S., 263, 327
Goldsmith, Oliver, 41
Good, Peter, 63
Gordon, Richard Francis, Jr., 238
Göttingen Magnetic Union, 86
Gould, John, 110
Graham, George, 16, 25, 32–33

Gray, Asa, 155
Gray, Hawthorne, 222
Great Surveys, 146–152
    40th Parallel Survey, 147–150
    See also Powell, John Wesley
Green, Charles, 59
Greenwood, Isaac, 39
Gregory, Herbert, 177
Grissom, Virgil, 229, 233
"Gulf Stream and Currents of the Sea,
    The" (Maury), 165
Gulliver's Travels (Goldsmith), 41
Gunpowder, 1

Haeckel, Ernst, 96
Hague, Arnold, 148
Hague, James, 148–149
Haise, Fred Wallace, Jr., 238
Hale, George Ellery, 241
Halley, Edmund, 15–18, 26, 32, 33, 54,
    55, 94
Halley's Comet, 259
Harmonices mundi libri V (Kepler), 10
Harrison, John, 16, 32–33, 58
Hasselquist, Frederick, 50
Hassler, Ferdinand, 136, 137
Hawkes, Graham, 206
Hayden, Ferdinand, 146
Hays, Harvey Cornelius, 177
Hegel, Georg, 127
Henry the Navigator, 6–7
Henry, John, 90–91
Hensen, Victor, 175
Henslow, John Stevens, 105, 106, 107
Herschel, John, 87, 89–90, 106
Herschel, William, 89
Hillary, Edmund, x
Himalayan Journals (J. Hooker), 113,
    115
Himalayas, 113
Histoire naturelle (Leclerc), 40, 47
Historia medicinal (Monardes), 43
Hiten, 262
Hjort, Johan, 175
Holland, 48
Hollister, Gloria, 209
Hooke, Robert, 15
Hooker, Joseph Dalton, 111–117, 114
    (photo), 119, 120, 130, 131, 309
Hooker, William Jackson, 111–112

Hortus cliffortianus (Linnaeus), 40
Hubble, Edwin, 241
Hubble Space Telescope, 241
Hubbs, Carl, 195
Hulbert, E. O., 209
Humanson, M. L., 241
Humboldt, Alexander von, 18, 46, 62, 68,
    72, 74–80, 78 (figure), 85–86, 87, 88,
    112, 115, 299
  and British imperialism, 80–82
Huxley, Thomas Henry, 116, 121–129,
    130, 131
  as "Darwin's bulldog," 123
Huygens, Christian, 15, 19, 26
Hydrological Office, 181

Imperialism, 72, 129, 131
Induction, 11–12
Instauratio Magna (Bacon)
  frontispiece of (figure), 13
Institut Oceanographique de Paris, 172
Institute of France, 83, 86
Institute of Marine Resources (IMR),
    180
Inter-Agency Solar-Terrestrial Program,
    263–264
International Council for the
    Exploration of the Sea (ICES), 175
International Space Station (ISS), 241
"Investigating Space with Rocket
    Devices" (Tsiolkovskiy), 246
Isabella (queen), 7
Iselin, Columbus, 178, 210
Isis, vii
Islamic scholars, at the Bait al-Hikmah
    (House of Wisdom), 5
Island Life (Wallace), 118

James II (king), 17
Japan, 258, 259
Jefferson, Thomas, 136, 139–140, 152
John II (king), 7
Johnson, Lyndon, 212, 226, 229–230
Johnson, Martin, 179
*Journal of Researches into the Geology
    and Natural History of the Various
    Countries Visited by H.M.S. Beagle*
    (Darwin), 105
Jupiter, 255, 256, 261
Jupiter rockets, 225, 255

Kalm, Pehr, 50
Kennedy, John F., 226, 228, 229–230
Kepler, Johannes, 10, 12
Khrushchev, Nikita, 228
King, Charles, 146, 147–150,153–154, 155
Kipling, Rudyard, 129
Komarov, 234
Korolev, Sergei, 224, 225, 226, 229, 230,
    231, 232, 233–234, 242, 246

L-1 (Earth-Sun Libration Point), 264
La figure de la terre (Maupertuis),
    23–24, 25
La Pérouse, 53, 65, 66, 67, 71
Lady with a Spear (E. Clark), 198
LaFond, Eugene, 179
Le Gentil, Guillaume Joseph Hyacynthe
    Jean-Baptiste de la Galaisiere,
    56–57
Leclerc, Georges Louis, 40, 47, 66
Leibniz, Gottfried Wilhelm, 21
Lewis, Meriwether, x, 74, 139–142, 141
    (painting)
Lewis and Clark Centennial Exposition,
    156
Lillie, Frank, 177
Lincoln, Abraham, 153
Link, Edward, 206
Linnaean Society of London, 41
Linnaeus, Carl, 40–41, 44, 47–52, 63
    "apostles" of, 49–52
Lloyd, John Augustus, 87
Löfling, Pehr, 50
Long, Stephen H., 142
Longitude Act/Prize, 29, 30
Louis XIII (king), 44
Louis XIV (king), 19, 21, 22, 24, 26
Louis XV (king), 22, 24, 47, 52–53
Lovell, James A., Jr., 235, 238
Lowell, Percival, 245
Lowell Observatory, 245
Lubbock, John William, 83–84
Luce, Henry, 226
Luna 1 (the Cosmic Rocket), 248
Luna 2, 248
Luna 3, 248
Luna 9, 249
Lunar Orbiters, 249
    Lunar Orbiter I, 249
Lunar Prospector, 263

Lunar Surveyors, 249
  Surveyor 1, 249
Lunar theory, the lunar-distance
  method, 28–31, 33
Lyell, Charles, 107, 109, 116, 119, 120
Lyman, Theodore, 96

Mackenzie, Alexander, 139
Magellan, Ferdinand, 7–8
Magellan, 260
Malaspina, Alejandro, 65
Malay Archipelago, The (Wallace), 118,
  120
Malthus, Thomas, 106, 110–111
Man and Nature (Marsh), 155
Man's Place in Nature (Huxley), 126
Manifest Destiny, 135
Marine Biological Laboratory (MBL),
  177
*Marine Mammals of the Northwest
  Coast of North America*
  (Scammon), 167
Mariner probes
Mariner 1, 249
Mariner 2, 249–250
Mariner 3, 251
Mariner 4, 251
Mariner 5, 250
Mariner 9, 253
Mars, 251–252, 253, 264, 267
Mars (Lowell), 245
Mars and Its Canals (Lowell), 245
Mars as the Abode of Life (Lowell), 245
Mars Climate Orbiter, 265–266
Mars Express, 267
Mars Global Surveyor, 264–265, 267
Mars Observer, 263
Mars Odyssey, 266–267
Mars Pathfinder, 265
Mars Polar Lander, 266
Mars probes
  Mars 2, 252
  Mars 3, 252
  Mars 5, 252
  Mars 6, 252
  Mars 7, 252–253
  Mars 8, 265
Marsh, George Perkins, 155
Maskelyne, Nevil, 26, 30–31, 33, 55
Mason, Charles, 56

Masson, Francis, 63
Mathematics, 6
Maupertuis, Pierre-Louis Moreau de,
  23–25
Maury, Matthew Fontaine, 90, 91, 163
  (figure), 162, 164–166
Mayer, Johann Tobias, 30
Medical Research Laboratory, 201
Mencke, Otto, 14
Merriam, John C., 177
Meteor Expedition, 173–174
Meteorology, 79, 89–91, 98
Millikan, Robert, 209
Miner, Roy Waldo, 192–193
Mir, 241
Mishin, Vasiliy, 234, 240
Monardes, Nicolas, 43
Montgolfier, Jacques, 219
Montgolfier, Joseph, 219
Moon, 263
  Apollo 11 mission to, 235–237
  Apollo 12 mission to, 237–238
  See also Lunar theory, the lunar-
    distance method
Mountaineering in the Sierra Nevada
  (King), 147, 153–154
Muir, John, 152, 154–155, 156
Munk, Walter, 179
Murray, John, 97, 171

N-1 rockets, 231, 232, 240
Nansen, Fridtjof, 173, 174
Napoleon Bonaparte, 74
Nares, George Strong, 97
*Narrative of Travels on the Amazon
  and Rio Negro, A* (Wallace), 117,
  118, 119
National Academy of Science (NAS)
  Committee on the Effects of Atomic
    Radiation on Oceanography and
    Fisheries, 184
  Committee on Oceanography, 177
National Academy of Sciences (St.
  Petersburg), 21
National Aeronautics and Space
  Administration (NASA), 182, 201,
  226, 230, 231, 240. *See also specific
  space projects*
National Defense Research Committee,
  179

National Museum (U.S.), 156

National Oceanic and Atmospheric Association, 182

National Research Council (NRC), 176

Natural history, 40, 72
  rise of, 47–49
  stimulus to, 42–44

Natural History of the Sperm Whale, The (Beale), 167

Natural selection, theory of. *See* Evolution, Darwin's theory of

Nautical Almanac and Astronomical Ephemeris (Maskelyne), 30–31

Naval Consulting Board, Committee on Submarine Detection, 176

Navigation, 3, 6, 8

and chronometers, 31–34
  determination of latitude, 25
  determination of longitude, 25–28
  and ephemeredes, 28

Nedelin, Mitrofan, 229

Nelson, David, 63

Neptune, 257

New Atlantis (Bacon), 2, 12, 289

New Guinea, 124–125

"New Method for Composing a Natural History of Meteors" (Greenwood), 39

New York Aquarium, 157

New York Zoological Society, 157

Newton, Isaac, 15, 16, 17, 21, 22, 28–29, 31, 32, 35, 47, 83, 104

Nicholson, William, 204

Nicks, Oran, 258

Nixon, Richard M., 226, 239

Northwest Passage, search for, 91–94

Notes on the State of Virginia (Jefferson), 152

Novum organum (Bacon), 2

Oberth, Herman, 242

Ocean ethic, 206

Oceanography, 94–98, 316, 317
  and the dynamic method, 173, 174, 175
  influence of World War II on, 178–179
  Scandinavian, 172–176
  U.S., 176–181

Oceanography (Bigelow), 177

Office of Naval Research (ONR), 181, 212

Office of Scientific Research and Development, 179

Oldenberg, Henry, 14, 15, 291

Olsen, Chris, 192–193

"On the Navigation of Cape Horn" (Maury), 164

On the Origin of Species (Darwin), 105, 107, 110, 115, 120, 125, 126

Operation Crossroads, 180, 181, 185

Orbiting Solar Observatories (OSOs), 247–248

Osborn, Henry Fairfield, 209

Owen, Richard, 122

Pacific Railroad Reports (Davis), 145–146

Paley, William, 106

Parallax, 54

Paris Observatory, 26

Parry, William Edward, 88, 92, 93

Pascal, Blaise, 19

Patterson, Robert, 140

Pax Britannica, 81

Peirce, Charles, 138

Personal Narrative (Humboldt), 75, 106, 117, 122, 299

Peter the Great, 21

Petersen, Johannes, 175

Pettersson, Otto, 173

Phillip II (king), 8

Philosophes, 42, 65, 67

Philosophia botanica (Linnaeus), 40

Philosophical Transactions of the Royal Society of London, 15, 16, 29

Physical geography, 79

Physical Geography of the Sea, The (Maury), 91, 162, 164, 319

Physical sciences, 72

Piccard, Auguste, 210, 222

Piccard, Jean, 222

Piccard, Joseph, 211

Pike, Zebulon, 142

Pingre, Alexandre-Gui, 56

Pioneer probes, 248, 249
  Pioneer III, 248
  Pioneer IV, 248
  Pioneer V, 248
  Pioneer VI, 248
  Pioneer 10, 254–255, 256
  Pioneer 11, 256

Pioneer Venus probes
  Pioneer Venus 1, 257–258
  Pioneer Venus 2, 258
Plankton Expedition, 175
Polar, 264
Polyka, Valeri, 241
Portugal, 6–8
Powell, John Wesley, 146, 147, 150–152, 155
Principia mathematica (Newton), 15, 16, 17, 28–29, 31
Principles of Geology (Lyell), 107
Printing, 1, 12
Project 5M, 260
Project Apollo, 230, 231–232, 233, 238–239
  Apollo 4, 233
  Apollo 7, 233
  Apollo 8, 235
  Apollo 9, 235
  Apollo 10, 235
  Apollo 11, 235–237
  Apollo 12, 237–238
  Apollo 13, 238
  Apollo 14, 238
  Apollo 15, 238
  Apollo 17, 239
Project FAMOUS, 213
Project Gemini, 230, 231, 232
Project Lunar Orbiter, 231
Project Man High, 224
Project Mercury, 226, 229, 231
Project Paperclip, 255
Project Ranger, 231, 248–249
  Ranger 7, 249
Project Stratolab, 224
Project Surveyor, 231
Ptolemy, Claudius, 6, 9
  chart of the new continents, 9 (figure)

Quadrant, 6

R-1 rockets, 224
R-7 rockets, 228
Racism, 128, 129, 131
Rae, John, 94
Raleigh, Walter, 11
Randall, John, 199
Reagan, Ronald, 324

Reflections on the Decline of Science in England (Babbage), 73
Renaissance, the, 5–6
Report on the Lands of the Arid Region (Powell), 152
Report on the Scientific Results of the Voyage of the H.M.S. Challenger, 171
Revelle, Roger, 179, 189
Rhea darwini, 109
Richardson, John, 92, 122
Richer, John, 17, 21, 22
Ridgeway, Robert, 148
Ritter, William Emerson, 178
Robert, Marie-Nöel, 220 (figure)
Robinson Crusoe (Stevenson), 41
Roemer, Ole, 27
Ross, James Clark, 73, 87, 88, 92, 93, 112, 302
Ross, John, 88, 92, 93, 302
Rossby, Carl Gustav, 178
Rothman, Johan, 40
Rousseau, Jean-Jacques, 41, 63
Royal Academy of Sciences in Paris, 2, 18–22, 23, 24–25, 44, 46, 56, 66
  and the Jardin du Roi, 44–47, 45 (figure)
Royal Academy of Sciences (Sweden), 41
Royal Academy of Uppsala, 48
Royal Botanic Gardens (Kew), 112, 115–116
Royal Geographical Society, 72, 74
Royal Greenwich Observatory, 26, 29
Royal Hydrographic Office, 81–82, 98, 166
Royal Society of London, 2, 14–15, 29, 32, 44, 46, 56, 58, 59, 74, 88, 170
  lack of professional scientists in, 73
  as modeled on Bacon's *New Atlantis*, 14
Rozier, Françoise Pilâtre de, 219–220
Rudolphine Tables (Kepler), 10
Rush, Benjamin, 140
Russia, 30, 88–89

Sabine, Edward, 87, 98
Sagan, Carl, 256
Sailing Directions (Maury), 165

Sakigake, 259

Sallo, Denis de, 14

Salyut, 240

Sars, Georg, 96

Sarton, George, vii

Satellite imaging, 182–184

Saturn, 256–257

   the moon Titan of, 268

Saturn rockets, 231–233, 239, 255

Scammon, Charles Melville, 167, 168

Schiaparelli, Giovanni, 245

Schiller, Friedrich, 76

Schirra, Walter, 233

Schmidt, Oscar, 96

Schmitt, Harrison, 239

Science

   Humboldtian science, 81

   internationalization of its practice,
      98–99

Scientific journals, 14–15. See also *Isis;*
     Philosophical Transactions of the
     Royal Society of London

Scientific method, 8

Scientific Lazzaroni, 137

Scientific revolution, the, 9–12

Scientific societies, 291

   See also specific societies

Scientists in the Sea program, 204, 206

Scoresby, William, 92, 166–167, 167–168,
     316

Scott, David, 237

Scripps, E. W., 178

Scripps Institution of Oceanography
     (SIO), 177–178, 179

Scuba diving, 196, 197, 200

Sea Change (Earle), 206

"Sea Power and the Sea Bed" (Craven),
     203

SEALAB program, 202

   SEALAB I, 202

   SEALAB II, 202–203, 204

   SEALAB III, 203–204

Second scientific revolution, 98

Sedgwick, Adam, 105, 116

Seward, William, 156

Shark Lady (E. Clark), 198

Sheffield Scientific School, 148

Shepard, Francis, 179

Shepherd, Alan, 229

Shoemaker, Eugene, 263

Sierra Club, 155

"Significance of the Frontier in
     American History, The" (Turner),
     155

Silliman, Benjamin, 154

Skylab, 240

Slocum, Kenneth, 212

Smith, Adam, 106

Smith, F. C. Walton, 195

Smith, James Edward, 41

Smithsonian Institution, 145, 156

Social Darwinism, 129

Social Statics (Spencer), 127

Solander, Daniel Carl, 50–51, 59, 292

Solar and Heliospheric Observatory
     (SOHO), 264

Sondes, 221, 222–223

Sound Surveillance System (SOSUS),
     182

South America, 76, 77, 109

Southern California Academy of
     Science, 184–185

Soviet Union, 228

Soyuz, 234–235

Space race, the, 223–239, 243, 245–246

Space Shuttles, 240, 241

   Atlantis, 260–261

   Challenger, 260

   Columbia, 242

Spacelab, 240

Spain, 6–8, 65

Sparrman, Anders, 50, 61

Special Projects Office, 203

Species plantarum (Linnaeus), 40

Spencer, Herbert, 127, 129

Sputnik, 225, 226, 227 (photo), 228, 250

Stanley, Owen, 122, 124, 125

Stevens, Samuel, 117

Stevenson, Robert Louis, 41

Submarine Signal Company, 176

Suisei, 259

"Survey of the Greenland Sea"
     (Scoresby), 167

Sverdrup, Harald, 179, 180

Svësdni Gorodok, 230

Sweden, 49

Swigert, John L., Jr., 238

System of Nature (Linnaeus), 40, 48–49

Systematic Geology (King), 149
Systeme de la nature (Maupertuis), 25

Tahiti, 53, 59, 67
Tärnström, Christopher, 50
Tektite Project, 204–205
    Tektite I, 205
    Tektite II, 205, 206
Tereshkova, Valentina, 231
Terra Australis Incognita, 60, 61
Terrestrial magnetism, 85–89, 98
    and the Magnetic Crusade, 87, 90
Thayer, Sylvanus, 137
Thomson, Charles Wyville, 95–96, 170, 171
Thoreau, Henry David, 115
Thouin, Andre, 66
Thunberd, Carl Peter, 50
Tides, 83–85, 84 (figure), 98
Titan rockets, 232
Torrey, John, 145, 146, 155
Tournefort, Joseph Pitton de, 40, 44–45
Townsend, Charles Haskins, 156–157
Trask, John B., 153
Travel literature, 41–42, 103, 131
Triangulation, 54
Tropical Ocean-Global Atmosphere (TOGA) program, 183–184
Tsiolkovskiy, Konstantin, 224, 242, 246
Turner, Frederick Jackson, 155
Two Lectures on the Parallax and Distance from the Sun, as Deducible from the Transit of Venus (Winthrop), 56
Tyndall, John, 155

Ulysses, 262
Underwater habitats, 200
Uniformitarianism, 107, 149
University of California Division of War Research, 179
University of Cambridge, 106
University of Göttingen, 75–76
University of Leiden, 48
University of Uppsala, 48
Uranus, 257
U.S. Coast and Geodetic Survey, 178
U.S. Coast Survey, 135, 136–139
U.S. Exploring Expedition (U.S. Ex Ex), 169

U.S. Geological Survey, 138–139, 147
U.S. Navy, and American oceanography, 176–181
U.S. Navy Hydrographic Office, 178
U.S.S.R., 222

V-2 rockets, 223–224, 255
Van Allen, James, 248
Van Dover, Cindy Lee, 190–191
Vancouver, George, 65, 140
Vanderbilt, Albert, 198
Vanderbilt, William, 198
Vaughan, Thomas Wayland, 177, 178, 180
Vega probes
    Vega 1, 259
    Vega 2, 259
Venera probes
    Venera 2, 250
    Venera 4, 250
    Venera 11, 258
    Venera 12, 258
    Venera 13, 258
    Venera 14, 258
Venus, 249, 251, 251, 257–258, 260
    transit of, 54–57
Venus physique (Maupertuis), 25
Verne, Jules, 245, 246, 255
Vespucci, Amerigo, 7
Vestiges of the Natural History of Creation (Anon.), 116, 119
Viking probes, 253
    Viking 1, 254
    Viking 2, I254
Volcanic Islands (Darwin), 105
Voltaire, 24, 41, 75
Voskhod, 231, 232
Vostok, 228–229
Voyage Made within the Inland Parts of South America (Condamine), 41, 46
Voyage of Discovery and Research in the Southern and Antarctic Regions during the Years 1839 to 1843, A (James Ross), 88, 92
Voyage of the Beagle (Darwin), 122
Voyage Round the World (Bougainville), 53, 75
Voyage Round the World, A (G. Forster), 62
Voyage to the Moon (Verne), 245

Voyager probes, 256
    Voyager 1, 256–257
    Voyager 2, 257
    Voyager Interstellar Mission, 257

Wallace, Alfred Russel, 105, 117–121,
    129, 130, 131, 307
Wallace's Line, 118
Wallis, John, 6
Wallis, Samuel, 53
Walsh, Don, 211
Warren, G. K., 145–146
Webb, James, 229
Weber, Wilhelm, 86
Wells, H. G., 255
Werner, Abraham, 76
Wheeler, George Montague, 146
Whewell, William, 83–85, 98, 106, 106
White, Edward, II, 233
Whitney, Josiah Dwight, 148, 152–154,
    155
Wilberforce, Samuel, 127

Wilkes, Charles, 95, 169
Wilkinson, James, 142
Willdonow, Karl Ludwig, 74–75
William IV (king), 93
Williamson, John Ernest, 192
Wind, 264
Winthrop, John, 55–56
Wistar, Caspar, 140
Woods Hole Oceanographic Institute
    (WHOI), 177–178, 181
World Climate Research Programme,
    184
World Ocean Circulation Experiment
    (WOCE), 184
Wren, Christopher, 14

Young, Brigham, 152

Zheng He, 4–5, 162
Zhu Di, 4, 5
Zond tests, 234
    Zond 1, 250